Statistics and Numerical Methods
in BASIC for Biologists

Statistics and Numerical Methods in BASIC for Biologists

J. D. LEE

Senior Lecturer in Chemistry and Computing
Loughborough University of Technology

and

T. D. LEE

Cambridge

VAN NOSTRAND REINHOLD COMPANY

New York — Cincinnati — Toronto — London — Melbourne

**Published by Van Nostrand Reinhold Company Ltd.,
Molly Millars Lane, Wokingham, Berkshire, England**

*Published in 1982 by Van Nostrand Reinhold Company,
135 West 50th Street, New York, NY 10020, USA*

*Van Nostrand Reinhold Limited,
1410 Birchmount Road, Scarborough, Ontario, M1P 2E7,
Canada*

*Van Nostrand Reinhold
480 Latrobe Street, Melbourne, Victoria 3000, Australia*

Library of Congress Cataloging in Publication Data

Lee, J. D. (John David), 1931–
    Statistics and numerical methods in BASIC for
biologists.

    Bibliography: p.
    Includes index.
    1. Biometry—Data processing. 2. Basic (Computer
program language) I. Lee, T. D. (Timothy D.)
II. Title
QH323.5.L4   574′.072   82-1954
ISBN 0-442-30476-5   AACR2
ISBN 0-442-30481-1 (pbk.)

Printed and bound in Great Britain by the Alden Press, Oxford

Preface

In the last decade there has been a trend to include statistics in an increasing number of biology and pharmacy courses at university level, and at all levels in schools. In general, biologists and pharmacists do much less numerical and mathematical work than chemistry, physics, engineering or mathematics students, and they often have difficulty using abstract concepts and working with symbols. There are essentially two different problems; firstly knowing which statistical test to apply, knowing when it is valid to use it, and drawing the correct inference from it, and secondly the lesser problem of getting the arithmetic right!

Learning how to apply statistical methods is a skill rather like learning to ride a bicycle—reading about it and thinking about it are not enough. It is only through the practical experience of actually doing it that one will gradually acquire understanding and practical skill. The computer revolution has minimised difficulties over arithmetic manipulation, provided that suitable computer programs are available. Use of such programs makes it easy to perform sophisticated statistical procedures, but programs should not be used without an understanding of the methods and data. Understanding is essential to prevent the mis-application and abuse of statistical methods. Time is well spent thinking out how to design and carry out an experiment before actually performing it, since no amount of fancy mathematics or statistics can counteract the effect of a badly planned experiment.

The wide variation in behaviour of living material make it essential that biologists understand and apply statistics to their experiments.

This book aims to cover a number of widely used topics involving statistics and probability in a simple, down to earth and readable manner. Jargon has been avoided wherever possible, and techniques are introduced by means of detailed worked examples. In a conflict between mathematical rigour and readability, we have chosen to err on the side of readability. A computer program is provided with each method. The programs are easy to use, and print out helpful instructions. They include numerous error checks against incorrect or pathological data, and are thought to be nearly 'student-proof'. The deliberate use of an elementary subset of BASIC for writing the programs is to make it easy to implement them with little or no change on a wide variety of computers. A trial run is also provided, to show what the program actually does, and to provide trial data for users to test the program on their own computer. Where appropriate, chapters include exercises, and solutions giving many of the intermediate steps are provided at the end of the book.

We are most grateful to a number of colleagues for help, advice and constructive criticism in the preparation of the book. These include Professor J. N. Miller, Dr A. G. Briggs, Mr J. R. Buxton, Mrs I. Calus, Mr J. Fernandez, Mr M. J. Hunt, Dr B. Negus, Mr S. Sherman, and Dr R. J. Stretton.

J. D. Lee   *January 1982*
T. D. Lee

Contents

Introduction

This book deals with statistical inference, that is methods of obtaining the facts from numerical data. The chapters are largely self contained, and it is not necessary to read all the chapters in order. However, Chapter 1 (Errors) and Chapter 2 (Average and spread of results) are widely used throughout the text, and should be read by those who are not familiar with this elementary statistical material. There is some interdependence of chapters, and Chapter 5 (Chi-Squared Test), Chapter 6 (Comparison of two samples) and Chapter 7 (Comparison of more than two samples) all use distributions, which are discussed in Chapter 4. Similarly, Chapter 9 (Straight line fitting) uses correlation coefficients, which are discussed in Chapter 8. In addition, Chapter 10 (Some biological applications of linear regression), Chapter 11 (Comparison of regression lines) and Chapter 12 (Polynomials) all use least squares, which is discussed in Chapter 8.

The computer programs provided have been run without difficulty on several different machines. They are written in an elementary subset of BASIC, and should be implemented with little difficulty on a wide variety of microcomputers and mainframes which support floating point BASIC and permit the printing of 72 columns of output on a line. The readability of the program listings has been improved by indenting the FOR . . . NEXT loops as is common in ALGOL. The programs are friendly to the user in that they print explanatory messages. Whilst long messages are essential for beginners, they become a time wasting irritation for experienced users, and the programs allow the user to choose between long or abbreviated messages. The programs include a number of checks on the data, in an attempt to make them robust towards incorrect or pathological data. The numerical methods have been chosen with care, to avoid loss of accuracy. Most of the programs include a subroutine to check, and if necessary amend, the input data before performing the calculations. This is invaluable for correcting wrongly typed data, particularly when there are quite a lot of values. It does, however, increase the size of the programs. The book makes no attempt to teach BASIC programming, but some implementation notes are given covering features not defined in the ANSI minimal BASIC standard specification. These should assist users to make minor changes necessary for their own particular computer.

Guide to the Implementation of Programs

The program listings use the following conventions to avoid confusion between symbols:

Number zero is crossed ∅, while letter O is not.
The international money symbol (dollar) is crossed $, whilst letter S is not.
Number one is not crossed 1, whilst letter I is crossed at the top and bottom.

The following points not covered by the ANSI Minimal BASIC standard specification may need attention by the user:

1. In these programs strings are DIMensioned in the first line of each program, e.g.

   10 DIM I$(3)

   This is intended to reserve space for a maximum of three characters in the string I$. On some computers it in fact reserves space for four strings: I$(0), I$(1), I$(2) and I$(3), each of which may contain a series of characters. On computers implementing the latter form it is better and will save space if the string declarations are removed from the DIMension statement.

2. Certain versions of BASIC including TRS-80 level II, RML 9K and Xitan disc BASIC should have the string declarations removed from the DIMension statements as above, and instead they require a CLEAR statement to reserve space for all the characters in all of the strings used.

3. The spacing of the printout produced by the programs may vary slightly from machine to machine because of slightly different implementations of TAB and , in PRINT statements.

4. Certain of the programs use an IF statement to jump out of a FOR . . . NEXT loop prematurely. This procedure works with most versions of BASIC, but some including Xitan disc BASIC and North Star BASIC may report a stack error. To cure this the offending IF statement should be changed to read

   IF . . . THEN EXIT line number .

   instead of IF . . .THEN line number

5. Sample runs are provided to give sample data for testing the programs. All user input is preceded by a question mark and a space.

6. Certain features such as IF . . . THEN . . . ELSE, PRINT USING, ON . . . GOTO, and multi-statement lines have been avoided because they are neither universally available, nor implemented the same way on all machines.

7. The word LET is included in all arithmetic statements, since this will always work. However, in many implementations of BASIC it may be omitted.

1

Errors

Accuracy and Precision

Experimental results are always subject to errors, and in assessing the reliability of the final result the accuracy and precision should be considered. Accuracy may be expressed as how closely the experimental result agrees with the true or most probable value. In contrast, precision may be defined as how closely a set of measurements of the same quantity agree with each other. Thus accuracy expresses the correctness of a measurement whereas precision describes the reproducibility of the measurement.

These two concepts are illustrated in Table 1.1 by the results of four separate students, who each determined the density of a piece of glass three times. The true density of the glass is 2600 kg m^{-3}. The results may be summar-

Table 1.1

Student	Reading 1	Reading 2	Reading 3	Average and spread
1	2580	2590	2600	2590 ± 10
2	2440	2460	2450	2450 ± 10
3	2610	2760	2460	2610 ± 150
4	2880	2620	2750	2750 ± 130

ised in Table 1.2. Two measures of precision are the standard deviation and the coefficient of variation, both of which are discussed in Chapter 2.

Table 1.2

	Accurate	Inaccurate
Precise	Student 1 2590 ± 10	Student 2 2450 ± 10
Imprecise	Student 3 2610 ± 150	Student 4 2750 ± 130

Significant Figures and Rounding

An understanding of significant figures is essential in order to avoid claiming an unreasonably high or low accuracy in a *final* result. The number of significant figures in a value is the total number of digits in the value excluding leading zeros (see Table 1.3).

Table 1.3

Value	Number of significant figures
3.142	4
0.020	2
12.020	5
0.012 020	5

An alternative way of working out the number of significant figures is to express the value in standard (scientific) form as used on many calculators. The number of significant

3

figures is the number of decimal places plus one (see Table 1.4).

Quantities which are measured should be recorded with only one uncertain figure. Thus if a measurement is subject to error in the third decimal place, then only three decimal figures may be claimed. Furthermore the implication of claiming a value of 3.142 (with three decimal figures), is that the true value is closer to 3.142 than either 3.141 or 3.143. Thus

Table 1.4

Value	Standard form	Decimal places	Significant figures
3.142	3.142×10^0	3	4
0.020	2.0×10^{-2}	1	2
12.020	1.2020×10^1	4	5
0.012 020	1.2020×10^{-2}	4	5

$3.1415 \leqslant$ true value < 3.1425. It is important that no rounding of intermediate answers is performed when making calculations using measured quantities, since this would introduce additional and unnecessary errors. In contrast, the final answer should always be rounded to give one uncertain figure. The purpose of rounding is to remove inaccurate digits from a final answer so that all of the digits quoted are meaningful. Thus if a three significant figure answer is required, the three digit value closest to the calculated value should be given, for example

Table 1.5

Number of significant figures	Rounded value of π
6	3.141 59
5	3.141 6
4	3.142
3	3.14
2	3.1
1	3

$\pi \simeq 3.141\ 592\ 653\ 6$ (Table 1.5). It can be seen that rounding a number amounts to truncating it to the appropriate number of digits, except when the first truncated digit is 5 or more then 1 is added to the least significant digit retained.

Classification of Errors

The types of errors which may affect an experimental result may be conveniently divided into three groups: gross accidental errors, systematic errors and random errors.

Gross Accidental Errors

These are large irregular errors which are caused by *incorrect* experimental technique or calculation. These include:

(i) weighing a damp sample rather than a dry sample;
(ii) weighing a sample which is still warm (the weight will be affected by convection currents);
(iii) using a 20 ml pipette inadvertently when requiring a 25 ml pipette;
(iv) using the wrong value for a shunt on a galvanometer;
(v) mismatching units, e.g. density in g per cc instead of $kg\ m^{-3}$, temperature in °C rather than K, or solutions whose concentrations are expressed as normal rather than molar;
(vi) transcription errors;
(vii) errors in calculation—particularly common with the use of electronic calculators.

Systematic Errors

These consistently give a result which is wrong by a fixed amount. Some examples are as follows:

(i) Instrumental errors such as the zero incorrectly adjusted, or calibration errors (particularly thermometers, pipettes, barometers, voltmeters).

4

(ii) Reagent errors. A sample used as a primary standard may be impure, or may have been made up to the wrong concentration of solution.

(iii) Personal errors, for example parallax if a voltmeter is always viewed in an identical way from the side, or if a burette is consistently viewed from beneath. Some experiments involve detecting a colour change (titrations or colour dyeing), and people observe colours differently.

Random Errors

If the same observer makes the same measurement repeatedly, under apparently identical conditions, it will be found that slight variations occur. These are random errors, and are caused by unknown factors changing without the observer being aware of this. Some examples of such factors are small changes in room temperature, pressure or humidity, draughts, fluctuations in the mains electricity supply, vibrations, stray magnetic fields or parallax errors arising from reading an instrument from differing angles. The slight variation in the level to which a pipette or a standard flask is filled with a solution is another source of random error.

If a sufficiently large number of measurements are taken, the average value obtained will be close to the true value since positive and negative errors are equally likely, and will on average cancel. Furthermore the distribution of random errors will be approximately *normal*. Such a distribution is shown in Fig. 1.1. An alternative name for this distribution is Gaussian.

Figure 1.1 shows that small errors occur much more frequently than large errors. Since there is an equal chance of obtaining a positive error and a negative error, the curve is symmetrical, and the mean error is zero.

The horizontal or error axis of Fig. 1.1 is calibrated in terms of σ, where σ is the standard deviation of the population of measurements. The standard deviation is a

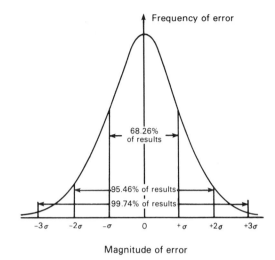

Fig. 1.1 *Normal* distribution curve.

measure of the spread of the measurements. Thus a wide spread of results yields a large standard deviation and closely grouped results give a small standard deviation. The standard deviation is estimated as

$$\text{Estimated } \sigma = \sqrt{\frac{\text{sum of errors squared}}{\text{number of measurements} - 1}}$$

This is fully described in Chapter 2.

Figure 1.1 also shows that 68.26% of the errors lie within $\pm$ one standard deviation, 95.46% of the errors lie within $\pm$ two standard deviations and 99.74% of the errors lie within $\pm$ three standard deviations for a normal distribution. The word normal has a precise mathematical meaning, and the area under a normal curve is tabulated in Appendix 4. The normal curve is used together with the standard deviation to calculate confidence limits as mentioned later.

If there are only a small number of measurements then the normal distribution is not adequate for this purpose. This problem was overcome by Student in 1908 by using the t-distribution when the number of measurements was small. ('Student' is a pseudonym for W. S. Gosset, *Biometrika*, 1908, **6**, 1.) The t-distribution tends towards the normal distribution as the number of terms becomes large,

5

as can be seen in Appendix 6. (In this context the number of degrees of freedom is equal to the number of measurements minus one.)

How to Avoid or Reduce the Effects of Errors

Gross accidental errors are completely avoidable by care in experimental technique. Transcription of numbers should be avoided wherever possible by recording directly into the laboratory notebook. Repeating the experiment generally indicates the erroneous value.

Systematic errors cannot be detected from a single experiment, no matter how often the experiment is replicated using the same equipment and apparatus. The only way of detecting these errors is to repeat the experiment using a different set of apparatus and different reagents.

Random errors arise from causes beyond the control of the experimenter, and consequently cannot be avoided. Since these errors are just as likely to give a low reading as a high reading, their effect can be reduced by taking many readings and averaging the results. The average result will then be close to the true result provided that gross accidental errors and systematic errors are insignificant.

Ways of Quantifying Errors

Two common ways of expressing the magnitude of the errors are (i) to give the range of results, and (ii) to quote the standard deviation of the results.

Range

The range of the results is obtained by subtracting the smallest value obtained from the largest value. Frequently the result is expressed as average value $\pm\frac{1}{2}$ range. In many cases, the final result depends on readings of several quantities, each of which has an error term. The total error in the final result

depends on the individual errors combined in an appropriate manner. The rules for combining errors are derived below.

Standard Deviation

The standard deviation of a set of results is harder to calculate than the range, but is more representative of the spread of results since it takes account of all of the results rather than just the upper and lower values as used for the range. It is possible to calculate 95% or 99% confidence limits (that is limits within which the true value is 95% or 99% certain to lie) from the standard deviation and either the normal distribution or the t-distribution. The way in which standard deviations and confidence limits are calculated is described in the next chapter. If the final result is based on more than one quantity then the standard deviation of the final result depends on the individual standard deviations combined in an appropriate manner. The rules for combining standard deviations are given in Chapter 2.

How to Combine Errors

Consider two terms x and y which are used to calculate the final result. The range of the x readings is $2\Delta x$, hence the value of x is quoted as $x \pm \Delta x$. Similarly the range of the y values is $2\Delta y$, hence the value of y is quoted as $y \pm \Delta y$. The way in which x and y are combined to give the final result determines the way in which the errors Δx and Δy are combined. The derivation of the four cases for addition, subtraction, multiplication and division is given below. Readers who are only interested in applying the rules should skip the derivations and move on to the summary.

Derivation of Rules for Combining Errors

(i) *Addition*   result $= x + y$

$$(x \pm \Delta x) + (y \pm \Delta y)$$

6

The largest possible result is

$$x+y+\Delta x+\Delta y = (x+y)+(\Delta x+\Delta y)$$

The smallest possible result is

$$x+y-\Delta x-\Delta y = (x+y)-(\Delta x+\Delta y)$$

The range of the results is thus $2(\Delta x+\Delta y)$. Hence the result should be expressed as

$$(x+y)\pm(\Delta x+\Delta y) \qquad (1)$$

(ii) *Subtraction*   result $=x-y$

$$(x\pm\Delta x)-(y\pm\Delta y)$$

The largest possible result is

$$x-y+\Delta x+\Delta y = (x-y)+(\Delta x+\Delta y)$$

The smallest possible result is

$$x-y-\Delta x-\Delta y = (x-y)-(\Delta x+\Delta y)$$

The range of the result is thus $2(\Delta x+\Delta y)$. Hence the result should be expressed as

$$(x-y)\pm(\Delta x+\Delta y) \qquad (2)$$

(iii) *Multiplication*   result $=x\cdot y$

$$(x\pm\Delta x)\cdot(y\pm\Delta y)$$
$$= xy\pm x\Delta y\pm y\Delta x\pm\Delta x\Delta y$$

The second order term $\Delta x\Delta y$ is ignored since it is very small

$$\simeq xy\pm x\Delta y\pm y\Delta x$$

The largest possible result is

$$\simeq xy+(|x\Delta y|+|y\Delta x|)$$

The smallest possible result is

$$\simeq xy-(|x\Delta y|+|y\Delta x|)$$

Hence the result should be expressed as

$$xy\pm(x\Delta y+y\Delta x) \qquad (3)$$

Rather than considering the *absolute* errors Δx and Δy it is convenient to consider the *relative* errors δx and δy, where

$$\delta x = |\Delta x/x| \quad \text{and} \quad \delta y = |\Delta y/y| \qquad (4)$$

Substituting δx and δy into Equation 3

$$= xy\pm(xy\delta y+xy\delta x)$$
$$= xy\pm xy(\delta x+\delta y) \qquad (5)$$

The relative error in the product is thus $(\delta x+\delta y)=$ the sum of the relative errors.

(iv) *Division*   result $=x/y$. To simplify the derivation, it is assumed that both x and y are positive.

$$(x\pm\Delta x)/(y\pm\Delta y)$$

The largest possible result is

$$\frac{x+\Delta x}{y-\Delta y}$$

Substituting for Δx and Δy using Equations 4,

$$\frac{x+x\delta x}{y-y\delta y} = \frac{x}{y}\left(\frac{1+\delta x}{1-\delta y}\right)$$

making the approximation that $1/(1-\delta y)\simeq 1+\delta y$ for small δy

$$\simeq \frac{x}{y}(1+\delta x)(1+\delta y)$$

$$= \frac{x}{y}(1+\delta x+\delta y+\delta x\delta y)$$

The second order term $\delta x\delta y$ is ignored since it is very small.

Largest possible result $\simeq \dfrac{x}{y}+\dfrac{x}{y}(\delta x+\delta y)$

Similarly,

Smallest possible result $\simeq \dfrac{x}{y}-\dfrac{x}{y}(\delta x+\delta y)$

Hence the result of the division can be expressed as

$$\frac{x}{y}\pm\frac{x}{y}(\delta x+\delta y) \qquad (6)$$

This result applies for all combinations of x and y both positive and negative. The relative error in the quotient is thus $(\delta x+\delta y)=$ the sum of the relative errors.

Summary of Equations for Combining Errors

If the *absolute* errors in two terms x and y are Δx and Δy and the *relative* errors are $\delta x=|\Delta x/x|$ and $\delta y=|\Delta y/y|$, then

the absolute error in $(x+y)$
    is $\pm(\Delta x+\Delta y)$ (see Equation 1)
the absolute error in $(x-y)$
    is $\pm(\Delta x+\Delta y)$ (see Equation 2)
the abolute error in $x \cdot y$
    is $\pm(x\Delta y+y\Delta x)$ (see Equation 3)
    or $\pm xy(\delta x+\delta y)$ (see Equation 5)
and the absolute error in x/y
    is $\pm\dfrac{x}{y}(\delta x+\delta y)$ (see Equation 6)

These equations can easily be memorised by remembering that the *absolute* error of a sum or a difference is the sum of the *absolute* errors. The *relative* error of a product or a quotient is the sum of the *relative* errors.

Standard Error

If an experiment is repeated n times, it is likely that the final result calculated from each experiment will vary. The spread of experimental readings in one experiment is measured by the standard deviation of the readings and is discussed in Chapter 2. In an analogous manner the spread of the final results from the replicate experiments is measured by the standard error of the final results.

The first step in calculating the standard error is to evaluate the mean $\bar{R}$ of all of the n final results. The differences Δ_i between each of the final results R_i and the mean $\bar{R}$ are then evaluated. The standard error is calculated:

Standard error of final results

$$= \sqrt{\frac{\text{Sum of differences squared}}{\text{Number of terms minus one}}}$$

$$= \sqrt{\frac{\Sigma\Delta_i^2}{n-1}}$$

It can be seen that this equation is essentially the same as that given earlier for calculating standard deviations. The difference between standard errors and standard deviations is that standard errors refer to the spread of calculated values such as means, slopes or final results whereas standard deviations refer to the spread of experimentally observed results.

Confidence Limits

Having calculated a final answer, which could be a slope, a mean or any other derived quantity, it is often useful to derive a range within which the true result is 95% or 99% likely to fall. This range is called the 95% or 99% confidence limit. Confidence limits are calculated:

Confidence limits
    = Answer $\pm t \cdot$ Standard error of answer

The constant t is based on the confidence level chosen (95% or 99%) and the number of points measured. The t-distribution is discussed in Chapter 4, and a table of values is given in Appendix 6. Applications of confidence limits are found in Chapter 2.

Example 1—Errors Involved in Making a Standard Solution

In many branches of physical science it is necessary to prepare standard solutions, that is solutions whose concentrations are exactly known. Generally this involves dissolving an exactly known weight of solute in the solvent, and making the volume of the resultant solution up to an exactly known amount in a graduated flask. Ignoring any personal (human) errors, and errors due to impurities in the solute and solvent, the accuracy of the standard solution depends on the accuracy of the weighing and the accuracy of calibration of the graduated flask.

The National Physical Laboratory permits the tolerances shown in Table 1.6 on Class A standard weights. (NPL recognises no other grade of weights.) Two grades of graduated flasks Class A and Class B are commonly used. The tolerances for Class B flasks are given in Table 1.7. (Class A have approximately half these tolerances.) Concentration is calculated as weight/volume. In this example 4.2501 g of silver nitrate were weighed, dissolved in water and made up to 250 cm³ of solution in a graduated flask. The concentration is $4.2501/250$ g/cm³ $= 0.017\ 00$ g/cm³. The

Table 1.6 Tolerance of Grade A Weights

Weight/g	Tolerance/g
100	0.000 5
50	0.000 25
30	0.000 15
20	0.000 1
10–0.1	0.000 05
0.05–0.01	0.000 02

Table 1.7 Tolerance of Grade B Flasks

Volume of flask/cm³	Tolerance/cm³
1000	0.80
250	0.30
100	0.15
25	0.06
5	0.04

weight of 4.2501 g was made up of two 2 g weights, a 0.25 g weight plus the rider. The tolerance on the weight of 4.25 is thus the sum of the tolerances of the three individual weights (ignoring any error due to the rider) =

$$0.000\ 05 + 0.000\ 05 + 0.000\ 05\ g$$
$$= 0.000\ 15\ g$$

The tolerance in a Class B 250 cm³ graduated flask is 0.30 cm³. The error in the concentration is calculated as shown in the summary of equations for combining errors. The relative errors in the weight and volume respectively are

$$\frac{0.000\ 15}{4.2501} = 0.000\ 035$$

and

$$\frac{0.30}{250} = 0.0012$$

The relative error of the quotient (mass/volume = concentration) is the sum of the two relative errors

$$= 0.001\ 235$$

The absolute error in the concentration is thus

$$0.001\ 235 \times 0.017\ 00 =$$
$$0.000\ 021\ 0\ g/cm^3$$

The concentration should thus be stated as $0.017\ 00 \pm 0.000\ 02\ g/cm^3$.

The error derived in this way is the maximum possible, and usually the actual error will be smaller than this, because the individual errors may be less than their maxima, and one error may partly cancel another.

It is apparent that the relative error from the glassware exceeds that from the weights by a factor of 34, and hence the error in the weights is almost insignificant. In the laboratory, weights can be determined much more accurately than volumes.

Example 2—Errors in Volumetric Analysis

Chemists and biologists commonly carry out titrations to estimate the concentration of a solution of unknown strength. Titrations are also used to estimate the purity of a sample by preparing a solution containing an accurately known weight of sample, and performing a titration to find the weight of active ingredient actually present. Two solutions are used which will react together—a standard solution and the unknown solution. A known volume of one solution is measured with a pipette and delivered into a conical flask. The other solution is added from a burette until the equivalence point is reached, which is usually detected by a colour change in an indicator which has been added to the conical flask. Ignoring gross personal errors and indicator errors, four main errors remain:

(i) The pipette may not deliver exactly the stated volume even though filled exactly up to the calibration mark.
(ii) The burette may not deliver exactly the volume indicated by the scale due to incorrect calibration.
(iii) It is not possible to deliver less than one drop of solution from the burette. Thus the volume delivered by the burette may be up to one drop beyond the true equivalence point.

9

(iv) Observational errors when reading the burette.

Pipettes and burettes are available to Class A and Class B specifications. Typical tolerances are given in Table 1.8 and it should be noted that the tolerance of Class B apparatus is approximately double that for Class A.

Table 1.8   Tolerance of Pipettes and Burettes

Volume/ cm^3	Class A pipettes (cm^3)	Class B pipettes (cm^3)	Class A burettes (cm^3)	Class B burettes (cm^3)
100	± 0.06	± 0.12	± 0.10	± 0.20
50	0.04	0.08	0.06	0.10
25	0.03	0.06	0.04	0.08
10	0.02	0.04	0.02	0.04
5	0.02	0.03	0.02	0.03

Consider a titration using a 25 cm^3 Class A pipette and a 50 cm^3 Class A burette. The magnitudes of the four errors given above are:

Pipette calibration	± 0.03 cm^3
Burette calibration	± 0.06 cm^3
One drop	0.04 cm^3
Error in reading burette	± 0.01 cm^3

Most burettes deliver between 20 and 30 drops per cm^3 hence the size of one drop is approximately 0.04 cm^3. The excess volume of solution added from the burette will vary between 0 and 0.04 cm^3. On average the excess will be 0.02 cm^3, and this should be subtracted from the amount delivered. This corrected delivered volume is subject to an error of ± 0.02 cm^3, that is $\pm$ half a drop.

Most burettes are calibrated in intervals of 0.1 cm^3, which corresponds to about 2 mm in height. The burette can usually be read to the nearest 0.02 cm^3, thus the maximum error from a single reading is ± 0.01 cm^3.

The equation for the titration may be written

$$\begin{array}{l}\dfrac{\text{Volume of}}{\text{standard}} \times \dfrac{\text{Concentration}}{\text{of standard}} \\[2ex] = \dfrac{\text{Volume}}{\text{of unknown}} \times \dfrac{\text{Concentration}}{\text{of unknown}}\end{array}$$

Hence

$$\begin{array}{l}\dfrac{\text{Concentration}}{\text{of unknown}} = \dfrac{\text{Volume of standard}}{\text{Volume of unknown}} \\[2ex] \times \text{Concentration of standard}\end{array}$$

Assuming that the standard solution was measured using the pipette, and the unknown using the burette

Volume of standard

$= 25 \pm 0.03$ cm^3

Initial burette reading

$= 0.53 \pm 0.01$ cm^3

Final burette reading

$= 22.36 \pm 0.01$ cm^3

Difference in burette readings

$= (23.36 \pm 0.01) - (0.53 \pm 0.01)$ cm^3

$= (23.36 - 0.53) \pm (0.01 + 0.01)$ cm^3

$= 22.83 \pm 0.02$ cm^3

(The errors have been combined using Equation 2.)

The observed volume from the burette is thus 22.83 ± 0.02 cm^3, but the tolerance to which the burette is calibrated is ± 0.06 cm^3. The delivered volume from the burette must lie in the range

$$22.83 \pm 0.02 \pm 0.06 \text{ cm}^3$$
$$= 22.83 \pm 0.08 \text{ cm}^3$$

This volume will on average be too large by half a drop $= 0.02$ cm^3, and is also subject to an error of $\pm$ half a drop $= \pm 0.02$ cm^3.

The volume of unknown used to reach the equivalence point is thus

$$(22.83 \pm 0.08) - (0.02 \pm 0.02) \text{ cm}^3$$
$$= 22.81 \pm 0.10 \text{ cm}^3$$

In working out the concentration of the unknown, the volume of standard is divided by the volume of unknown. It can be seen from Equation 6 that the relative error of a quotient is the sum of the relative errors of numerator and denominator.

Relative error of numerator

= relative error of volume of standard

= 0.03/25 = 0.0012

Relative error of denominator

= Relative error of volume of unknown

0.10/22.81 = 0.0044

Total relative error

= 0.0012 + 0.0044 = 0.0056

= 0.56%

The final answer for the concentration of the unknown is thus subject to an error of 0.56% regardless of the units used. It is worth noting that the burette introduces most of the error.

Exercises

1.1 (a) Explain what is meant by the standard error of the mean.
   (b) In the two examples (i) and (ii) given below the mean of a large population is given, together with three possible values for the standard deviation.

(i) Boxes of matches are stated to contain 50 matches. The mean number per box is 51.2 matches, which standard deviation is the most reasonable: 0.5, 5.0 or 50?

(ii) The average height of adult males is 175 cm. Which standard deviation is the most reasonable: 0.7, 7.0 or 70 cm?

1.2 Calculate the density of a ball-bearing, and give the accuracy of the answer using the following data: mass = 3.8251 ± 0.001 g; diameter = 1.126 ± 0.0005 cm; volume of sphere = $\frac{4}{3}\pi r^3$.

1.3 A machine to produce bars of chocolate can be adjusted to give any average mass required, but the standard deviation is always 5 g. What average mass should the machine be set to so that fewer than

(a) 15.87%
(b) 2.27%
(c) 0.13%

of the bars have a mass less than 250 g?

2

Average and Spread of Results

Arithmetic Mean

Suppose we have a set of numbers x_1, x_2, x_3, ..., x_n which for example correspond to the total number of GCE subjects passed by all the n pupils taking the examination in the whole country. The arithmetic mean (usually called just the mean) is denoted by the Greek symbol μ and is the sum of the number of subjects passed by each student divided by the total number of students. Thus

$$\mu = \frac{x_1 + x_2 + x_3 + \ldots + x_n}{n}$$

$$= \frac{\sum_{i=1}^{i=n} x_i}{n} \qquad (1)$$

This gives an exact result because we are using all the numbers in the population rather than taking just a sample.

Median

Another measure of the pupils' achievements is given by the median value. This is calculated by arranging the n values in either descending order (largest first) or ascending order (smallest first), and choosing the middle value, that is the $\frac{1}{2}(n+1)$th value in the ordered list. If there are an even number of values, then there is no middle value in the list and the average of the middle two values is taken. Sorting a large list of numbers into order by hand is laborious and time consuming, and for these reasons some automatic sorting process is necessary.

Various methods of sorting are known. An extremely fast method known as a Shell sort is used in a computer program for calculating the median.

The median is generally much less affected by a freak value than the mean. This is illustrated by the following example.

A company has 1000 shareholders, 999 of whom have invested £10 each, and the remaining shareholder has invested £90 010. The mean value invested by each shareholder calculates as £100, but the median value is £10. Clearly in this case the median is more representative of a 'typical' shareholder.

It is found that the median tends to vary more between different samples from the same data than does the mean. The consequence of this is that the median is less reliable than the mean, hence statistical inferences are usually based on the mean rather than the median.

It is sometimes useful to compare the values for the mean and the median for a set of data. Whether these two values are nearly the same or considerably different indicates whether the data are symmetrical or skewed (lopsided) as shown in Figs. 2.1 and 2.2. A measure of how much a distribution is skewed is given by Pearson's coefficient of skewness later in the chapter.

Description of Program to Calculate the Median (see Program 2.1)

After printing a heading (lines 20–30) the user is asked whether full instructions are required (line 50). The reply, which must be YES or NO

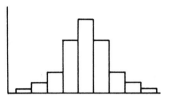 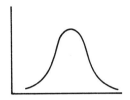

Fig. 2.1 Symmetrical distributions.

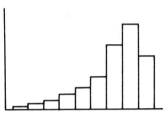 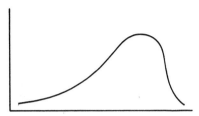

Fig. 2.2 Skewed distributions.

is checked in a subroutine (lines 490–560). The instructions given in lines 80–120 are printed only on the first run if the answer is YES.

A message (line 140) requests the number of data values, and the value typed is checked (lines 160–190) to ensure that it is an integer, and between 4 and 100 inclusive. The lower limit is necessary since the inter-quartile range is subsequently calculated, and the upper limit is imposed by the DIMension of the X array in line 10. The requested number of data values are input (lines 220–240).

Next a subroutine is called (line 260) to sort the data values into descending order. The particular method used is the 'Shell sort'.

The median value is calculated and printed (lines 270–290). If there is an odd number of data values then the middle value is selected as the median, but if there is an even number of values then the median is taken as the average of the two middle values.

The range of the numbers is printed (line 290) using the first and last of the values in the sorted list.

The inter-quartile range is calculated and printed (lines 310–360) by taking the two values corresponding to values $\frac{1}{4}$ and $\frac{3}{4}$ down the sorted list. The way in which this is done is illustrated by an example:

Values

12.5	7.9	6.3	5.5	4.3	2.8

Position in sorted list

1	2	3	4	5	6

Middle values

$$\text{Median} = \tfrac{1}{2}(6.3 + 5.5) = 5.9$$

$$\text{Range} \; = 12.5 - 2.8 = 9.7$$

Inter-quartile range = 2.25th value and 4.75th value

2.25th value

$$= \frac{(3 \times \text{second value}) + \text{third value}}{4}$$

$$= \frac{3 \times 7.9 + 6.3}{4} = 7.5$$

4.75th value

$$= \frac{\text{fourth value} + (3 \times \text{third value})}{4}$$

$$= \frac{5.5 + 3 \times 4.3}{4} = 4.6$$

Inter-quartile range

$$= 4.6 \text{ to } 7.5$$

Following this the user is offered the option of printing an ordered list of the data values (lines 380–430) and finally the option of another run with new data is given (lines 450–470).

Mode

A third measure of the pupils' achievements is given by the modal value. The mode is defined as the most frequently occurring value. The

Program 2.1   Trial run.

```
          MEDIAN, RANGE AND INTERQUARTILE RANGE
          ======  ===== === ============= =====

WOULD YOU LIKE FULL INSTRUCTIONS
TYPE YES OR NO AND PRESS RETURN.
? YES
THIS PROGRAM CALCULATES THE MEDIAN, RANGE AND INTERQUARTILE
RANGE FROM A SET OF DATA BY FIRST SORTING YOUR VALUES INTO
DESCENDING ORDER.  YOU ARE REQUIRED TO SPECIFY HOW MANY
VALUES YOU HAVE AND PRESS RETURN, AND THEN TYPE IN THE
VALUES ONE AT A TIME, PRESSING RETURN AFTER EACH VALUE.




TYPE NUMBER OF DATA VALUES
? 9
TYPE THE DATA VALUES, ONE AT A TIME.
PRESS RETURN AFTER EACH VALUE
? 6
? 4
? 3
? 7
? 5
? 8
? 1
? 9
? 2
MEDIAN VALUE = 5
RANGE OF VALUES = 1 TO 9 EQUALS 8
INTERQUARTILE RANGE = 3 TO 7

WOULD YOU LIKE A LIST OF YOUR VALUES IN DESCENDING ORDER.
TYPE YES OR NO AND PRESS RETURN.
? YES
 9
 8
 7
 6
 5
 4
 3
 2
 1

WOULD YOU LIKE ANOTHER RUN
TYPE YES OR NO AND PRESS RETURN.
? NO
```

14

```
10  DIM X(100), Q$(9)
20  PRINT TAB(15); "MEDIAN, RANGE AND INTERQUARTILE RANGE"
30  PRINT TAB(15); "====== ===== === ============= ====="
40  PRINT
50  PRINT "WOULD YOU LIKE FULL INSTRUCTIONS"
60  GOSUB 500
70  IF Q$ = "NO" THEN 130
80  PRINT "THIS PROGRAM CALCULATES THE MEDIAN, RANGE AND INTERQUARTILE"
90  PRINT "RANGE FROM A SET OF DATA BY FIRST SORTING YOUR VALUES INTO"
100 PRINT "DESCENDING ORDER.  YOU ARE REQUIRED TO SPECIFY HOW MANY"
110 PRINT "VALUES YOU HAVE AND PRESS RETURN, AND THEN TYPE IN THE"
120 PRINT "VALUES ONE AT A TIME, PRESSING RETURN AFTER EACH VALUE."
130 PRINT
140 PRINT "TYPE NUMBER OF DATA VALUES"
150 INPUT N
160 IF N <> INT(N) THEN 180
170 IF (N - 4) * (N - 100) <= 0 THEN 200
180 PRINT "THE NUMBER OF VALUES MUST BE AN INTEGER BETWEEN 4 & 100"
190 GOTO 140
200 PRINT "TYPE THE DATA VALUES, ONE AT A TIME."
210 PRINT "PRESS RETURN AFTER EACH VALUE"
220 FOR I = 1 TO N
230    INPUT X(I)
240 NEXT I
250 REM CALL SHELL SORT SUBROUTINE
260 GOSUB 1000
270 LET M = INT((N + 1.001) / 2)
280 LET M1 = (N + 1) / 2 - M
290 PRINT "MEDIAN VALUE ="; (1 - M1) * X(M) + M1 * X(M + 1)
300 PRINT "RANGE OF VALUES ="; X(N); "TO"; X(1); "EQUALS"; X(1) - X(N)
310 LET Q = INT((3 * N + 1.001) / 4)
320 LET Q1 = (3 * N + 1) / 4 - Q
330 PRINT "INTERQUARTILE RANGE ="; (1 - Q1) * X(Q) + Q1 * X(Q + 1);
340 LET Q = INT((N + 3.001) / 4)
350 LET Q1 = (N + 3) / 4 - Q
360 PRINT "TO"; (1 - Q1) * X(Q) + Q1 * X(Q + 1)
370 PRINT
380 PRINT "WOULD YOU LIKE A LIST OF YOUR VALUES IN DESCENDING ORDER."
390 GOSUB 500
400 IF Q$ = "NO" THEN 440
410 FOR I = 1 TO N
420 PRINT X(I)
430 NEXT I
440 PRINT
450 PRINT "WOULD YOU LIKE ANOTHER RUN"
460 GOSUB 500
470 IF Q$ = "YES" THEN 130
480 STOP
490 REM *** SUBROUTINE TO CHECK YES/NO ANSWERS
500 PRINT "TYPE YES OR NO AND PRESS RETURN."
510 INPUT Q$
520 IF Q$ = "YES" THEN 560
530 IF Q$ = "NO" THEN 560
540 PRINT "REPLY '"; Q$; "' NOT UNDERSTOOD. RE-";
550 GOTO 500
560 RETURN
990 REM *** SUBROUTINE TO PERFORM SHELL SORT
1000 LET L = N
1010 LET L = INT(L / 2)
1020 IF L = 0 THEN 1180
```

```
1030 LET M = N - L
1040 FOR I = 1 TO M
1050   LET J = I
1060   LET J2 = J + L
1070   IF X(J) >= X(J2) THEN 1160
1080   LET T = X(J)
1090   LET X(J) = X(J2)
1100   LET X(J2) = T
1140   LET J = J - L
1150   IF J > 0 THEN 1060
1160 NEXT I
1170 GOTO 1010
1180 RETURN
1190 END
```

relation between the mean, median and mode is shown in Fig. 2.3.

If the distribution is symmetrical then the mean, median and mode all coincide, but if the distribution is skewed the values differ. The mean and the centre of gravity of the distribution coincide, the median bisects the area of the distribution while the mode cuts the highest point on the distribution. The value obtained for the mode is not very

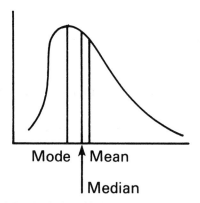

Mode   Mean

Median

Fig. 2.3   Relationship between mean, median and mode.

reliable unless a continuous distribution is used, or if a very large number of discrete data values are used which approximate to a continuous distribution. For this reason an approximation is often used to calculate the mode for discrete data.

$$\text{Mean} - \text{Mode} \simeq 3\,(\text{Mean} - \text{Median})$$

Hence

$$\text{Mode} \simeq \text{Mean} - 3\,(\text{Mean} - \text{Median})$$
$$\simeq (3 \times \text{Median}) - (2 \times \text{Mean})$$

Skewness

Skewed distributions are shown in Fig. 2.2. Two features describe how skewed a distribution is:

(i) the direction of the skew;
(ii) the magnitude of the skew.

The direction of the skew may be either negative as in Fig. 2.2, or positive as in Fig. 2.3. It can be seen that the positive skew has the longer tail on the right while a negative skew has the longer tail on the left.

There are many methods of expressing the magnitude of the skew, and the most commonly used is Pearson's coefficient of skewness:

$$\text{Skewness} = \frac{\text{Mean} - \text{Mode}}{\text{Standard deviation}}$$

(An explanation of what standard deviations are, and how they are calculated is given later in this chapter.)

The drawback of this equation for skewness is that the calculated value for the mode is often unreliable. For this reason the mode is replaced using the empirical relationship

$$\text{Mean} - \text{Mode} \simeq 3\,(\text{Mean} - \text{Median})$$

16

Thus skewness is often calculated as

$$\text{Skewness} = \frac{3\,(\text{Mean} - \text{Median})}{\text{Standard deviation}}$$

Not only does the magnitude of the number indicate the degree of skew, but the sign (+ or −) of the number indicates the direction (positive or negative) of the skew.

Range, Variance and Standard Deviation of a Set of Values

While the mean and median are useful measures of the students' achievements, they give no indication as to the spread of results, and some students will have passed in 10 subjects whereas others will have passed in only one.

The simplest way of expressing the spread or precision of a set of results is to quote the range of the numbers, that is the largest value and the smallest value, or alternatively the difference between them. This quantity is not representative since it only takes account of two values from the set and may be badly distorted by a single abnormal value. A method of avoiding this problem is to use the 'inter-quartile range'. For this one arranges the values in ascending order and chooses the values one-quarter and three-quarters down the list. (For a normal distribution the inter-quartile range represents 1.35 standard deviations. The meaning of this should become apparent later.)

A better measure of the spread or precision of the set of results is given by the variance s^2, and the standard deviation s. The variance is defined as the sum of the squared differences between each of the terms and the mean value. Thus:

$$s^2 = \frac{(x_1-\mu)^2+(x_2-\mu)^2+(x_3-\mu)^2+\ldots+(x_n-\mu)^2}{n}$$

$$s^2 = \frac{\sum_{i=1}^{i=n}(x_i-\mu)^2}{n} \qquad (2)$$

It should be noted that each of the differences is squared before it is summed. This is necessary since some of the differences are positive and some are negative, and if just the differences were added then their total would be zero as a direct consequence of the definition of the mean.

The standard deviation s is given by

$$s = \sqrt{\left[\frac{\Sigma(x_i-\mu)^2}{n}\right]} \qquad (3)$$

Clearly a large variance or a large standard deviation corresponds to a wide spread of results, and small values for s^2 and s correspond to closely grouped results. Consider a very much simplified example of five pupils who obtained passes in 6, 9, 3, 7 and 5 subjects respectively. The mean

$$\mu = \frac{6+9+3+7+5}{5} = \frac{30}{5} = 6$$

The variance is evaluated

$$s^2 = \frac{\Sigma\,(6-6)^2+(9-6)^2+(3-6)^2+(7-6)^2+(5-6)^2}{5}$$

$$= \frac{0+9+9+1+1}{5} = 4$$

hence the standard deviation

$$s = \sqrt{4} = 2$$

This method of evaluating the variance is long-winded, particularly if the number of terms is large, and the calculation may be simplified. Starting with Equation 2

$$s^2 = \frac{\sum_{i=1}^{i=n}(x_i-\mu)^2}{n}$$

The term

$$\begin{aligned}
\Sigma(x_i-\mu)^2 &= \Sigma\,[(x_i-\mu)\cdot(x_i-\mu)] \\
&= \Sigma\,[x_i^2-2\,x_i\mu+\mu^2] \\
&= \Sigma x_i^2 - 2\mu\Sigma x_i + \Sigma\mu^2 \\
&= \Sigma x_i^2 - 2\mu\,n\mu + n\mu^2 \\
&= \Sigma x_i^2 - n\mu^2
\end{aligned}$$

17

hence

$$s^2 = \frac{\Sigma x_i^2 - n\mu^2}{n}$$

or

$$s^2 = \frac{\Sigma x_i^2}{n} - \mu^2 \qquad (4)$$

$$s^2 = \frac{\Sigma x_i^2}{n} - \frac{(\Sigma x_i)^2}{n^2} \qquad (5)$$

Equations 4 and 5 are both much simpler for the calculation of the variance by hand, but Equation 4 squares any rounding error produced in calculating the mean μ, hence Equation 5 is to be preferred. This equation is always used on calculators because the individual values need not be stored as they are entered, but only the sum of the values Σx_i, the sum of the values squared Σx_i^2 and the number of terms must be collected. However, with a computer Equation 2 should be used because it is less prone to numerical rounding errors. The use of this equation requires that the numbers are stored in an array, the mean is calculated, and then the individual differences are calculated, squared and summed. Equation 2 gives the most accurate answer. Equations 4 and 5 are potentially worse since they involve the subtraction of two numbers which are approximately equal for small variances. This always results in a loss of numerical accuracy, and most acute when a limited number of significant figures are carried.

Mean, Variance and Standard Deviation Using a Sample of Readings

In most statistical work the population is very large, and it is usual to work with a sample of readings rather than the entire population. Similarly in scientific experiments one performs a limited number of readings such as titrations to find the concentration of a solution. The readings obtained are random samples from the entire population, and the mean of the sample, the variance of the sample and the standard deviation *of the sample* can be

calculated as before. From these one would like to obtain the best estimate of the mean, variance and standard deviation *of the whole population*.

The mean of the sample is denoted by $\bar{x}$, and this gives the best estimate of the mean of the whole population μ

$$\bar{x} = \frac{\Sigma x_i}{n} = \text{Estimated value for } \mu$$

However, the sample variance s^2 underestimates the variance of the entire population σ^2, and for a sample of n terms the best estimate of the population variance is obtained using $(n-1)$ as the divisor rather than n as in Equation 2.

$$\text{Estimated } \sigma^2 = \frac{\Sigma (x_i - \bar{x})^2}{n-1} \qquad (6)$$

and in a similar way the standard deviation of the population is estimated by:

$$\text{Estimated } \sigma = \sqrt{\left[\frac{\Sigma (x_i - \bar{x})^2}{n-1} \right]} \qquad (7)$$

This equation is generally used with a computer, but for hand calculations it is often simplified:

$$\text{Estimated } \sigma = \sqrt{\left[\frac{\Sigma x^2}{(n-1)} - \frac{(\Sigma x)^2}{n(n-1)} \right]} \qquad (8)$$

When the sample size is large, the estimate of σ using the divisor $(n-1)$ is almost the same as that obtained using the divisor n. With a sample of 50 the difference between the true σ and the estimated σ is less than 1%, and for a sample size of 100 the difference is less than 0.5%. For large samples of perhaps 30 or more it is customary to use the divisor n.

Explanation of the Divisors n and $(n-1)$

Equation 2 for the standard deviation s of a set of numbers uses the divisor n. Equation 7 for the estimated standard deviation σ based on a random sample from the whole population uses the divisor $(n-1)$. The difference between these equations may be explained (rather than proved) in the following manner.

If the whole population is used, then the true mean μ is known, and the standard deviation can be calculated using the n divisor. If the true mean is not known (as is the case when using a sample of readings) then the estimated mean $\bar{x}$ must be used instead. However, the sum of all the differences $(x_i - \bar{x})$ is zero by the definition of $\bar{x}$. It follows that if $(n-1)$ of the differences are given, then the last $(n$th$)$ difference can be calculated. There are thus only $(n-1)$ degress of freedom when using a sample to estimate the standard deviation of the entire parent population. Degrees of freedom are further discussed in Chapter 5.

Calculation of Confidence Limits

The probability that the mean of the whole population μ lies outside the 95% confidence limits is 5%. Similarly the probability that μ lies outside the 99% confidence limits is 1%. When there is only a small sample of data values, the confidence limits are calculated using the t-distribution as follows:

$$\text{Confidence limit} = \bar{x} \pm \frac{t \cdot \sigma}{\sqrt{n}}$$

where $\bar{x}$ is the mean of the sample, t is a constant based on the confidence level chosen and the number of points, σ is the estimated standard deviation of the sample and n is the number of data values. The value of t can be obtained from the table in Appendix 6. To use this one must select the appropriate confidence level and also the number of degrees of freedom v. In this case the number of degrees of freedom is $(n-1)$. Degrees of freedom are fully explained in Chapter 5. It should be noted that as v becomes large the t-distribution tends towards the normal distribution. For this reason it is common to use the individual t values for a small number of points but to use the infinity (normal) value for a large number of points (i.e. $n > = 30$).

Description of Standard Deviation Program (Program 2.2)

After printing a heading, the user is asked if full instructions are required. The reply must be YES or NO, and is checked by a subroutine (lines 1890–1980). No other reply is accepted. Full or abbreviated instructions are printed accordingly, but if the program is re-run then abbreviated instructions will always be given.

The user is prompted to type in the data values one at a time followed by pressing RETURN. The X array in the program can hold up to 100 values. Normally there will be fewer entries than this, and a dummy value of 999999 is typed to indicate the end of data input.

A check is then performed to ensure that at least one valid line of data has been typed (lines 230–250). Next a subroutine (lines 1140–1880) is called to print and edit the data if necessary. The user is asked if the data are correct, and if not the data are listed and instructions given to allow insertion and deletion of lines, replacement of existing values and re-listing the current data. Extensive checks are carried out to ensure that changes are valid. This subroutine is described more fully in Chapter 8.

A check is performed to ensure that at least two data values remain after editing. The average value is calculated and printed (lines 330–410). The sum of the differences (between the values and their average value) squared is accumulated, and the variance and standard deviation are calculated and printed (lines 420–570). It is worth mentioning that the divisor $(n-1)$ is used when there are less than 30 points—otherwise the divisor is n.

The user is then asked if confidence limits are required. If they are needed, they are calculated in the following manner:

(i) First the confidence limits are calculated (lines 630–650) using the infinity values corresponding to the 95% and 99% values for the normal distribution.

(ii) Thirty t values for the 95% confidence

limit are read from a table of DATA values stored in the program (lines 660–750). If there are more than 30 readings then the infinity value is retained, otherwise a new confidence limit is calculated from the t value appropriate for the number of readings (line 700).

(iii) The 99% confidence limit is calculated in a similar way (lines 760–850).

(iv) The confidence limits are printed (lines 860–910).

The user is then asked if another run is required. If so, a choice is given between typing in a completely new set of data, or editing the old (existing) data (lines 920–1090). If another run is not required then the run is terminated and a finishing message is printed (lines 1100–1130).

Program 2.2 Trial run.

```
STANDARD DEVIATION CALCULATION
======== ========= ===========

WOULD YOU LIKE FULL INSTRUCTIONS?
 TYPE YES OR NO & PRESS RETURN.

? YES

TYPE IN DATA VALUES. PRESS RETURN AFTER EACH TERM.
YOU WILL HAVE CHANCE TO CORRECT TYPING ERRORS LATER
TERMINATE DATA WITH THE VALUE 999999
? 25.00
? 24.99
? 25.01
? 999999
ARE THE DATA VALUES ENTERED CORRECT?   TYPE YES OR NO & PRESS RETURN.

? YES

NUMBER OF READINGS = 3

AVERAGE VALUE = 25

VARIANCE = 0.000100005

STANDARD DEVIATION = 0.0100002

WOULD YOU LIKE THE CONFIDENCE LIMITS?
 TYPE YES OR NO & PRESS RETURN.

? YES

WITH 95% CONFIDENCE THE MEAN OF THE PARENT POPULATION IS IN
THE RANGE 25 ,PLUS OR MINUS 0.024844   THAT IS 24.9752 TO 25.0248

WITH 99% CONFIDENCE THE MEAN OF THE PARENT POPULATION IS IN
THE RANGE 25 ,PLUS OR MINUS 0.0573033   THAT IS 24.9427 TO 25.0573

WOULD YOU LIKE ANOTHER RUN?
 TYPE YES OR NO & PRESS RETURN.

? NO

JOB COMPLETED.
```

20

```
10  DIM X(100), Q$(10), I$(3)
20  PRINT "STANDARD DEVIATION CALCULATION"
30  PRINT "======== ========= ==========="
40  PRINT
50  PRINT "WOULD YOU LIKE FULL INSTRUCTIONS?"
60  GOSUB  1910
70  LET I$ = Q$
80  PRINT
90  IF I$ = "YES" THEN 120
100 PRINT "INPUT DATA"
110 GOTO  140
120 PRINT "TYPE IN DATA VALUES. PRESS RETURN AFTER EACH TERM."
130 PRINT "YOU WILL HAVE CHANCE TO CORRECT TYPING ERRORS LATER"
140 PRINT "TERMINATE DATA WITH THE VALUE 999999"
150 LET N = 0
160 FOR I = 1 TO 100
170    INPUT X(I)
180    IF X(I) = 999999 THEN 230
190    LET N = N + 1
200 NEXT I
210 PRINT
220 PRINT "PROGRAM CAN ONLY HANDLE 100 VALUES."
230 IF N > 0 THEN 270
240 PRINT "PLEASE TYPE IN SOME DATA"
250 GOTO 160
260 REM CALL SUBROUTINE TO CHECK & EDIT DATA
270 GOSUB 1150
280 REM ABANDON RUN IF LESS THAN 2 TERMS
290 PRINT
300 IF N >= 2 THEN 340
310 PRINT "WITH ONLY ONE DATA POINT THE STANDARD DEVIATION MUST BE ZERO"
320 GOTO  920
330 REM WORK OUT AVERAGE VALUE
340 LET A = 0
350 FOR I = 1 TO N
360    LET A = A + X(I)
370 NEXT I
380 LET A1 = A / N
390 PRINT "NUMBER OF READINGS ="; N
400 PRINT
410 PRINT "AVERAGE VALUE ="; A1
420 REM WORK OUT THE DIFFERENCES SQUARED BETWEEN EACH MARK AND AVERAGE,
430 REM AND COLLECT TOTAL IN D.
440 LET D = 0
450 FOR I = 1 TO N
460    LET D1 = X(I) - A1
470    LET D = D + (D1 * D1)
480 NEXT I
490 REM WORK OUT VARIANCE V & STANDARD DEVIATION S
500 LET V = D / (N - 1)
510 IF N < 30 THEN 530
520 LET V = D / N
530 PRINT
540 PRINT "VARIANCE ="; V
550 PRINT
560 LET S = SQR(V)
570 PRINT "STANDARD DEVIATION ="; S
580 REM DECIDE WHETHER TO WORK OUT CONFIDENCE LIMITS
590 PRINT
600 PRINT "WOULD YOU LIKE THE CONFIDENCE LIMITS?"
610 GOSUB  1900
```

```
620 IF Q$ = "NO" THEN 920
630 REM CALCULATE CONFIDENCE LIMITS USING INFINITY VALUES
640 LET C = 1.95996 * S / SQR(N)
650 LET C1 = 2.57582 * S / SQR(N)
660 REM RE-CALCULATE 95% CONFIDENCE LIMIT IF N<= 30
670 FOR I = 1 TO 30
680    READ T
690    IF I <> N THEN 710
700    LET C = T * S / SQR(N)
710 NEXT I
720 DATA 0, 12.706, 4.303, 3.182, 2.776, 2.571, 2.447, 2.365
730 DATA 2.306, 2.262, 2.228, 2.201, 2.197, 2.160, 2.145, 2.131
740 DATA 2.120, 2.110, 2.101, 2.093, 2.086, 2.080, 2.074, 2.069
750 DATA 2.064, 2.060, 2.056, 2.052, 2.048, 2.045
760 REM RE-CALCULATE 99% CONFIDENCE LIMIT IF N <= 30
770 FOR I = 1 TO 30
780    READ T
790    IF I <> N THEN 810
800    LET C1 = T * S / SQR(N)
810 NEXT I
820 DATA 0, 63.657, 9.925, 5.841, 4.604, 4.032, 3.707, 3.499
830 DATA 3.355, 3.250, 3.169, 3.106, 3.055, 3.012, 2.977, 2.947
840 DATA 2.921, 2.898, 2.878, 2.861, 2.845, 2.831, 2.819, 2.807
850 DATA 2.797, 2.787, 2.779, 2.771, 2.763, 2.756
860 PRINT
870 PRINT "WITH 95% CONFIDENCE THE MEAN OF THE PARENT POPULATION IS IN"
880 PRINT "THE RANGE";A1;",PLUS OR MINUS";C;"  THAT IS";A1-C;"TO";A1+C
890 PRINT
900 PRINT "WITH 99% CONFIDENCE THE MEAN OF THE PARENT POPULATION IS IN"
910 PRINT "THE RANGE";A1;",PLUS OR MINUS";C1;" THAT IS";A1-C1;"TO";A1+C1
920 PRINT
930 PRINT "WOULD YOU LIKE ANOTHER RUN?"
940 GOSUB 1900
950 IF Q$ = "NO" THEN 1110
960 RESTORE
970 LET I$ = "NO"
980 PRINT "TYPE NEW FOR A RUN WITH COMPLETELY NEW DATA"
990 PRINT "  OR OLD TO EDIT AND RERUN THE EXISTING DATA"
1000 INPUT Q$
1010 IF Q$ = "NEW" THEN 1050
1020 IF Q$ = "OLD" THEN 1080
1030 PRINT "REPLY '"; Q$; "' NOT UNDERSTOOD"
1040 GOTO 980
1050 PRINT "TYPE IN A NEW SET OF DATA"
1060 PRINT "==== == = === === == ===="
1070 GOTO 140
1080 GOSUB 1200
1090 GOTO 290
1100 REM TERMINATE JOB
1110 PRINT
1120 PRINT "JOB COMPLETED."
1130 STOP
1140 REM SUBROUTINE TO CHECK THAT DATA ARE CORRECT & ALTER IF NECESSARY
1150 PRINT "ARE THE DATA VALUES ENTERED CORRECT?";
1160 REM A4 SHOULD BE SET TO THE NUMBER OF LINES ON THE VDU
1170 LET A4 = 20
1180 GOSUB 1900
1190 IF Q$ = "YES" THEN 1880
1200 PRINT "HERE IS A LIST OF THE CURRENT DATA"
1210 PRINT "LINE NUMBER", "X"
1220 FOR I = 1 TO N
```

```
1230    PRINT I, X(I)
1240    IF INT(I / (A4 - 1)) * (A4 - 1) <> I THEN 1280
1250    PRINT "WOULD YOU LIKE TO CONTINUE LISTING";
1260    GOSUB 1900
1270    IF Q$ = "NO" THEN 1290
1280 NEXT I
1290 PRINT "TYPE R TO REPLACE";
1300 IF I$ = "NO" THEN 1320
1310 PRINT " AN EXISTING LINE OF DATA"
1320 IF N = 100 THEN 1370
1330 PRINT TAB(5); " A TO ADD";
1340 IF I$ = "NO" THEN 1360
1350 PRINT " AN EXTRA LINE"
1360 IF N = 1 THEN 1400
1370 PRINT TAB(5); " D TO DELETE";
1380 IF I$ = "NO" THEN 1400
1390 PRINT " AN EXISTING LINE"
1400 PRINT TAB(5); " L TO LIST";
1410 IF I$ = "NO" THEN 1430
1420 PRINT " THE DATA"
1430 PRINT "   OR C TO CONTINUE";
1440 IF I$ = "NO" THEN 1460
1450 PRINT " THE CALCULATION"
1460 INPUT Q$
1470 IF Q$ = "R" THEN 1570
1480 IF N = 100 THEN 1510
1490 IF Q$ = "A" THEN 1690
1500 IF N = 1 THEN 1520
1510 IF Q$ = "D" THEN 1740
1520 IF Q$ = "L" THEN 1200
1530 IF Q$ = "C" THEN 1880
1540 PRINT "REPLY '"; Q$; "' NOT UNDERSTOOD."
1550 GOTO 1290
1560 REM REPLACE LINE
1570 PRINT "TYPE THE LINENUMBER OF THE LINE TO BE REPLACED";
1580 INPUT I
1590 IF I <> INT(I) THEN 1610
1600 IF (I - 1) * (I - N) <= 0 THEN 1640
1610 PRINT "LINENUMBER MUST BE AN INTEGER IN THE RANGE 1 -"; N
1620 PRINT "RE-";
1630 GOTO 1570
1640 PRINT "TYPE THE CORRECT LINE TO REPLACE THE ONE WHICH IS WRONG:"
1650 PRINT "X"
1660 INPUT X(I)
1670 GOTO 1720
1680 REM ADD A NEW LINE
1690 LET N = N + 1
1700 PRINT "TYPE THE ADDITIONAL LINE OF DATA AS SHOWN:    X"
1710 INPUT X(N)
1720 PRINT "OK"
1730 GOTO 1290
1740 REM DELETE A LINE
1750 PRINT "TYPE THE LINENUMBER OF THE LINE TO BE DELETED"
1760 INPUT J
1770 IF (J - 1) * (J - N) > 0 THEN 1790
1780 IF J = INT(J) THEN 1810
1790 PRINT "LINENUMBER MUST BE AN INTEGER IN THE RANGE 1 -"; N
1800 GOTO 1750
1810 FOR I = J + 1 TO N
1820    LET X(I - 1) = X(I)
1830 NEXT I
```

```
1840 LET N = N - 1
1850 PRINT "OK"
1860 IF J > N THEN 1290
1870 GOTO 1200
1880 RETURN
1890 REM SUBROUTINE TO CHECK REPLIES
1900 IF I$ = "NO" THEN 1920
1910 PRINT " TYPE YES OR NO & PRESS RETURN."
1920 PRINT
1930 INPUT Q$
1940 IF Q$ = "YES" THEN 1980
1950 IF Q$ = "NO" THEN 1980
1960 PRINT "REPLY '"; Q$; "' NOT UNDERSTOOD.";
1970 GOTO 1910
1980 RETURN
1990 END
```

How to Combine Standard Deviations

Consider two independent terms x and y with standard deviation σ_x and σ_y. If x and y are used to calculate the final result z, then the standard deviation of z is σ_z.

Addition $z = x+y$

$$\text{then } \sigma_z = \sqrt{(\sigma_x^2 + \sigma_y^2)}$$

This is obtained directly from Appendix 8, Equation 4.

Subtraction $z = x-y$

$$\text{then } \sigma_z = \sqrt{(\sigma_x^2 + \sigma_y^2)}$$

This is obtained directly from Appendix 8, Equation 5.

Multiplication $z = x \cdot y$

$$\text{then } \frac{\sigma_z}{z} = \sqrt{\left(\frac{\sigma_x^2}{x^2} + \frac{\sigma_y^2}{y^2}\right)}$$

This is obtained directly from Appendix 8, Equation 6.

Division $z = x/y$

$$\text{then } \frac{\sigma_z}{z} = \sqrt{\left(\frac{\sigma_x^2}{x^2} + \frac{\sigma_y^2}{y^2}\right)}$$

This is obtained directly from Appendix 8, Equation 7.

The equations for addition and subtraction are the same, and the equations for multiplication and division are also the same.

Coefficient of Variation

The precision of a final result depends on the spread of the measurements. One common measure of precision is called the coefficient of variation, and compares the spread of the measurements with their magnitude:

Coefficient of variation =

$$\frac{\text{Standard deviation of measurements}}{\text{Mean value of measurements}} \times 100$$

This is arguably a more useful measure of precision than the standard deviation, because it takes into acount the magnitude of the numbers. As with the standard deviation, a low value for the coefficient of variation corresponds to high precision while a high value corresponds to low precision.

Mean, Variance and Standard Deviation Using Grouped Data

In many experiments which yield a large number of data values it is convenient to record the results in groups, each of which covers a range of values, rather than discrete values. For example a survey of blood alcohol

24

readings measured in a forensic laboratory over a large period of time might be grouped into bands of 10 or 20 mg per 100 ml of sample. In other cases the large number of readings involved makes grouping almost unavoidable—for example a national survey of the annual salaries of the 50 million inhabitants of Great Britain.

It is usual practice to collect the data in groups rather than as the exact values.

Consider a simple example where the Forestry Commission measured the girths of a set of trees. Data are collected into groups using tally marks (Table 2.1). To calculate the mean girth one uses the formula:

$$\text{Mean } \bar{x} = \frac{\Sigma\,[x_i \cdot f(x_i)]}{n} \qquad (9)$$

where x_i is the mid-point of the range of the girth of the tree and $f(x_i)$ is the frequency (number of occurrences) of the girth x_i. The number of readings n is equal to the sum of the frequencies

$$n = \Sigma\,f(x_i)$$

Using Equation 9 the mean $\bar{x}$ is evaluated:

$$\bar{x} = \frac{[(30 \times 3) + (50 \times 13) + (70 \times 22) +}{3 + \quad 13 + \quad 22 +}$$

$$\frac{+ (90 \times 5) + (110 \times 7)]}{+ \quad 5 + \quad 7} \text{ cm}$$

$$\bar{x} = \frac{3500}{50} \text{ cm} = 70 \text{ cm}$$

On average in the grouping, there will be as many values rounded up to the mid-point of the range as there are values which are rounded down. The positive and negative errors thus introduced should cancel each other.

The appropriate equation for the standard deviation for a grouped sample differs from Equation 7 only in that it allows for the frequency in each band.

$$\text{Estimated } \sigma = \sqrt{\left\{ \frac{\Sigma\,[(x_i - \bar{x})^2 \cdot f(x_i)]}{n-1} \right\}} \qquad (10)$$

Equation 7 may be considered as a special case for Equation 10 where all the frequencies are one. With grouped data, Equation 10 gives too high a value for the estimated standard deviation. This is because the error terms $(x_i - \bar{x})$ are squared, making all of the rounding errors, produced by grouping, into positive numbers. On average each $(x_i - \bar{x})^2$ term is too large by an amount of $c^2/12$ for each point, where c is the range of the group. In the case where all the groups have the same range (as in the example of tree girths) the equation becomes:

$$\text{Estimated } \sigma$$

$$= \sqrt{\left\{ \frac{\Sigma\,[(x_i - \bar{x})^2 \cdot f(x_i)]}{n-1} - \frac{c^2}{12} \right\}} \qquad (11)$$

This is known as Sheppard's correction.

Table 2.1

Range of girths (cm)	Mid-point of range x_i (cm)	Tally	Frequency $f(x_i)$
20 to 40	30	111	3
40 to 60	50	╫╫ ╫╫ 111	13
60 to 80	70	╫╫ ╫╫ ╫╫ ╫╫ 11	22
80 to 100	90	╫╫	5
100 to 120	110	╫╫ 11	7
			Σ50 = number of readings n

Using Equation 11 the standard deviation of tree girths is

Estimated σ

$$= \sqrt{\left[\frac{(30-70)^2 \cdot 3}{50} + \frac{(50-70)^2 \cdot 13}{50} + \right.}$$

$$+ \frac{(70-70)^2 \cdot 22}{50} + \frac{(90-70)^2 \cdot 5}{50} +$$

$$\left. + \frac{(110-70)^2 \cdot 7}{50} - \frac{20^2}{12}\right]$$

$$= \sqrt{\left(\frac{4800}{50} + \frac{5200}{50} + \frac{0}{50} + \frac{2000}{50} + \right.}$$

$$\left. + \frac{11\ 200}{50} - \frac{400}{12}\right)$$

$$= \sqrt{\frac{23\ 200}{50} - \frac{400}{12}} = 20.75\ \text{cm}$$

It should be noted that the number of terms n has been used as the denominator rather than $n-1$. This is commonly done when the number of terms is large as discussed previously. In this book 'large' is taken to mean 30 or more. Without Sheppard's correction the estimated value for $\sigma = 21.54$ cm, hence the correction decreases the value by almost 4%. It is worth mentioning that the change in divisor only changes the estimated standard deviation by about 1%.

Description of Standard Deviation Program for Grouped Data (Program 2.3)

First a heading is printed and the user is asked if full instructions are required. The reply, which must be YES or NO is checked by a subroutine (lines 2100–2190), and either long or short instructions are printed.

A message invites the user to input data in the form: lower limit of range, higher limit of range, number of readings in the range (frequency of range)—followed by pressing RETURN. The data input loop extends from lines 190–230. A number of checks are performed in a subroutine (lines 2200–2330) to ensure that the values typed are reasonable:

(i) The frequency must be a whole number.
(ii) The frequency must be positive or zero.
(iii) The high limit must be greater than or equal to the low limit.

The end of data input is signalled by typing a dummy line of 0, 0, 0 and pressing RETURN. The arrays L, H and F are dimensioned at 100 and restrict the maximum number of lines of data to 100. A check is performed to ensure that some valid data have been entered before the terminator. When data input is complete a subroutine is called (lines 1330–2090) which asks if the data entered are correct, and if necessary it permits the addition, deletion or replacement of lines as well as the option of listing the current data. This subroutine is described more fully in Chapter 8.

If at this stage there is only one data range, the program reports that the standard deviation must be zero, and then branches to the end of the program. Usually there will be more than one data range, and the program works out the average reading (lines 360–440). A check is performed to ensure that there are at least two readings since the estimated standard deviation cannot be calculated from a single reading.

The standard deviation is then calculated in two ways (lines 610–750), that is without and with Sheppard's group correction. Both results are printed. If the number of readings is less than 30 then the divisor used is the number of readings minus one, otherwise it is the number of readings. In the unlikely case of the correction exceeding the sum of differences squared the standard deviation is set to zero and a warning message is printed.

Provided that the standard deviation is greater than zero, the user is given the option of calculating confidence limits based on the t-distribution. This is fully discussed in the description of the (non-grouped) standard deviation program.

Finally the user is asked if another run is required, and if so whether completely new data are to be input or whether the existing data are to be edited and re-run.

Programm 2.3 Trial run.

```
STANDARD DEVIATION CALCULATION FOR GROUPED DATA
======== ========= =========== === ======= ====

WOULD YOU LIKE FULL INSTRUCTIONS?
 TYPE YES OR NO & PRESS RETURN.

? YES

TYPE THREE VALUES ON ONE LINE SEPARATED BY COMMAS.  THESE
ARE:  LOW LIMIT OF RANGE, HIGH LIMIT OF RANGE, FREQUENCY
THEN PRESS RETURN & TYPE THE NEXT LINE, RETURN ETC.
YOU WILL HAVE THE CHANCE TO CORRECT TYPING ERRORS LATER
TERMINATE DATA WITH THE DUMMY VALUES  0, 0, 0
LOW LIMIT, HIGH LIMIT, FREQUENCY
? 20, 40, 3
? 40, 60, 13
? 60, 80, 22
? 80, 100, 5
? 100, 120, 7
? 0, 0, 0
ARE THE DATA VALUES ENTERED CORRECT?  TYPE YES OR NO & PRESS RETURN.

? YES

NUMBER OF GROUPS = 5

NUMBER OF READINGS = 50

AVERAGE VALUE = 70

STANDARD DEVIATION = 21.5407

WITH THE GROUP CORRECTION THE BEST ESTIMATE FOR
THE STANDARD DEVIATION IS 20.7525

WOULD YOU LIKE THE CONFIDENCE LIMITS?
 TYPE YES OR NO & PRESS RETURN.

? YES

WITH 95% CONFIDENCE THE MEAN OF THE PARENT POPULATION IS IN
THE RANGE 70 PLUS OR MINUS 5.75218  THAT IS 64.2478 TO 75.7522

WITH 99% CONFIDENCE THE MEAN OF THE PARENT POPULATION IS IN
THE RANGE 70 PLUS OR MINUS 7.55964  THAT IS 62.4404 TO 77.5596

WOULD YOU LIKE ANOTHER RUN?
 TYPE YES OR NO & PRESS RETURN.

? NO

JOB COMPLETED.
```

```
10 DIM F(100), H(100), L(100), Q$(10), I$(3)
20 PRINT "STANDARD DEVIATION CALCULATION FOR GROUPED DATA"
30 PRINT "======== ========= =========== === ======= ===="
40 PRINT
50 PRINT "WOULD YOU LIKE FULL INSTRUCTIONS?"
60 GOSUB 2120
70 LET I$ = Q$
80 PRINT
90 IF I$ = "YES" THEN 120
100 PRINT "INPUT DATA"
110 GOTO 160
120 PRINT "TYPE THREE VALUES ON ONE LINE SEPARATED BY COMMAS.  THESE"
130 PRINT "ARE:  LOW LIMIT OF RANGE, HIGH LIMIT OF RANGE, FREQUENCY"
140 PRINT "THEN PRESS RETURN & TYPE THE NEXT LINE, RETURN ETC."
150 PRINT "YOU WILL HAVE THE CHANCE TO CORRECT TYPING ERRORS LATER"
160 PRINT "TERMINATE DATA WITH THE DUMMY VALUES  0, 0, 0"
170 PRINT "LOW LIMIT, HIGH LIMIT, FREQUENCY"
180 LET N = 0
190 FOR I = 1 TO 100
200    GOSUB 2220
210    IF ABS(L(I)) + ABS(H(I)) + F(I) = 0 THEN 260
220    LET N = N + 1
230 NEXT I
240 PRINT
250 PRINT "PROGRAM CAN ONLY HANDLE 100 VALUES."
260 IF N > 0 THEN 300
270 PRINT "PLEASE ENTER SOME DATA VALUES"
280 GOTO 120
290 REM CALL SUBROUTINE TO CHECK & EDIT DATA
300 GOSUB 1340
310 REM ABANDON RUN IF LESS THAN 2 TERMS
320 PRINT
330 IF N >= 2 THEN 370
340 PRINT "WITH ONLY ONE DATA RANGE THE STANDARD DEVIATION MUST BE ZERO"
350 GOTO 1110
360 REM WORK OUT AVERAGE VALUE & CORRECTION FACTOR C
370 LET A = 0
380 LET N1 = 0
390 LET C = 0
400 FOR I = 1 TO N
410    LET A = A + (L(I) + H(I)) / 2 * F(I)
420    LET C = C + (H(I) - L(I)) * (H(I) - L(I)) * F(I) / 12
430    LET N1 = N1 + F(I)
440 NEXT I
450 IF N1 >= 2 THEN 480
460 PRINT "THERE MUST BE AT LEAST TWO READINGS"
470 GOTO 1110
480 LET A1 = A / N1
490 PRINT "NUMBER OF GROUPS ="; N
500 PRINT
510 PRINT "NUMBER OF READINGS ="; N1
520 PRINT
530 PRINT "AVERAGE VALUE ="; A1
540 REM WORK OUT THE DIFFERENCES SQUARED BETWEEN EACH MARK AND AVERAGE,
550 REM AND COLLECT TOTAL IN D.
560 LET D = 0
570 FOR I = 1 TO N
580    LET D1 = (H(I) + L(I)) / 2 - A1
590    LET D = D + (D1 * D1) * F(I)
600 NEXT I
610 REM WORK OUT STANDARD DEVIATION S1
```

```
620 LET S1 = SQR(D / (N1 - 1))
630 LET S = (D - C) / (N1 - 1)
640 IF N1 < 30 THEN 670
650 LET S1 = SQR(D / N1)
660 LET S = (D - C) / N1
670 PRINT
680 PRINT "STANDARD DEVIATION ="; S1
690 PRINT
700 IF S >= 0 THEN 730
710 PRINT "SHEPPARDS CORRECTION IS SO LARGE THAT"
720 LET S = 0
730 LET S = SQR(S)
740 PRINT "WITH THE GROUP CORRECTION THE BEST ESTIMATE FOR"
750 PRINT "THE STANDARD DEVIATION IS"; S
760 REM DECIDE WHETHER TO WORK OUT CONFIDENCE LIMITS
770 PRINT
780 IF S = 0 THEN 1110
790 PRINT "WOULD YOU LIKE THE CONFIDENCE LIMITS?"
800 GOSUB 2110
810 IF Q$ = "NO" THEN 1110
820 REM CALCULATE CONFIDENCE LIMITS USING INFINITY VALUES
830 LET C = 1.95996 * S / SQR(N1)
840 LET C1 = 2.57582 * S / SQR(N1)
850 REM RE-CALCULATE 95% CONFIDENCE LIMIT IF N <= 30
860 FOR I = 1 TO 30
870    READ T
880    IF I <> N1 THEN 900
890    LET C = T * S / SQR(N1)
900 NEXT I
910 DATA 0, 12.706, 4.303, 3.182, 2.776, 2.571, 2.447, 2.365
920 DATA 2.306, 2.262, 2.228, 2.201, 2.197, 2.16, 2.145, 2.131
930 DATA 2.12, 2.11, 2.101, 2.093, 2.086, 2.08, 2.074, 2.069
940 DATA 2.064, 2.06, 2.056, 2.052, 2.048, 2.045
950 REM RE-CALCULATE 99% CONFIDENCE LIMIT IF N <= 30
960 FOR I = 1 TO 30
970    READ T
980    IF I <> N1 THEN 1000
990    LET C1 = T * S / SQR(N1)
1000 NEXT I
1010 DATA 0, 63.657, 9.925, 5.841, 4.604, 4.032, 3.707, 3.499
1020 DATA 3.355, 3.25, 3.169, 3.106, 3.055, 3.012, 2.977, 2.947
1030 DATA 2.921, 2.898, 2.878, 2.861, 2.845, 2.831, 2.819, 2.807
1040 DATA 2.797, 2.787, 2.779, 2.771, 2.763, 2.756
1050 PRINT
1060 PRINT "WITH 95% CONFIDENCE THE MEAN OF THE PARENT POPULATION IS IN"
1070 PRINT "THE RANGE";A1;"PLUS OR MINUS";C;" THAT IS";A1-C;"TO";A1+C
1080 PRINT
1090 PRINT "WITH 99% CONFIDENCE THE MEAN OF THE PARENT POPULATION IS IN"
1100 PRINT "THE RANGE";A1;"PLUS OR MINUS";C1;" THAT IS";A1-C1;"TO";A1+C1
1110 PRINT
1120 PRINT "WOULD YOU LIKE ANOTHER RUN?"
1130 GOSUB 2110
1140 IF Q$ = "NO" THEN 1300
1150 RESTORE
1160 LET I$ = "NO"
1170 PRINT "TYPE NEW FOR A RUN WITH COMPLETELY NEW DATA"
1180 PRINT "  OR OLD TO EDIT AND RERUN THE EXISTING DATA"
1190 INPUT Q$
1200 IF Q$ = "NEW" THEN 1240
1210 IF Q$ = "OLD" THEN 1270
1220 PRINT "REPLY '"; Q$; "' NOT UNDERSTOOD"
```

```
1230 GOTO 1170
1240 PRINT "TYPE IN A NEW SET OF DATA"
1250 PRINT "==== == = === === == ===="
1260 GOTO 160
1270 GOSUB 1390
1280 GOTO 320
1290 REM TERMINATE JOB
1300 PRINT
1310 PRINT "JOB COMPLETED."
1320 STOP
1330 REM SUBROUTINE TO CHECK THAT DATA ARE CORRECT & ALTER IF NECESSARY
1340 PRINT "ARE THE DATA VALUES ENTERED CORRECT?";
1350 REM A4 SHOULD BE SET TO THE NUMBER OF LINES ON THE VDU
1360 LET A4 = 20
1370 GOSUB 2110
1380 IF Q$ = "YES" THEN 2090
1390 PRINT "HERE IS A LIST OF THE CURRENT DATA"
1400 PRINT "LINE NUMBER", "LOW LIMIT", "HIGH LIMIT", "FREQUENCY"
1410 FOR I = 1 TO N
1420    PRINT I, L(I), H(I), F(I)
1430    IF INT(I / (A4 - 1)) * (A4 - 1) <> I THEN 1470
1440    PRINT "WOULD YOU LIKE TO CONTINUE LISTING";
1450    GOSUB 2110
1460    IF Q$ = "NO" THEN 1480
1470 NEXT I
1480 PRINT "TYPE R TO REPLACE";
1490 IF I$ = "NO" THEN 1510
1500 PRINT " AN EXISTING LINE OF DATA"
1510 IF N = 100 THEN 1560
1520 PRINT TAB(5); " A TO ADD";
1530 IF I$ = "NO" THEN 1550
1540 PRINT " AN EXTRA LINE"
1550 IF N = 1 THEN 1590
1560 PRINT TAB(5); " D TO DELETE";
1570 IF I$ = "NO" THEN 1590
1580 PRINT " AN EXISTING LINE"
1590 PRINT TAB(5); " L TO LIST";
1600 IF I$ = "NO" THEN 1620
1610 PRINT " THE DATA"
1620 PRINT "   OR C TO CONTINUE";
1630 IF I$ = "NO" THEN 1650
1640 PRINT " THE CALCULATION"
1650 INPUT Q$
1660 IF Q$ = "R" THEN 1760
1670 IF N = 100 THEN 1700
1680 IF Q$ = "A" THEN 1860
1690 IF N = 1 THEN 1710
1700 IF Q$ = "D" THEN 1930
1710 IF Q$ = "L" THEN 1390
1720 IF Q$ = "C" THEN 2090
1730 PRINT "REPLY '"; Q$; "' NOT UNDERSTOOD."
1740 GOTO 1480
1750 REM REPLACE LINE
1760 PRINT "TYPE THE LINENUMBER OF THE LINE TO BE REPLACED";
1770 INPUT I
1780 IF I <> INT(I) THEN 1800
1790 IF (I - 1) * (I - N) <=0 THEN 1830
1800 PRINT "LINENUMBER MUST BE AN INTEGER IN THE RANGE 1 -"; N
1810 PRINT "RE-";
1820 GOTO 1760
1830 PRINT "TYPE THE CORRECT LINE TO REPLACE THE ONE WHICH IS WRONG:"
```

```
1840 GOTO 1890
1850 REM ADD A NEW LINE
1860 LET N = N + 1
1870 LET I = N
1880 PRINT "TYPE THE ADDITIONAL LINE OF DATA AS SHOWN:"
1890 PRINT "LOW LIMIT, HIGH LIMIT, FREQUENCY"
1900 GOSUB 2220
1910 PRINT "OK"
1920 GOTO 1480
1930 REM DELETE A LINE
1940 PRINT "TYPE THE LINENUMBER OF THE LINE TO BE DELETED"
1950 INPUT J
1960 IF (J - 1) * (J - N) >0 THEN 1980
1970 IF J = INT(J) THEN 2000
1980 PRINT "LINENUMBER MUST BE AN INTEGER IN THE RANGE 1 -"; N
1990 GOTO 1940
2000 FOR I = J + 1 TO N
2010    LET L(I - 1) = L(I)
2020    LET H(I - 1) = H(I)
2030    LET F(I - 1) = F(I)
2040 NEXT I
2050 LET N = N - 1
2060 PRINT "OK"
2070 IF J > N THEN 1480
2080 GOTO 1390
2090 RETURN
2100 REM SUBROUTINE TO CHECK REPLIES
2110 IF I$ = "NO" THEN 2130
2120 PRINT " TYPE YES OR NO & PRESS RETURN."
2130 PRINT
2140 INPUT Q$
2150 IF Q$ = "YES" THEN 2190
2160 IF Q$ = "NO" THEN 2190
2170 PRINT "REPLY '"; Q$; "' NOT UNDERSTOOD.";
2180 GOTO 2120
2190 RETURN
2200 REM SUBROUTINE TO INPUT FREQUENCY & CHECK THAT IT IS NOT NEGATIVE
2210 REM & THAT HIGH LIMIT OF RANGE >= LOW LIMIT
2220 INPUT L(I), H(I), F(I)
2230 IF F(I) = INT(F(I)) THEN 2260
2240 PRINT "FREQUENCY MUST BE A WHOLE NUMBER"
2250 GOTO 2310
2260 IF F(I) >= 0 THEN 2290
2270 PRINT "FREQUENCY MUST BE POSITIVE OR ZERO"
2280 GOTO 2310
2290 IF H(I) >= L(I) THEN 2330
2300 PRINT "THE UPPER LIMIT MUST BE GREATER THAN THE LOWER LIMIT"
2310 PRINT "LAST LINE OF DATA REJECTED - RETYPE CORRECTLY"
2320 GOTO 2220
2330 RETURN
2340 END
```

Exercises

2.1 The following are the speeds of a sample of 24 cars in m.p.h. along a certain stretch of road at about the same time on the same day.

60, 49, 77, 66, 60, 66, 55, 65, 48, 50, 57, 63, 55, 60, 70, 51, 68, 67, 53, 60, 60, 62, 57, 61.

Calculate the mean and standard deviation of these speeds and estimate the mean and standard deviation of all the cars on this road at that time. (For simplicity Sheppard's correction should be ignored.)

2.2 The heights of 100 women are given in Table 2.2. Calculate the mean and standard deviation of these values. (Remember to use Sheppard's correction since the data are grouped.)

2.3 Two random sets of values x_1 and x_2 are drawn from a parent population of mean μ and standard deviation σ. What is the mean and standard deviation of each of the following distributions:

(a) $x_1 + x_2$ (b) $x_1 - x_2$ (c) $\frac{1}{2}(x_1 + x_2)$
(d) $\frac{1}{2}(x_1 - x_2)$

2.4 A student's class grades in Biology and maths are both distributed with means of 55 and standard deviations of 10. Write down the mean and standard deviation of:

(a) his total mark in Biology and maths;
(b) the difference between his marks in Biology and maths;
(c) his average mark in Biology and maths.

2.5 Define the mean, median and mode of a distribution. Calculate the mean and median weekly wage for a set of people using the data in Table 2.3 and use these to estimate the mode and skewness.

Table 2.2

Height (cm)	Frequency
Less than 147.5	4
147.5–152.5	9
152.5–157.5	21
157.5–162.5	32
162.5–167.5	20
167.5–172.5	11
More than 172.5	3

Table 2.3

Weekly wage (£)	Number of employees
35–45	11
45–55	35
55–65	58
65–75	64
75–85	50
85–95	33
95–105	22
105–115	12
115–125	9
125–135	6

3

Central Limit Theorem

The central limit theorem may be stated:

> If samples of size n_1 are drawn at random from a parent population of mean μ and standard deviation σ, the sample means constitute a population of mean μ and the standard deviation tends to $\sigma/\sqrt{n_1}$ as n_1 tends to infinity. Regardless of whether the parent population was normal or not, the sample means tend to a normal distribution as n_1 tends to infinity. The approximation may be quite good with n_1 as small as 5, and is generally acceptable for $n_1 \geqslant 15$.

One application of the central limit theorem is to estimate the mean and standard deviation of a parent population, given only a set of sample means. The procedure is similar to that for standard deviations described in the previous chapter. The difference in the two procedures is that the calculation of standard deviations requires *all* of the individual data values, whereas the central limit theorem uses only a set of sample means collected from the original data. (A sample mean is the mean of n_1 values drawn from the original data.) The central limit theorem should be used only if the original data are not available. If the original data are available it would be pointless to draw samples of size n_1, calculate the sample means and then use central limit theorem to estimate the mean and standard deviation of the original data, since these values can be obtained directly by the standard deviation procedure. Furthermore the standard deviation procedure is more reliable.

Example

A headmaster asked eight of his teachers to find the number of hours that their pupils watched television in a week. Each teacher questioned 12 children, and gave the headmaster the average hours of television viewing. The headmaster does not have the original data from the 96 children—he only has eight sample means. He can use the central limit theorem to estimate the mean and standard deviation of hours of television viewing by his pupils.

The eight sample means that the headmaster has are:

> 20.6, 18.8, 17.5, 19.6, 16.3, 17.1, 20.4, 17.8 hours.

The mean of these values is 18.5 hours and their standard deviation is 1.6 hours. These values can be used as *estimates* for μ and $\sigma/\sqrt{n_1}$, hence the estimated value of μ is 18.5 hours, and the estimated value of $\sigma = 1.6 \times \sqrt{12} = 5.5$ hours. It is interesting to compare these values with those obtained by the standard deviation procedure using all the viewing data collected by the teachers (see Table 3.1). (If the headmaster had all of these data he should not have used the central limit theorem.)

Using all of these figures to estimate μ and σ, the mean and standard deviation of viewing time by pupils were:

> mean viewing time $\mu = 18.5$ hours
> standard deviation of viewing time $\sigma = 4.4$ hours

Table 3.1

Teacher	Individual viewing times of pupils in hours	Sample mean
1	15, 18, 17, 17, 28, 17, 19, 19, 24, 23, 27, 23	20.6
2	19, 18, 23, 19, 13, 20, 23, 13, 25, 18, 15, 19	18.8
3	23, 18, 10, 16, 20, 18, 21, 12, 16, 22, 20, 14	17.5
4	18, 22, 20, 23, 19, 11, 21, 26, 27, 17, 14, 17	19.6
5	10, 23, 18, 15, 16, 17, 18, 20, 22, 14, 13, 10	16.3
6	11, 22, 18, 16, 23, 22, 12, 21, 13, 19, 10, 18	17.1
7	17, 16, 20, 23, 14, 25, 26, 22, 19, 25, 21, 17	20.4
8	13, 15, 18, 17, 21, 14, 27, 18, 23, 21, 16, 10	17.8

These values should be compared with the estimates obtained using the central limit theorem ($\mu = 18.5$ hours, $\sigma = 5.5$ hours). It should be noted that the means are the same, but the estimate for the standard deviation is wrong by 25%. This emphasises two points:

(i) The central limit theorem provides an *estimate* and should only be used when the full data are not available.

(ii) The sample size of 12 is rather small, and closer results would probably be obtained with a larger sample.

Outline of Central Limit Theorem

Three points emerge from the central limit theorem:

(i) The mean of the sample means tends to the mean of the parent population μ.

(ii) The standard deviation of the sample means tends to $\sigma/\sqrt{n_1}$.

(iii) The distribution of the sample means is approximately normal for large values of n_1.

These three points are discussed in turn:

(i) It is self-evident that if only a small number of samples is taken from a large parent population, it is possible that the samples are not representative of the whole population, and thus the mean of the sample means may differ appreciably from μ the true mean of the parent population. Clearly as the number of samples is increased it becomes less likely that the samples are not representative. (For example the chance of obtaining 3 'freak' samples is much larger than the chance of obtaining 30 'freak' samples.) Thus provided the number of samples is fairly large, the mean of the sample means will approximate to μ.

(ii) Consider a parent population which comprises a large number of loaves of bread produced by a bakery. A sample of two loaves is taken and each loaf is weighed giving weights x_1 and x_2. If the sampling is performed several times then different values for x_1 and x_2 will be obtained. The standard deviation of these readings will be σ_{x_1} and σ_{x_2}, and their variances will be $\sigma_{x_1}^2$ and $\sigma_{x_2}^2$ respectively. Using Appendix 8 Equation 4 the variance of the sum of the weights of the two loaves is $\sigma_{x_1}^2 + \sigma_{x_2}^2$.

If, however, each sample comprises n_1 loaves, the variance of the sum of the weights will be

$$\sigma_{x_1}^2 + \sigma_{x_2}^2 + \sigma_{x_3}^2 + \ldots + \sigma_{x_{n_1}}^2$$

Since each loaf weighed is from the parent population which has a mean weight μ and standard deviation σ, the estimated standard deviation for the x_1 weights will tend to σ. Similarly the estimated standard deviation for the $x_1, x_2, x_3, \ldots, x_{n_1}$ will also tend to σ. Thus the variance of the sum of the weights tends to $\sigma^2 + \sigma^2 + \sigma^2 + \ldots + \sigma^2$. Since there are n_1

terms being added, the variance of the sum of the weights is $n_1\sigma^2$.

Thus the standard deviation of the sum of the weights

$$= \sqrt{(n_1\sigma^2)}$$

$$= \sigma\sqrt{n_1}$$

Since the mean sample weight or the 'sample mean' is equal to the sum of the weights divided by the sample size n_1

Standard deviation of sample mean

$$= \text{Standard deviation of}\left(\frac{\text{sum of weights}}{n_1}\right)$$

$$= \frac{1}{n_1}\cdot\text{standard deviation of sum of weights}$$

$$= \frac{1}{n_1}\sigma\sqrt{n_1}$$

$$= \frac{\sigma}{\sqrt{n_1}}$$

(iii) In one sample of loaves it is highly likely that some loaves will weigh more than the mean weight of the parent population, and others will weigh less. Calculating a mean weight for each sample tends to cancel out these differences. However, the differences do not totally disappear, although small differences occur more frequently than large differences and the mean difference is zero. Random differences of this kind generally follow an approximately normal distribution provided that the number of terms n_1 is large. (See Chapter 1.)

Description of the Central Limit Theorem Program (Program 3.1)

The program first prints a heading, and then asks if full instructions are required. The answer is checked in a subroutine (lines 1640–1730), and must be either YES or NO. Full or abbreviated instructions are printed as requested for the first run, but shortened instructions are always given on subsequent runs.

Next the user is invited to type in up to a maximum of 100 sample means (lines 90–250). Generally there will be fewer than 100 sample means, and the user indicates the end of data input by typing the dummy value of 999999. A check is performed (lines 270–290) to ensure that at least one sample mean has been entered before the terminator. Though at least two sample means are required to perform a meaningful calculation, it is possible to add additional data in the checking/editing subroutine described in the next paragraph. The check performed here is essential to ensure the correct functioning of this subroutine.

A subroutine (lines 890–1630) is then entered which asks if the data are correct. The answer must be either YES or NO and is checked in another subroutine (lines 1640–1730). If the data are correct then the editing subroutine is exited, but otherwise the current data are listed. Instructions are then given for replacement of incorrect sample means, addition of new sample means or deletion of incorrect sample means, together with commands for re-listing the data and continuing the calculation. The operation of this subroutine is described more fully in Chapter 8.

On returning from the checking subroutine a check is performed (lines 320–350) to ensure that at least two sample means exist. Should there be only one sample mean then a warning message is printed explaining that the standard deviation must be zero. In this case no calculations are performed, and the user is asked if another run is required.

Generally there are two or more sample means, and the program next asks for the number of values which were used to calculated the sample means (i.e. the number of values in each sample). A number of checks (lines 380–480) are performed on the value typed in:

(i) The value typed must be an integer.
(ii) The value must be less than or equal to 1000. This is an empirical limit.
(iii) The value must be 5 or more since

sample means with fewer than 5 values are unreliable.

(iv) If the value is less than 15, a warning message is printed as the calculated standard deviation may not be reliable.

(v) A special case is made if the value is 1. This corresponds to typing in all of the individual readings, and the results obtained will be identical to those from the standard deviation program.

Next the mean of the samples mean is calculated and printed (lines 490–560). Then the sum of the squared differences from the mean is accumulated (lines 570–610) and the variance is evaluated and printed (lines 620–660). In common with the standard deviation procedure described previously, the denominator used is $(n-1)$ if there are less than 30 sample means, or n if there are 30 or more sample means, where n is the number of

Program 3.1   Trial run.

```
        CENTRAL LIMIT THEOREM
        ======= ===== =======
WOULD YOU LIKE FULL INSTRUCTIONS
  TYPE YES OR NO & PRESS RETURN.

? YES
THIS PROGRAM ESTIMATES THE VALUES FOR THE MEAN AND STANDARD
DEVIATION OF A PARENT POPULATION, FROM WHICH A SAMPLE OF
READINGS HAVE BEEN TAKEN.   THE READINGS WERE COLLECTED INTO
GROUPS OF N VALUES.   THE ORIGINAL N VALUES ARE NO LONGER
NEEDED (OR AVAILABLE) BUT THEIR MEAN IS REQUIRED.   THIS IS
CALLED THE SAMPLE MEAN.
TYPE IN THE SAMPLE MEANS & PRESS RETURN AFTER EACH VALUE
YOU WILL BE GIVEN THE CHANCE TO EDIT INCORRECT DATA LATER
TERMINATE THE DATA WITH A DUMMY VALUE OF 999999
? 20.6
? 18.8
? 17.5
? 19.6
? 16.3
? 17.1
? 20.4
? 17.8
? 999999
ARE THE DATA VALUES ENTERED CORRECT?   TYPE YES OR NO & PRESS RETURN.

? YES
TYPE THE NUMBER OF VALUES USED TO OBTAIN EACH SAMPLE MEAN
? 12
TAKE CARE - THE STANDARD DEVIATION MAY BE UNRELIABLE
SINCE THE SAMPLE SIZE IS LESS THAN 15

THE AVERAGE OF THE SAMPLE MEANS TYPED IN IS 18.5125
AND THIS IS THE BEST ESTIMATE FOR MEAN OF PARENT POPULATION

VARIANCE OF SAMPLE MEANS = 2.51553
ESTIMATED STANDARD DEVIATION OF SAMPLE MEANS = 1.58604

BEST ESTIMATE FOR STANDARD DEVIATION OF PARENT POPULATION   IS 5.49421

WOULD YOU LIKE ANOTHER RUN
  TYPE YES OR NO & PRESS RETURN.

? NO
END OF JOB
```

```
10 DIM M(100), Q$(10), I$(3)
20 PRINT TAB(10); "CENTRAL LIMIT THEOREM"
30 PRINT TAB(10); "======= ===== ======="
40 PRINT
50 PRINT "WOULD YOU LIKE FULL INSTRUCTIONS"
60 GOSUB 1660
70 LET I$ = Q$
80 IF I$ = "NO" THEN 180
90 PRINT "THIS PROGRAM ESTIMATES THE VALUES FOR THE MEAN AND STANDARD"
100 PRINT "DEVIATION OF A PARENT POPULATION, FROM WHICH A SAMPLE OF"
110 PRINT "READINGS HAVE BEEN TAKEN.  THE READINGS WERE COLLECTED INTO"
120 PRINT "GROUPS OF N VALUES.  THE ORIGINAL N VALUES ARE NO LONGER"
130 PRINT "NEEDED (OR AVAILABLE) BUT THEIR MEAN IS REQUIRED.  THIS IS"
140 PRINT "CALLED THE SAMPLE MEAN."
150 PRINT "TYPE IN THE SAMPLE MEANS & PRESS RETURN AFTER EACH VALUE"
160 PRINT "YOU WILL BE GIVEN THE CHANCE TO EDIT INCORRECT DATA LATER"
170 GOTO 190
180 PRINT "TYPE THE SAMPLE MEANS"
190 PRINT "TERMINATE THE DATA WITH A DUMMY VALUE OF 999999"
200 LET N = 0
210 FOR I = 1 TO 100
220    INPUT M(I)
230    IF M(I) = 999999 THEN 270
240    LET N = N + 1
250 NEXT I
260 PRINT "PROGRAM CAN ONLY HANDLE 100 SAMPLE MEANS"
270 IF N > 0 THEN 310
280 PRINT "PLEASE ENTER SOME SAMPLE MEANS"
290 GOTO 210
300 REM CHECK THAT DATA ARE CORRECT
310 GOSUB 900
320 IF N > 1 THEN 360
330 PRINT "SINCE THERE IS ONLY ONE SAMPLE MEAN THE STANDARD DEVIATION"
340 PRINT "MUST BE ZERO"
350 GOTO 720
360 PRINT "TYPE THE NUMBER OF VALUES USED TO OBTAIN EACH SAMPLE MEAN"
370 INPUT N1
380 IF N1 <> INT(N1) THEN 410
390 IF N1 = 1 THEN 430
400 IF (N1 - 5) * (N1 - 1000) <= 0 THEN 460
410 PRINT "RETYPE AN INTEGER NUMBER IN THE RANGE 5 - 1000"
420 GOTO 370
430 PRINT "WITH A SAMPLE OF SIZE 1 THE RESULTS ARE THE SAME AS"
440 PRINT "FROM THE STANDARD DEVIATION PROGRAM."
450 GOTO 490
460 IF N1 >= 15 THEN 490
470 PRINT "TAKE CARE - THE STANDARD DEVIATION MAY BE UNRELIABLE"
480 PRINT "SINCE THE SAMPLE SIZE IS LESS THAN 15"
490 LET S = 0
500 FOR I = 1 TO N
510    LET S = S + M(I)
520 NEXT I
530 PRINT
540 LET M1 = S / N
550 PRINT "THE AVERAGE OF THE SAMPLE MEANS TYPED IN IS"; M1
560 PRINT "AND THIS IS THE BEST ESTIMATE FOR MEAN OF PARENT POPULATION"
570 LET S = 0
580 FOR I = 1 TO N
590    LET S1 = M(I) - M1
600    LET S = S + S1 * S1
```

```
610 NEXT I
620 LET V = S / (N - 1)
630 IF N < 30 THEN 650
640 LET V = S / N
650 PRINT
660 PRINT "VARIANCE OF SAMPLE MEANS ="; V
670 LET S = SQR(V)
680 PRINT "ESTIMATED STANDARD DEVIATION OF SAMPLE MEANS ="; S
690 PRINT
700 PRINT "BEST ESTIMATE FOR STANDARD DEVIATION OF PARENT POPULATION";
710 PRINT " IS"; S * SQR(N1)
720 PRINT
730 PRINT "WOULD YOU LIKE ANOTHER RUN"
740 GOSUB 1650
750 LET I$ = "NO"
760 IF Q$ = "NO" THEN 870
770 PRINT "WOULD YOU LIKE TO USE COMPLETELY NEW DATA"
780 PRINT "OR EDIT AND RE USE THE OLD EXISTING DATA"
790 PRINT "TYPE NEW OR OLD & PRESS RETURN"
800 INPUT Q$
810 IF Q$ = "NEW" THEN 180
820 IF Q$ = "OLD" THEN 850
830 PRINT "REPLY '"; Q$; "' NOT UNDERSTOOD.   RE-";
840 GOTO 790
850 GOSUB 950
860 GOTO 320
870 PRINT "END OF JOB"
880 STOP
890 REM SUBROUTINE TO CHECK THAT DATA ARE CORRECT AND ALTER IF NECESSARY
900 PRINT "ARE THE DATA VALUES ENTERED CORRECT?";
910 REM A4 SHOULD BE SET TO THE NUMBER OF LINES ON THE VDU
920 LET A4 = 20
930 GOSUB 1650
940 IF Q$ = "YES" THEN 1630
950 PRINT "HERE IS A LIST OF THE CURRENT DATA"
960 PRINT "LINE NUMBER", "SAMPLE MEAN"
970 FOR I = 1 TO N
980    PRINT I, M(I)
990    IF INT(I / (A4 - 1)) * (A4 - 1) <> I THEN 1030
1000    PRINT "WOULD YOU LIKE TO CONTINUE LISTING";
1010    GOSUB 1650
1020    IF Q$ = "NO" THEN 1040
1030 NEXT I
1040 PRINT "TYPE R TO REPLACE";
1050 IF I$ = "NO" THEN 1070
1060 PRINT " AN EXISTING SAMPLE MEAN"
1070 IF N = 100 THEN 1120
1080 PRINT TAB(5); " A TO ADD";
1090 IF I$ = "NO" THEN 1110
1100 PRINT " AN EXTRA SAMPLE MEAN"
1110 IF N = 1 THEN 1150
1120 PRINT TAB(5); " D TO DELETE";
1130 IF I$ = "NO" THEN 1150
1140 PRINT " AN EXISTING SAMPLE MEAN"
1150 PRINT TAB(5); " L TO LIST";
1160 IF I$ = "NO" THEN 1180
1170 PRINT " THE DATA"
1180 PRINT "   OR C TO CONTINUE";
1190 IF I$ = "NO" THEN 1210
1200 PRINT " THE CALCULATION"
1210 INPUT Q$
```

```
1220 IF Q$ = "R" THEN 1320
1230 IF N = 100 THEN 1260
1240 IF Q$ = "A" THEN 1440
1250 IF N = 1 THEN 1270
1260 IF Q$ = "D" THEN 1490
1270 IF Q$ = "L" THEN 950
1280 IF Q$ = "C" THEN 1630
1290 PRINT "REPLY '"; Q$; "' NOT UNDERSTOOD."
1300 GOTO 1040
1310 REM REPLACE LINE
1320 PRINT "TYPE THE LINENUMBER OF THE LINE TO BE REPLACED";
1330 INPUT I
1340 IF I <> INT(I) THEN 1360
1350 IF (I - 1) * (I - N) <= 0 THEN 1390
1360 PRINT "LINENUMBER MUST BE AN INTEGER IN THE RANGE 1 -"; N
1370 PRINT "RE-";
1380 GOTO 1320
1390 PRINT "TYPE THE CORRECT VALUE TO REPLACE THE ONE WHICH IS WRONG:"
1400 PRINT "SAMPLE MEAN"
1410 INPUT M(I)
1420 GOTO 1470
1430 REM ADD A NEW LINE
1440 LET N = N + 1
1450 PRINT "TYPE THE ADDITIONAL SAMPLE MEAN"
1460 INPUT M(N)
1470 PRINT "OK"
1480 GOTO 1040
1490 REM DELETE A LINE
1500 PRINT "TYPE THE LINENUMBER OF THE LINE TO BE DELETED"
1510 INPUT J
1520 IF (J - 1) * (J - N) > 0 THEN 1540
1530 IF J = INT(J) THEN 1560
1540 PRINT "LINENUMBER MUST BE AN INTEGER IN THE RANGE 1 -"; N
1550 GOTO 1500
1560 FOR I = J + 1 TO N
1570    LET M(I - 1) = M(I)
1580 NEXT I
1590 LET N = N - 1
1600 PRINT "OK"
1610 IF J > N THEN 1040
1620 GOTO 950
1630 RETURN
1640 REM SUBROUTINE TO CHECK REPLIES
1650 IF I$ = "NO" THEN 1670
1660 PRINT " TYPE YES OR NO & PRESS RETURN."
1670 PRINT
1680 INPUT Q$
1690 IF Q$ = "YES" THEN 1730
1700 IF Q$ = "NO" THEN 1730
1710 PRINT "REPLY '"; Q$; "' NOT UNDERSTOOD.";
1720 GOTO 1660
1730 RETURN
1740 END
```

sample means. The estimated standard deviation of the sample means is calculated as the square root of the variance, and is printed (lines 670–680). The best estimate for the standard deviation of the parent population,

i.e. the estimate for σ is calculated and printed (lines 700–710).

The user is then asked if another run is required. If the answer is NO, then the run is finished, but if the answer is YES the user

vited to choose between typing in com-
~~ctely new data, or editing and re-running
using the old (existing) data.

Exercises

3.1 On five consecutive days the weight of fish
landed by a trawler was 500, 700, 600,
800, 650 lbs. Each day's catch was
packaged into 25 boxes. Use the central
limit theorem to estimate the mean and
standard deviation of the weight of fish
in a box.

3.2 A lift is designed to carry a maximum load
of 1 tonne (1000 kg). The manufacturers
claim that it will carry up to 13 people. If
people have a mean weight of 70 kg with a
standard deviation of 10 kg is the claim
fair, and what is the chance of overloading the lift with 13 people?

3.3 The weights of men travelling by air from
London to Edinburgh were found to have

a mean value of 75 kg and a standard
deviation of 8 kg. What is the probability
that 64 men travelling on this aircraft
have a combined weight greater than
4925 kg?

3.4 A factory produces washers which should
be of uniform thickness. Quality control
consists of taking 100 washers at random
from each day's production, measuring
their total thickness, and calculating the
average thickness of these 100. Estimate
the mean and standard deviation of the
thickness of a washer from the 20 days'
records given in Table 3.2.

Table 3.2

Thickness of 100 washers (mm)				
253.6	246.8	249.3	251.7	252.4
251.4	250.8	249.5	248.3	245.4
254.5	251.1	248.9	248.1	246.5
247.2	250.4	250.1	251.8	248.5

4

Data and Distributions

Discrete and Continuous Data

Some experiments and all opinion polls yield discrete data. For example the result of tossing a coin is either 'heads' or 'tails', while a poll on political affiliations would yield data on the numbers of Conservative, Labour, Liberal and other supporters. The essential feature of such data is that there is only a limited number of possible outcomes or results.

In contrast other experiments yield data on a continuous scale. Examples of such data include the speed of motor cars, the heights of men or the weights of laboratory animals. The essential feature of this type of data is that there is a continuum of possibilities for each of the examples.

These two different types of data must be handled differently. This is because when using discrete data it is possible to give the probability of an event occurring—for example the chance that a tossed coin will land 'heads' uppermost. In contrast, the probability that a laboratory animal weighed *exactly* 3 kg is infinitesimally small. Instead of quoting the probability for an exact value, when using continuous data probabilities are associated with intervals or ranges. One might reasonably ask for the probability of a laboratory animal weighing between 2.5 kg and 3.5 kg, or a car travelling at a speed of between 35 m.p.h. and 40 m.p.h. The difference between these two types of data is shown in Figs. 4.1 and 4.2. Figure 4.1 shows the results obtained by tossing a coin 10 times, recording the number of 'heads', and repeating the proce-

dure many times. Graphs such as this are called bar charts. Figure 4.2 shows the distribution of the weights of a large number of dogs. Curves such as this are called probability density curves and represent continuous distributions.

The shaded area on the bar chart (Fig. 4.1) represents the number of occurrences of 'heads' appearing exactly 3 times in 10 tosses, while the shaded area in the graph (Fig. 4.2) gives the probability of a dog weighing between 2.5 and 3.5 kg. For a continuous distribution, the probability of obtaining a value between x_1 and x_2 is the area under the probability density curve between the limits x_1

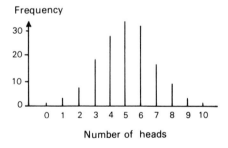

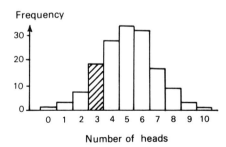

Fig. 4.1   Bar charts showing the number of 'heads'.

41

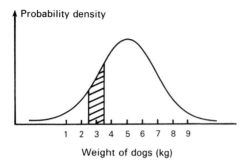

Fig. 4.2 Graph showing the distribution of the weights of dogs.

and x_2. The total area under the probability density curve from minus infinity to plus infinity is one since this represents every possible outcome.

The area under the bar chart in Fig. 4.1 increases with the number of times that the procedure (in this case tossing a coin 10 times), is repeated. This introduces difficulties which are avoided by plotting the relative frequency rather than frequency and obtaining a histogram.

Relative frequency of result

$$= \frac{\text{Frequency of result}}{\text{Number of times procedure is repeated}}$$

A histogram of the results is shown in Fig. 4.3, and the important difference from the bar chart in Fig. 4.1 is that the area under the histogram is always one.

The shaded area on the histogram (Fig. 4.3) represents the relative frequency of 'heads' appearing 3 times in 10 tosses. Figures 4.2 and 4.3 are similar in that areas represent probabilities, but they differ in two ways. Firstly, Fig. 4.2 uses probability density whereas Fig. 4.3 uses relative frequency. Secondly, Fig. 4.2 is a smooth (continuous) curve, while Fig. 4.3 is a stepped (discrete) curve.

The distinction between discrete and continuous data becomes less important if the number of discrete outcomes and the number of times the procedure was repeated are both large. This case is illustrated by tossing 250 coins at a time, counting the number of

'heads', and repeating the procedure 2000 times. The results are shown in Fig. 4.4, and it can be seen that the discrete data of the histogram approximate quite closely to a continuous curve.

The sharp distinction between discrete and continuous data is often blurred for another reason. The speeds of motor cars travelling along a road clearly form a continuous distribution. However, if the instrument measuring the speeds is only accurate to the nearest 1 m.p.h., then the continuous speeds are actually recorded as discrete data. Plotting such data would produce a histogram

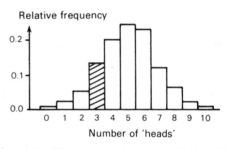

Fig. 4.3 Histogram showing relative frequency of number of 'heads'.

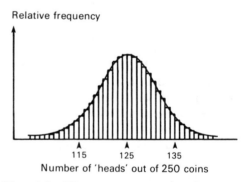

Fig. 4.4 Histogram approximating to a continuous distribution.

approximating to a continuous curve as in Fig. 4.4.

In this chapter, four different distributions are discussed: normal distribution, t-distribution, Poisson distribution and binomial distribution, and the first two of these are used extensively elsewhere in the book.

Elementary Probability

The concept of probability, that is the chance that an event will occur, is used in many places in this book. Some of the simple ideas and laws are outlined.

The probability that an event will occur is a number in the range 0 to 1. A probability of 0 means that the event will never occur, whereas a probability of 1 means that it is certain to occur. A probability of $\frac{1}{2}$ means that there is an equal chance of the event occurring or not occurring, for example the chance of getting 'heads' when tossing an unbiased coin or of getting an even number when rolling a fair die. The probability of rolling any one particular number in the range 1 to 6 (for example a 4) is $\frac{1}{6}$. These probabilities are written symbolically as:

$$P(\text{heads}) = \frac{1}{2}$$

$$P(\text{even}) = \frac{1}{2}$$

$$P(4) = \frac{1}{6}$$

If all the possible outcomes of an event are listed, then one of these outcomes must happen, so the sum of the individual probabilities must be one (otherwise there would be a chance of some other outcome). In coin tossing experiments there are two possible outcomes: heads and tails, hence

$$P(\text{heads}) + P(\text{tails}) = 1$$

Similarly in rolling a die there are six possible outcomes:

1, 2, 3, 4, 5 or 6

hence

$$P(1) + P(2) + P(3) + P(4) + P(5) + P(6) = 1$$

If the die is fair then each of these individual probabilities is equal to $\frac{1}{6}$, but the sum of the individual probabilities is always 1 regardless of whether the die is fair or not.

Consider the probability of obtaining a score of 5 *or* 6 when rolling a die. Clearly

$$P(5 \text{ or } 6) = P(5) + P(6) \qquad (1)$$

Let A represent getting a score of 5 or 6

then $P(A) = P(5) + P(6)$

Let B represent getting an even score

then $P(B) = P(2) + P(4) + P(6)$

From Equation 1 it would appear that the probability of A or B is equal to the probability of A plus the probability of B. Thus

$P(A \text{ or } B)$ would appear to be
$P(5) + P(6) + P(2) + P(4) + P(6)$.

Intuitively this is wrong since the left-hand side is the probability of being even, or 5 or 6 that is the probability of shaking a 2, 4, 6, 5 or 6 and the number 6 appears twice in the list. $P(A \text{ or } B) = P(A) + P(B)$ *only* when the outcomes A and B are mutually exclusive (complementary). When A and B are not mutually exclusive, any outcomes which occurs in both A and B is counted twice. In the above example 6 is counted twice, and

$$P(A \text{ or } B) = P(A) + P(B) - P(6)$$

Generally $P(A \text{ or } B) = P(A) + P(B) - P(A \text{ and } B)$. This is illustrated in the Venn diagram (Fig. 4.5). Consider the probability of tossing a 'head' *and* rolling a 6 in an experiment where one fair coin is tossed and one fair die is rolled. Intuitively the probability is one-twelfth. Thus it would appear that

$$P(A \text{ and } B) = P(A) \times P(B) \qquad (2)$$

This result is true provided that the events A and B are independent, that is the probability of one is unaffected by the occurrence or non-occurrence of the other. An example of this result failing is the probability of a coin being simultaneously heads and tails. Plainly the probability is zero, yet $P(\text{heads}) = \frac{1}{2}$ and $P(\text{tails}) = \frac{1}{2}$ so $P(\text{heads and tails}) = \frac{1}{2} \times \frac{1}{2} = \frac{1}{4}$ from Equation 2. Equation 2 is only valid if A and B are independent.

Normal Distribution

The normal distribution is one of the most widely used continuous distributions in statistics (see Fig. 4.6). It is sometimes called the 'Gaussian distribution', or 'the normal curve of errors', because the frequencies of observed random errors in the repeated measurement

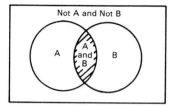

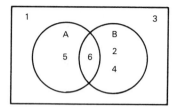

Fig. 4.5   Venn diagram.

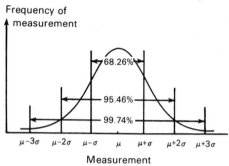

Fig. 4.6   Graph of a normal distribution.

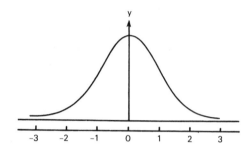

Fig. 4.7   Graph of standard normal curve.

of the same quantity approximate to this distribution, as discussed in Chapter 1.

One important feature of the normal distribution is that it is symmetrical about μ. The curve is bell-shaped, and extends to infinity in both directions, coming closer and closer to the horizontal axis without ever touching it. The horizontal axis is calibrated in terms of σ; the standard deviation of the measurement. It is not necessary to extend the tails of the curve very far, and because the area under the curve beyond ± 4 standard deviations is only 0.000 06 this is ignored for most practical purposes.

Normal distributions can have different shapes by stretching the x axis (and squashing the y axis to keep the area equal to one), or vice versa. The whole curve may be translated (moved) to the left or right, but there is only one normal curve with a given mean μ and standard deviation σ. The most commonly used normal curve has $\mu = 0$ and $\sigma = 1$. This is called the standard normal curve (Fig. 4.7).

The equation of the standard normal curve is

$$y = \frac{1}{\sqrt{2\pi}}e^{-\frac{1}{2}x^2} \tag{3}$$

where x is the number of standard deviations. The integral of Equation 3 is used to calculate the area under the standard normal curve, and is tabulated in Appendix 4. The area tabulated is the area from minus infinity up to the number of standard deviations specified x, as shown in Fig. 4.8.

If the area from the given number of standard deviations to plus infinity is required, this is calculated as one minus the table value, since the total area under the curve is by definition one.

The table does not give the areas for negative standard deviations (Fig. 4.9a).

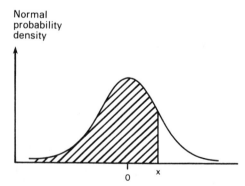

Fig. 4.8   Area under standard normal curve tabulated in Appendix 4.

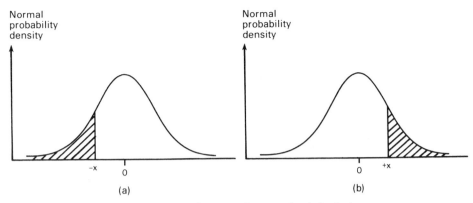

Fig. 4.9   Areas for negative standard deviations.

Clearly the shaded area in Fig. 4.9a is equal to the shaded area in Fig. 4.9b, which is equal to one minus the shaded area in Figure 4.8.

Area below $(-x)$ standard deviations
$= 1 -$ area below $(+x)$ standard deviations

The normal distribution is widely used for comparing two large samples, and for calculating confidence limits.

t-Distribution

For comparing small samples (number of readings less than 30), or for calculating confidence limits for small sets of data, the normal distribution is not adequate, and the t-distribution should be used. Its properties are similar to the normal distribution (symmetrical, total area equals one, and mean equals zero), but its shape varies depending on the number of degrees of freedom. In many situations the number of degrees of freedom is the number of readings minus one, but this may vary in individual cases, and is defined with each example. A table of t values for varying numbers of degrees of freedom is given in Appendix 6. It should be noted that the t-distribution tends to the normal distribution as the number of degrees of freedom becomes large. Care should be exercised to avoid confusion between one- and two-tailed possibilities.

Figure 4.10a shows the one-tailed probability of 10%, while Fig. 4.10b shows the two-tailed probability of 20%. Two-tailed tests are used to establish if the samples are significantly different, regardless of which is the larger, while one-tailed tests are used to establish either if sample 1 is significantly greater than sample 2, or if sample 1 is significantly smaller than sample 2, but *NOT* both.

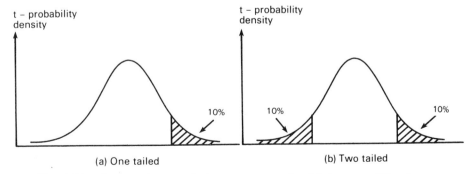

Fig. 4.10   One- and two-tailed probabilities for 25 degrees of freedom.

Poisson Distribution

The Poisson distribution is used to calculate the number of occurrences of a particular event for discrete data, for example the number of defective transistors in a packet. The Poisson distribution assumes that the events occur randomly and independently. It can be derived from the binomial distribution, and may be stated:

Probability of n occurrences of event in a given sample

$$= \frac{e^{-\mu}.\mu^n}{n!}$$

where μ is the mean number of occurrences of the event per sample, and $n!$ is factorial $n = 1 \cdot 2 \cdot 3 \ldots n$. Three points should be borne in mind:

(i) The derivation of the Poisson distribution assumes that the probability of the event occurring is small, hence large samples are required.

(ii) The variance of the Poisson distribution is equal to the mean μ.

(iii) The Poisson distribution tends to the normal distribution with mean μ and variance μ for large values of n.

A Poisson distribution is calculated in Chapter 5, Example 6.

Binomial Distribution

The binomial distribution relates to discrete data. In a single trial, an event either occurs with a probability p, or does not occur with a probability q. Clearly

$$p + q = 1$$

The probability of i occurrences of the event in n independent trials is given by the coefficient of t^i in $(q + pt)^n$.

Exercises

4.1 The weights of individual apples from a particular tree are found to be normally distributed. In one year, the crop constituted 150 apples, of which 15 were below 71 g, in mass and 30 were above 103 g. Estimate the mean and standard deviation of the crop, and calculate how many apples are likely to weigh above 96 g.

4.2 What is meant by the inter-quartile range of a distribution? Verify for a 'normal' distribution with mean = 0 and standard deviation = 1 that the quartiles are ± 0.675. (See Chapter 2.)

4.3 An intelligence quotient (IQ) test is standardised to give a 'normal' distribution with mean 100 and standard deviation 16. Calculate the probability of an individual having an IQ in the following ranges:

(a) less than 70
(b) 70–80
(c) 80–90
(d) 90–100
(e) 100–110
(f) 110–120
(g) 120–130
(h) greater than 130

4.4 In a particular town the number of houses struck by lightning over a period of 100 years had the following distribution:

Number of houses struck	0	1	2	3 or more
Number of years	36	37	17	10

Calculate the mean and variance for this distribution, and calculate the Poisson distribution with the same mean. Without performing calculations, do the data appear to fit a Poisson distribution, and could this have been guessed from the mean and variance?

5

Chi-Squared Test

In many experiments the results obtained are compared with those predicted by a theory which is under test. For example if a die is tossed 600 times, then in theory one would expect 100 occurrences of each of the numbers 1 to 6. If this experiment is carried out (or simulated on a computer) it is extremely unlikely that the exact theoretical result will be obtained. One must therefore ask how far the observed frequencies must differ from the theoretical frequencies before one can reasonably claim that the die is biased or the method of throwing is unfair.

A measure of how far the observed and expected frequencies differ is given by the chi-squared χ^2 statistic

$$\chi^2 = \Sigma \left[\frac{(O-E)^2}{E} \right]$$

where O are the observed frequencies and E the expected frequencies.

Example 1—Die Throwing Experiment and Chi-Squared

The way in which χ^2 is calculated is illustrated with the results from the die throwing experiment (Table 5.1). Thus the value of χ^2 is calculated as 1.96. Given the number of degrees of freedom v, this value of χ^2 can be looked up in χ^2 tables for various probability levels. In this example there are six pairs of O and E values (1, 2, 3, 4, 5 and 6) that is there are six classes. There is one restriction in calculating the expected frequencies in that the expected frequencies have the same total

as the observed frequencies. The number of degrees of freedom v is calculated:

$$v = \text{Number of classes} - \text{Number of restrictions}$$
$$= 6 - 1 = 5$$

Mathematicians have calculated for a random sample with five degrees of freedom the probabilities of obtaining values of χ^2 greater than or equal to certain values:

Probability $P =$	99%	95%	90%	80%
Chi-squared $\chi^2 >$	0.55	1.15	1.61	2.34

It can be seen that the χ^2 value of 1.96 corresponds to a probability of between 80% and 90%. This means that there is between an 80% and 90% chance that results as bad (different from the predicted) or worse could have arisen by chance. Since one would only expect a better result (agreeing more closely with the theoretical) between 10% and 20% of the time there are no statistical grounds for suspecting the fairness of the die.

If the value of χ^2 evaluated as a large number (for example 12), then the probability of a deviation greater than or equal to χ^2 is less than 5%. Since there is less than a 1 in 20 chance of obtaining this χ^2 result by chance one would suspect that the die was unfairly balanced. Moreover if the value of χ^2 exceeded 15, there is less than a 1% chance of a fair die giving this result hence the die used is highly suspect.

Conversely if the value of χ^2 evaluates as a very small number (for example 1) then the

47

Table 5.1

Score	Observed frequency O	Expected frequency E	$O-E$	$(O-E)^2$	$(O-E)^2/E$
1	105	100	5	25	0.25
2	92	100	-8	64	0.64
3	103	100	3	9	0.09
4	95	100	-5	25	0.25
5	97	100	-3	9	0.09
6	108	100	8	64	0.64
	$\Sigma\ 600$	$\Sigma\ 600$			$\Sigma\ 1.96 = \chi^2$

probability of a deviation greater than or equal to χ^2 is greater than 95%. Since there is less than a 1 in 20 chance of obtaining a result this good one starts to suspect that the data are not random in that 'only good results' have been selected by the experimenter and 'bad' results discarded. If χ^2 is smaller than 0.5 which corresponds to a probability of 99% then the data are almost too good to be true and serious doubts exist as to whether the data are truly random or may even be fictitious.

A complete table of χ^2 values for certain probabilities and varying numbers of degrees of freedom is given in Appendix 5. This should be used to verify the examples given in this chapter.

Example 2—Chi-Squared Applied to Blood Groups

There are four main blood groups in man—groups A, B, AB and O. The genes responsible for types A and B are dominant while the gene for type O is recessive. It follows that a person belonging to group A (phenotype A) may be either homozygous with two A genes, or heterozygous with an A gene and an O gene. In an identical way a person who is phenotype B may have either the genotype BB or BO. People who are phenotype AB must be genotype AB, and similarly phenotype O must be

genotype OO (see Table 5.2). Simple genetic theory predicts that the blood groups of children of two AB parents should be in ratio 1 AA:2 AB:1 BB.

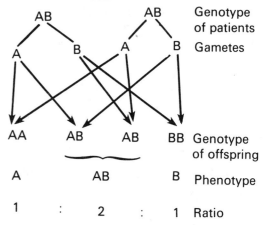

In a survey of 100 children born to such parents the following results were obtained A = 18, AB = 53 and B = 29. Do these results constitute evidence that the theory is false?

Table 5.2

Phenotype	Possible genotypes	
A	AA	AO
B	BB	BO
AB	AB	
O	OO	

48

Table 5.3

Phenotype (observed blood group)	Observed frequency O	Expected frequency E	$O-E$	$(O-E)^2$	$(O-E)^2/E$
A	18	25	-7	49	1.96
AB	53	50	3	9	0.18
B	29	25	4	16	0.64
	Σ 100	Σ 100			Σ 2.78 $= \chi^2$

Chi-squared is evaluated as shown in Table 5.3. There are three classes A, AB, B, and one restriction (the totals of O and E have been made to agree). The number of degrees of freedom v is given by

$$v = \text{Number of classes}$$
$$- \text{Number of restrictions}$$
$$= 3 - 1 = 2$$

Reference to χ^2 significance tables (Appendix 5) shows that

Probability P = 50% 30% 20% 10%
Chi-squared $\chi^2 > 1.39$ 2.41 3.22 4.61

The calculated value of χ^2 of 2.78 lies between the 20% and 30% probability levels, and it follows that assuming the genetic theory is correct there is between a 20% and 30% chance that results as bad or worse than this would be obtained because of random errors. Though the observed frequencies differ appreciably from the predicted ones, one would expect results as bad about every fourth survey hence there is no statistical evidence to suspect the genetic theory.

Example 3—Chi-Squared and Contingency Tables

A doctor has 200 patients who suffer from the same disease. The patients may be treated with an old drug, a new drug, or no drug at all. The disease may result in considerable pain, slight pain, or no pain, and the number of patients in each category is shown in the contingency table (Table 5.4). The doctor would like to know if the treatment has any statistical effect on the degree of pain experienced. First, one makes the assumption that the treatment has no effect on the level of pain. Using this assumption one calculates how many people would be expected in each of the categories, that is one produces a contingency table of expected frequencies. This is done as follows.

The proportion of people experiencing considerable pain is $70/200 = 35\%$. The proportion of people having no drug is $85/200 = 42.5\%$. The expected proportion of people having both considerable pain and no drug is $35\% \times 42.5\% = 14.875\%$ which corresponds to 29.75 people out of the 200 sampled. It is quite valid in this context to have a fraction of a person since expected frequencies

Table 5.4

Observed frequencies	No drug	New drug	Old drug	Row totals
Considerable pain	40	16	14	70
Slight pain	30	17	18	65
No pain	15	32	18	65
Column totals	85	65	50	200

do not have to be integers. The other entries in the expected contingency table are generated in a similar manner. The results are shown in Table 5.5. Chi-squared is then calculated in the usual way taking each of the nine pairs of observed and expected frequencies (see Table 5.6). To calculate the number of degrees of freedom one requires the number of classes and the number of restrictions. Since there are nine pairs of O and E frequencies the number of classes is nine. The number of restrictions is slightly more difficult. The total of each horizontal row in the expected table equals that in the observed table. Thus each row constitutes one restriction (in this case there are three restrictions). In addition the total of

Table 5.5

Expected frequencies	No drug	New drug	Old drug	Row totals
Considerable pain	29.75	22.75	17.5	70
Slight pain	27.625	21.125	16.25	65
No pain	27.625	21.125	16.25	65
Column totals	85	65	50	200

Table 5.6

Observed frequency O	Expected frequency E	$(O-E)^2/E$
40	29·75	3·53
16	22.75	2.00
14	17.5	0.70
30	27.625	0.20
17	21.125	0.81
18	16.25	0.19
15	27.625	5.77
32	21.125	5.60
18	16.25	0.19
		$\Sigma\ 18.99 = \chi^2$

each vertical row is the same in both expected and observed tables. Though this would seem to suggest a further three restrictions in the example given, it in fact only adds a further two restrictions. This is because by adding the row totals (70, 65 and 65) one obtains the total number of patients (200). Since the sum of the column totals (85, 65 and 50) must also equal the number of patients, one can easily calculate one column total from three row totals and two column totals, hence in this case there are $3+2=5$ restrictions. In general for an $m \times n$ contingency table there are $m+n-1$ restrictions and $m \cdot n$ classes. The number of degrees of freedom v is given by

$$\text{or} \quad \begin{aligned} v &= (m \cdot n) - (m+n-1) = 9-5 \\ v &= (m-1) \cdot (n-1) \quad\quad = 2 \times 2 \end{aligned} \Bigg\} = 4$$

Reference to the χ^2 distribution table (Appendix 5) shows that for four degrees of freedom a χ^2 value of 18.99 is greater than the value (18.46) for a probability $P=0.1\%$. This means that there is less than 0.1% chance (that is less than 1 chance in 1000) that results which differ from the theory as badly as these could have arisen by chance. This provides extremely strong evidence that the hypothesis assumed is incorrect. The doctor should therefore reject the hypothesis that the treatment has no effect on the level of pain. Conversely this indicates that it is 99.9% certain that one or both of the drugs are beneficial. This does not resolve whether the new, the old or both drugs are beneficial, and this is left as a challenge to the reader. (Hint—calculate χ^2 from a 3×2 contingency table comprising no drug, old drug, and three pain levels and look up significance. Then repeat for no drug and new drug.) Should a problem require a 2×2 contingency table, then it is necessary to use Yates's correction.

Example 4—Use of Yates's Correction when $v=1$

In all cases when the number of degrees of freedom $v=1$ it is necessary to apply Yates's correction for continuity. This states that when $v=1$ the absolute values of the $(O-E)$ differences must each be reduced by 0.5. This will always be necessary with a 2×2 contingency table and also in other cases. Though the reason for this correction is beyond the scope of this book, it is sufficient to observe that χ^2 is a continuous distribution and it is being used with discrete results. The correction is unnecessary when there are several observed and expected frequencies (i.e. $v > 1$).

For example if a penny is tossed 100 times, one would predict that it would land showing 'heads' on 50/100 occasions and 'tails' on 50/100 occasions provided that the coin was not biased or the method of tossing unfair. In practice the result may differ slightly from the theoretical value, and the chi-squared test is used to find if the differences between experimental and theoretical frequencies are significant or not. Suppose the result of the experiment was 'heads' 55 times and 'tails' 45 times—can one conclude that the coin is biased, or is the result reasonable for an unbiased coin? (See Table 5.7.) Reference to the χ^2 distribution table (Appendix 5) for $\chi^2 = 0.81$ and $v = 1$ gives a probability of a large deviation P between 30% and 50% which implies a good chance of such a result occurring by chance with a fair coin.

How to Handle Continuous Data

In each of the four examples considered so far there have been only a limited number of possible outcomes. For example when tossed a coin must land with either a 'head' or a 'tail'

uppermost, and a die can show only one of the digits 1–6. Such data are usually termed discrete by mathematicians or discontinuous by biologists.

Many experiments yield different types of data with continuous variation. Some examples of these sorts of experimental data are:

  (i) the heights of a group of men;
  (ii) the weights of individual hen's eggs;
  (iii) the girths of tree trunks;
  (iv) a set of titration results;
  (v) a set of amplification factors from apparently identical transistors;
  (vi) the speeds of cars travelling along a particular road.

With such continuous data it is necessary to group the data into a number of discrete classes when performing statistical tests. Grouping of data is commonly performed, for example, when plotting a bar chart. When data are grouped in preparation for the chi-squared test the following points should be observed:

  (i) A large number of measurements should be made. A minimum of at least 30 measurements is essential.
  (ii) To obtain reasonable discrimination (resolution) there should be at least 6–8 classes (i.e. groups of data).
  (iii) No class or group should be so small that it contains less than five expected values, that is the expected frequency must be five or more in each class. This limitation is a direct result of the derivation of the chi-squared distribution.

Table 5.7

	Observed frequencies O	Expected frequencies E	$(O-E)$	$Y = |O-E|-0.5$	Y^2/E
Heads	55	50	5	4.5	0.405
Tails	45	50	-5	4.5	0.405
					$\Sigma\ 0.81 = \chi^2$

(iv) Though it is common to group data into bands which cover an equal range, this is not essential. If one considers the traffic speed data in Example 5, it is seen that most of the classes cover a speed range of 10 m.p.h., but one class covers a range of 5 m.p.h. and another class covers an infinite range. Should a class contain less than five expected values (restriction iii) then two classes are merged into one with a greater range than before.

It should be noted that unless the number of measurements is large, the value of χ^2 will vary appreciably depending on how the measurements are arranged into different groups. It is partly for this reason that a very high value of χ^2 corresponding to a low probability P is required before a hypothesis is rejected. In general if $P < 5\%$ the hypothesis is suspect, if $P < 1\%$ the hypothesis is highly suspect, and if $P < 0.1\%$ the hypothesis is almost certainly false. These rejection limits are severe compared with the limit of one standard deviation from the mean which is used in other branches of statistics (31.7% of a normal population lie beyond a limit of one standard deviation).

Example 5—Comparison of Observed Frequencies with a Normal Distribution

A civil engineer has monitored the speeds of cars travelling along a main road to decide whether road improvements are necessary. The results are shown in Table 5.8, and by plotting a bar chart of the frequency of various speeds the results appear to be approximately normally distributed (see Fig. 5.1). To enable reasonable predictions the engineer requires a model of traffic speeds, and the problem is to establish whether the normal distribution model is realistic. The observed results are shown in Table 5.8. The mean speed of the cars is 40 m.p.h. and the standard deviation of the speeds is 14.49 m.p.h. These figures can be obtained from the standard deviation program described in

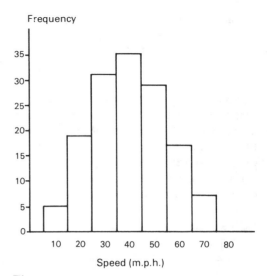

Fig. 5.1

Table 5.8

Traffic speed (m.p.h.)	Observed frequencies O
0–5	0
5–15	5
15–25	19
25–35	31
35–45	35
45–55	29
55–65	17
65–75	7
over 75	0

Σ 143 = number of cars recorded

Chapter 2. It should be noted that this value for the standard deviation involves Sheppard's correction because the speeds have been grouped into intervals of 10 m.p.h. resulting in a slightly smaller standard deviation than the value of 14.77 m.p.h. which would be obtained if grouping was ignored.

A normal population is generated which has the same mean speed, standard deviation of speed and also the same number of cars.

52

Once this has been done the observed and expected frequencies may be compared to give chi-squared.

Method of Generating Normal Population

(i) A table is prepared showing the various speed ranges.

(ii) The speeds are then expressed as the number of standard deviations from the mean speed. Speeds below the mean give rise to negative values while speeds above the mean give positive values.

(iii) Each of these values (standard deviations from mean) is looked up in a table showing the area under a normal probability curve starting at the left-hand side up to the number of standard deviations specified. (For example looking up the value of one standard deviation in the normal table given in Appendix 4 gives a value of 0.841. This means that 84.1% of any normal population are less than one standard deviation above the mean.) Tables do not usually give negative values for the standard deviation. This is discussed in Chapter 4 and may be handled as follows: firstly, the normal curve is symmetrical. In consequence the area below a given negative value is identical to the area above the same positive value. Secondly, the total area under the normal curve represents all possible outcomes and must by definition be equal to one. The area above a certain value (which corresponds to the area below the required negative value) can be evaluated as $1 -$ the area below the positive value which can be obtained directly from the tables. Thus:

Area below $(-x)$ standard deviations $= 1 -$ Area below $(+x)$ standard deviations (see Fig. 5.2).

(iv) To obtain the area under the normal curve in each speed range one subtracts

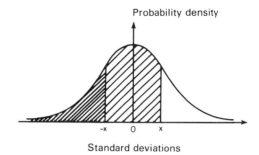

Probability density

Standard deviations

Fig. 5.2

the area below the low speed from the area below the high speed (see Fig. 5.3).

(v) The expected frequencies are derived by multiplying the area (probability) in each speed range by the total number of cars (143 in this example).

This is illustrated in Table 5.9. It must be noted that the sum of the expected frequencies is not exactly 143 cars as would be expected. There are two reasons for this:

(a) Small arithmetic rounding errors have accumulated.

(b) More important the sum of all the areas is 0.997 rather than 1.000. This is caused by the lower speed limit of 0 m.p.h. While this is a perfectly reasonable physical restriction, mathematically with a normal distribution there is a small probability (0.003) that cars travel at less than 0 m.p.h. since the normal probability curve extends to minus infinity. This problem can be hidden and overcome by re-labelling the first speed range as less than 5

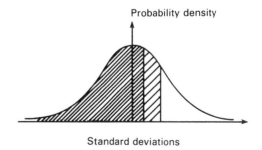

Probability density

Standard deviations

Fig. 5.3

Table 5.9

Speed m.p.h.	Speeds as standard deviations from mean	Area of normal curve below this standard deviation	Area within this speed range	Expected frequencies E
0–5	−2.76 to −2.42	0.003 to 0.008	0.005	0.72
5–15	−2.42 to −1.73	0.008 to 0.042	0.034	4.86
15–25	−1.73 to −1.04	0.042 to 0.149	0.107	15.30
25–35	−1.04 to −0.35	0.149 to 0.363	0.214	30.60
35–45	−0.35 to +0.35	0.363 to 0.637	0.274	39.18
45–55	+0.35 to +1.04	0.637 to 0.851	0.214	30.60
55–65	+1.04 to +1.73	0.851 to 0.958	0.107	15.30
65–75	+1.73 to +2.42	0.958 to 0.992	0.034	4.86
over 75	over +2.42	0.992 to 1.000	0.008	1.14
			Σ 0.997	Σ 142.56

m.p.h. rather than 0–5 m.p.h., and this results in an extra 0.003×143 cars in this speed range thus changing the expected frequency for this range from 0.72 to 1.15.

It is now possible to perform the chi-squared test in the usual way (see Table 5.10). The value of χ^2 is thus 1.96. The number of degrees of freedom v is calculated

v = Number of classes

− Number of restrictions

The number of classes was originally nine, but merging of classes has reduced this to seven. The number of restrictions is three (same total of 143, same mean speed of 40 m.p.h. and same standard deviation of 14.49 m.p.h). Thus

$v = 7 - 3 = 4$

Reference to the χ^2 distribution table in Appendix 5 for $\chi^2 = 1.96$ and $v = 4$ gives a probability P between 0.70 and 0.80. This means that there is between a 70% and 80% chance of obtaining worse results than this by chance if the traffic is normally distributed. χ^2 can never prove a hypothesis but in this case considerable support has been given to the hypothesis that the traffic flow is normally distributed.

Table 5.10

Observed frequencies O	Expected frequencies E		$(O-E)^2/E$
0 5 } 5	1.15 4.86 } 6.01		0.17
19	15.30		0.89
31	30.60		0.01
35	39.18		0.45
29	30.60		0.08
17	15.30		0.19
7 0 } 7	4.86 1.14 } 6.00		0.17
			Σ $1.96 = \chi^2$

Example 6—Comparison of Observed Frequencies and Poisson Distribution

A Poisson distribution is useful for analysis of discontinuous (discrete) data (see Chapter 4) and is applicable when the chance of a particular result is small, and the chance is independent of the previous results. Because the chance of the result occurring is small, a large number of tests must be performed.

If one considers the number of defective transistors found in each packet of 100, then the chance that any one transistor is working or defective is totally independent of whether the one previously examined was working or defective, or of how many have already been found to be defective. The Poisson distribution is only applicable if the chance of any transistor being faulty is small and the number of transistors tested is large. The Poisson distribution is very easy to calculate, and this is a great advantage over the normal distribution. It should be noted that for *all* Poisson distributions the mean value is always equal to the variance. The Poisson distribution is calculated using the equation:

$$\text{Probability of } n \text{ occurrences} = \frac{e^{-\mu} \cdot \mu^n}{n!}$$

where μ is the mean and $n!$ is factorial n (i.e. $1 \cdot 2 \cdot 3 \ldots n$). To convert the probabilities obtained in this way into expected frequencies E (for comparison with the observed frequencies O in the chi-squared test), one multiplies the calculated probability by the number of readings, i.e. the number of packets opened.

A quality control department examined the data for the number of defective transistors in a bag. They wished to establish whether faulty components were produced at random, or if the production process tended to have 'bad runs'. If the process produces bad components at random the distribution of defective transistors should approximate to Poisson whereas with 'bad runs' the results will not fit the Poisson distribution.

Eighty bags each containing 100 transistors were examined. The number of transistors in each bag is irrelevant to calculation provided all bags contain the same number, and provided the proportion of faulty transistors is small. The results are as shown in Table 5.11. The mean number of defective transistors per bag is two (160 defective transistors in 80 bags). The Poisson frequencies are calculated as shown in Table 5.12.

The following points should be noted:

(i) By definition factorial 0 (0!) equals one so the Poisson probability for no defective transistors ($n=0$) reduces to $e^{-\mu}$.

Table 5.11

Defective transistors n	Frequency (number of bags) O
0	8
1	18
2	25
3	24
4	5
over 4	0
	—
	Σ 80 = number of bags opened

Table 5.12

Defective transistors n	Poisson probability $e^{-\mu} \cdot \mu^n/n!$		Expected frequencies E
0	e^{-2}	$= 0.135$	10.8
1	$e^{-2} \cdot 2/1$	$= 0.271$	21.7
2	$e^{-2} \cdot 2^2/(2 \cdot 1)$	$= 0.271$	21.7
3	$e^{-2} \cdot 2^3/(3 \cdot 2 \cdot 1)$	$= 0.180$	14.4
4	$e^{-2} \cdot 2^4/(4 \cdot 3 \cdot 2 \cdot 1) = 0.090$		7.2
Over 4		0.053	4.2
	Total 1.000	Total	80.0

(ii) The next and subsequent lines in the table can be generated by multiplying the probability on the previous line by μ (2 in this case) and dividing by n. With the exception of the last line, the entire table can be speedily generated in this way.

(iii) The probability for the last line ($n > 4$ in this case) is obtained indirectly by subtracting all the previously calculated probabilities from 1, since the sum of all probabilities must be 1 by definition.

The chi-squared statistic is calculated in a similar manner to the previous examples. It should be noted that expected frequency for more than four defective transistors per bag is less than five. The mathematical derivation of χ^2 requires that all the expected frequencies should be large. In practice large means at least five. To overcome this the last two classes (that is the lines where $n = 4$ and $n > 4$) have been combined to make a single class (see Table 5.13). The value of χ^2 is thus 11.85. The number of degrees of freedom v is calculated:

$$v = \text{Number of classes}$$

$$- \text{Number of restrictions}$$

The number of classes is five. Though there were originally six classes corresponding to $n = 0, 1, 2, 3, 4$ and $n > 4$ defective transistors, $n = 4$ and $n > 4$ have been combined into a single class. The number of restrictions is two

Table 5.13

n	Observed frequency O	Expected frequency E	$(O-E)^2/E$
0	8	10.8	0.73
1	18	21.7	0.63
2	25	21.7	0.50
3	24	14.4	6.40
4	5 } 5	7.2 } 11.4	3.59
>4	0	4.2	
	Σ 80	Σ 80.0	Σ 11.85 $= \chi^2$

(the totals of both O and E are 80 bags, and the mean number defective per bag is two for both O and E). Thus

$$v = 5 - 2 = 3$$

Reference to the chi-squared distribution table in Appendix 5 for $\chi^2 = 11.85$ and three degrees of freedom gives a probability P between 0.01 and 0.001 that is between 1% and 0.1%. If the observed distribution is indeed Poisson then there is less than a 1% chance that results which disagree as badly or worse than this could have arisen by chance. It is therefore reasonable to conclude that the observed results have not come from a Poisson distribution, and that the production of defective transistors is not random.

Finally it is worth checking that the mean and variance (standard deviation squared) of the observed data are nearly equal, since the expected (Poisson) frequencies will by definition have these two terms identical. It is plainly unreasonable to expect two sets of values to agree if their variances are very different. In this example the variance for the observed frequencies is 1.18 compared with the value of 2 for the expected frequencies. It is therefore not surprising that the Poisson distribution is not found to fit!

Interpretation of the Probability Results

The χ^2 calculation together with the number of degrees of freedom eventually yield a probability P. This probability corresponds to the chance that results as different from the expected values (or even more different) could have arisen by chance assuming that the hypothesis is correct. The P value gives an indication of how likely it is that the hypothesis is false. It should be noted that χ^2 may be used as evidence to disprove a hypothesis if the observed and expected values differ significantly. Conversely agreement shows that the observed results could have arisen from the hypothesis, but this does not prove that the hypothesis is true (see Tables 5.14–5.16). Table 5.15 shows the usual range where the

Table 5.14   When to Reject the Hypothesis

Value of χ^2	Difference between observed and expected values	The hypothesis
Between $P=5\%$ and $P=1\%$ values	Significant	Probably untrue
Between $P=1\%$ and $P=0.1\%$ values	Highly significant	Most probably untrue
Greater than $P=0.1\%$ value	Exceedingly significant	Almost certainly untrue

Table 5.15   When the Hypothesis May be True

Value of χ^2	Difference between observed and expected values	Evidence to reject the hypothesis
Between $P=10\%$ and $P=5\%$ values	Barely significant	Very slight
Between $P=95\%$ and $P=10\%$ values	Not significant	None

Table 5.16   When to Suspect the Data

Value of χ^2	*Agreement* between observed and expected values	Data
Between $P=99\%$ and 95% values	Unbelievably good	Suspect
Less than $P=99\%$ values	Too good to be true	Highly suspect

hypothesis is accepted, but one must remember that this does not prove that the hypothesis is true.

If χ^2 is in the range shown in Table 5.16 the agreement is so good that it is improbable that such results could have arisen by chance. In these circumstances one should suspect:

(i) that the data are totally fictitious or have been 'improved' by the experimenter removing 'bad' results;

(ii) that the sample is not random, and the sampling technique should be examined;

(iii) that the hypothesis has been designed to fit the data too perfectly.

Null Hypothesis

Some books refer to the 'null hypothesis'. The null hypothesis states that there is no significant difference between the observed and expected values. Chi-squared provides a method of rejecting or not rejecting the null hypothesis. Since confusion may arise between the scientific hypothesis under test and the null hypothesis, the foregoing examples and discussion refer only to the scientific hypothesis under test.

Details of the Computer Program (Program 5.1 and Trial Runs 5.1–5.6)

First the program prints the title (line 20), then asks if the user requires full instructions. A subroutine (lines 2560–2650) checks that an answer YES or NO has been given, and continues to prompt the user if any other reply

is given. Depending on the reply, long or shortened instructions are printed during the first run with the program, but a second or subsequent run always gets shortened messages.

The user is instructed to input the data values, and up to 100 pairs of observed and expected frequencies may be typed in (lines 190–290). Each pair of values is checked to ensure that the numbers typed are valid using a subroutine (lines 1600–1790). The observed frequency must be an integer and positive or

Trial run 5.1

```
        CHI-SQUARED TEST
        === ======= ====
WOULD YOU LIKE FULL INSTRUCTIONS?   TYPE YES OR NO & PRESS RETURN.

? YES
THE PURPOSE OF THE CHI-SQUARED TEST IS TO COMPARE A SET OF
OBSERVED FREQUENCY READINGS WITH THOSE PREDICTED BY THE
HYPOTHESIS UNDER TEST.   A HIGH VALUE SUGGESTS THAT THE
HYPOTHESIS IS INCORRECT, WHILST A LOW VALUE INDICATES GOOD
AGREEMENT WITH THE HYPOTHESIS.
TYPE IN A PAIR OF OBSERVED AND EXPECTED DATA FREQUENCIES
SEPARATED BY A COMMA.
PRESS RETURN AFTER EACH PAIR OF VALUES
YOU WILL BE GIVEN THE CHANCE TO CORRECT TYPING ERRORS LATER

INPUT DATA
TERMINATE THE DATA WITH 999999, 999999
OBS FREQ, THEOR FREQ
? 105, 100
? 92, 100
? 103, 100
? 95, 100
? 97, 100
? 108, 100
? 999999, 999999
ARE THE DATA VALUES ENTERED CORRECT?   TYPE YES OR NO & PRESS RETURN.

? YES
VALUE OF CHI-SQUARED = 1.96
YOU HAVE PROVIDED 6 CLASSES (DATA PAIRS)
TYPE NUMBER OF DEGREES OF FREEDOM, OR 0 FOR AN EXPLANATION
? 5

PROBABILITY THAT EXPERIMENTAL DATA WHICH AGREES AS BADLY
OR WORSE THAN THIS COULD HAVE ARISEN BY CHANCE IS 85   %

THERE IS NO EVIDENCE TO REJECT THE HYPOTHESIS

WOULD YOU LIKE ANOTHER RUN?   TYPE YES OR NO & PRESS RETURN.

? NO
END OF JOB
```

Trial run 5.2

```
       CHI-SQUARED TEST
       === ======= ====
WOULD YOU LIKE FULL INSTRUCTIONS?  TYPE YES OR NO & PRESS RETURN.

? NO

INPUT DATA
TERMINATE THE DATA WITH 999999, 999999
OBS FREQ, THEOR FREQ
? 18, 2500
? 18, 25
? 53, 50
? 29, 25
? 999999, 999999
ARE THE DATA VALUES ENTERED CORRECT?
? NO
HERE IS A LIST OF THE CURRENT DATA
LINE NUMBER   OBS FREQ      THEOR FREQ
1             18            2500
2             18            25
3             53            50
4             29            25
TYPE R TO REPLACE  A TO ADD  D TO DELETE  L TO LIST    OR C TO CONTINUE ? D
TYPE THE LINENUMBER OF THE LINE TO BE DELETED
? 1
OK
HERE IS A LIST OF THE CURRENT DATA
LINE NUMBER   OBS FREQ      THEOR FREQ
1             18            25
2             53            50
3             29            25
TYPE R TO REPLACE  A TO ADD  D TO DELETE  L TO LIST    OR C TO CONTINUE ? C
VALUE OF CHI-SQUARED = 2.78
YOU HAVE PROVIDED 3 CLASSES (DATA PAIRS)
TYPE NUMBER OF DEGREES OF FREEDOM, OR 0 FOR AN EXPLANATION
? 2

PROBABILITY THAT EXPERIMENTAL DATA WHICH AGREES AS BADLY
OR WORSE THAN THIS COULD HAVE ARISEN BY CHANCE IS 25  %

THERE IS NO EVIDENCE TO REJECT THE HYPOTHESIS

WOULD YOU LIKE ANOTHER RUN?
? NO
END OF JOB
```

zero. The theoretical frequency is checked to ensure that it is positive, and also that it is sufficiently large (greater than or equal to five). If any of these checks fail, the data pair is rejected and the user is told to re-type the correct values. In all of these cases except where the theoretical frequency is less than five the line of data is incorrect, but in the case of a theoretical frequency below five the user must decide whether to combine the data with that from another group, or omit it altogether. The number of valid data pairs is counted, and the end of data input is signalled by the user typing the terminator 999999, 999999.

Next a subroutine is entered to check the data which have been typed in, and to alter them if necessary (lines 1800–2550). The user is asked if the data entered are correct, and the

Trial run 5.3

```
        CHI-SQUARED TEST
        === ======= ====
WOULD YOU LIKE FULL INSTRUCTIONS?   TYPE YES OR NO & PRESS RETURN.

? NO

INPUT DATA
TERMINATE THE DATA WITH 999999, 999999
OBS FREQ, THEOR FREQ
? 40, 29.75
? 16, 22.75
? 14, 17.5
? 30, 27.625
? 17, 21.125
? 18, 16.25
? 15, 27.625
? 32, 21.125
? 18, 16.25
? 999999, 999999
ARE THE DATA VALUES ENTERED CORRECT?
? YES
VALUE OF CHI-SQUARED = 18.989
YOU HAVE PROVIDED 9 CLASSES (DATA PAIRS)
TYPE NUMBER OF DEGREES OF FREEDOM, OR 0 FOR AN EXPLANATION
? 4

PROBABILITY THAT EXPERIMENTAL DATA WHICH AGREES AS BADLY
OR WORSE THAN THIS COULD HAVE ARISEN BY CHANCE IS 0.1  %

THE HYPOTHESIS IS ALMOST CERTAINLY UNTRUE

WOULD YOU LIKE ANOTHER RUN?
? NO
END OF JOB
```

YES/NO subroutine (lines 2560–2650) is used to obtain an answer of YES or NO. If the data are correct the program returns from the check subroutine, but otherwise the following procedure is adopted. The data are listed in three columns—line number, observed frequency and calculated frequency. It should be noted that if there are a lot of data, the program stops the listing when the screen contains 20 lines. This is to allow the user to examine the data before they scroll off the top of the screen on a visual display unit. The user is asked if he would like to continue listing, and if so the next 20 lines are displayed. The value of 20 is chosen by setting the variable $A4$ equal to 20 in line 1830, and this value was chosen since many visual display units show 20 lines of 80 characters. For VDUs which display 16 lines of 64 characters the value of $A4$ should be changed to 16, and for continuous listing of all the data on a printing terminal $A4$ should be set to 999.

After listing the data, instructions are printed to explain how the user may modify an existing line of data, add an extra line, delete an existing line, list the data or continue the calculation. New data are checked to ensure that they are reasonable by calling the check subroutine (lines 1600–1790). In general any number of alterations may be made to the data except that one is not allowed to add data if the arrays already hold 100 data pairs, and one is not allowed to delete all the data. If a line other than the last line is deleted, the current data are automatically listed since the deletion changes the line numbers. Extensive

Trial run 5.4

```
         CHI-SQUARED TEST
         === ======= ====
WOULD YOU LIKE FULL INSTRUCTIONS?   TYPE YES OR NO & PRESS RETURN.

? NO

INPUT DATA
TERMINATE THE DATA WITH 999999, 999999
OBS FREQ, THEOR FREQ
? 55, 50
? 45, 50
? 999999, 999999
ARE THE DATA VALUES ENTERED CORRECT?
? NO
HERE IS A LIST OF THE CURRENT DATA
LINE NUMBER    OBS FREQ       THEOR FREQ
1                55            50
2                45            50
TYPE R TO REPLACE   A TO ADD   D TO DELETE   L TO LIST    OR C TO CONTINUE ? C
VALUE OF CHI-SQUARED = 1
YOU HAVE PROVIDED 2 CLASSES (DATA PAIRS)
TYPE NUMBER OF DEGREES OF FREEDOM, OR 0 FOR AN EXPLANATION
? 1

SINCE THERE IS ONLY ONE DEGREE OF FREEDOM
CHI-SQUARED WILL BE RECALCULATED USING YATES CORRECTION
RECALCULATED VALUE FOR CHI-SQUARED = 0.81

PROBABILITY THAT EXPERIMENTAL DATA WHICH AGREES AS BADLY
OR WORSE THAN THIS COULD HAVE ARISEN BY CHANCE IS 37  %

THERE IS NO EVIDENCE TO REJECT THE HYPOTHESIS

WOULD YOU LIKE ANOTHER RUN?
? NO
END OF JOB
```

checks prevent the user altering lines which do not exist. Eventually the user will continue with the calculation.

A check is performed to ensure that there are at least two data pairs (lines 360–380), and the value of chi-squared is calculated (lines 390–420) and printed (line 430).

The user is then asked to type the number of degrees of freedom. The value typed is checked to ensure that it is an integer, less than or equal to the number of data pairs, and greater than zero. If any of these tests fail, full instructions explaining how to calculate the number of degrees of freedom are printed out, and the user is requested to retype the correct value. Instructions may be obtained by typing 0 for the number of degrees of freedom.

If the number of degrees of freedom is 1 then the value of the chi-squared is recalculated using Yates's correction (lines 700–830).

In the unlikely event that the observed and theoretical frequencies agree exactly the run is abandoned. Otherwise an empirical polynomial equation is used to calculate the probability P that results with as bad or worse agreement as the experimental frequencies could have arisen by chance (lines 890–1150). This is equivalent to manually looking up a table of probabilities for the appropriate value of chi-squared and the particular number of degrees of freedom. With tables one might say, for example, that the value was within the 95% limit. The program evaluates the exact probability and rounds the answer to give

Trial run 5.5

```
      CHI-SQUARED TEST
      === ======= ====
WOULD YOU LIKE FULL INSTRUCTIONS?  TYPE YES OR NO & PRESS RETURN.

? NO

INPUT DATA
TERMINATE THE DATA WITH 999999, 999999
OBS FREQ, THEOR FREQ
? 5, 6.01
? 19, 15.3
? 31, 30.6
? 35, 39.18
? 29, 30.6
? 17, 15.3
? 7, 6.00
? 999999, 999999
ARE THE DATA VALUES ENTERED CORRECT?
? YES
VALUE OF CHI-SQUARED = 1.9549
YOU HAVE PROVIDED 7 CLASSES (DATA PAIRS)
TYPE NUMBER OF DEGREES OF FREEDOM, OR 0 FOR AN EXPLANATION
? 4

PROBABILITY THAT EXPERIMENTAL DATA WHICH AGREES AS BADLY
OR WORSE THAN THIS COULD HAVE ARISEN BY CHANCE IS 74  %

THERE IS NO EVIDENCE TO REJECT THE HYPOTHESIS

WOULD YOU LIKE ANOTHER RUN?
? NO
END OF JOB
```

either an integer or one decimal figure in the printed probability. The program then prints a comment on the significance of the value and the validity of the hypothesis.

Finally the user is asked if he would like another run (lines 1410–1570), and the YES/NO subroutine is used to check the reply. If another run is required, the user is offered the choice of typing in completely new data, or editing the data already typed in prior to another run.

Exercises

5.1 (a) Use the chi-squared test to establish if the results from a die throwing experiment suggest that the die is fair.

Score	1	2	3	4	5	6
Frequency	47	44	69	48	45	47

(b) Test whether the die gives an abnormal proportion of 3s.

Score	3	others
Frequency	69	231

Explain the results.

5.2 The values below are extracted from 100 one digit numbers taken from a table claiming to be random numbers:

98133	55804	84863	08022	96684
22037	11087	40257	33483	10143
94299	66246	29286	92984	67425
22266	86541	77191	64578	17755

Test whether the frequencies of the digits differ significantly from random.

5.3 A laboratory assistant is responsible for growing batches of 100 pea seeds, and has to record the number of dwarf plants from each batch. From theoretical considerations it is expected that on average 25 out of each 100 grown will be dwarf. The number of dwarfs claimed for 8 batches of 100 plants were

17, 31, 21, 29, 30, 16, 20, 36

Does this support the suspicion that the laboratory assistant is careless or is manufacturing results?

5.4 A seaside resort has three beaches. A survey of the three beaches was made at the same time on the same day, with the following results:

Beach	A	B	C
Number of males	71	82	87
Number of females	79	68	113

Test whether the proportion of females differs significantly between the beaches.

5.5 A roulette wheel is suspected of being unfair. There are 37 possible scores which are labelled 0 to 36 respectively, and the gaming house wins on the score of zero. Of 999 trial spins, the house won (that is zero came up) on 44 occasions. Does this constitute evidence that the roulette wheel is biased? Would it be fair to suspect the wheel if zero had been the outcome on 35 occasions? (Remember to use Yates's correction since $v = 1$.)

5.6 Explain why the χ^2 statistic calculated from a 2×2 contingency table requires Yates's correction. 100 guinea pigs were fed with diet 1, and another 100 similar guinea pigs were fed with diet 2. Vitamin deficiency was observed in 33 animals fed with diet 1 and 45 animals fed with diet 2. Does the proportion of animals suffering vitamin deficiency differ significantly?

Trial run 5.6

```
      CHI-SQUARED TEST
      === ======= ====
WOULD YOU LIKE FULL INSTRUCTIONS?  TYPE YES OR NO & PRESS RETURN.

? NO

INPUT DATA
TERMINATE THE DATA WITH 999999, 999999
OBS FREQ, THEOR FREQ
? 8, 10.8
? 18, 21.7
? 25, 21.7
? 24, 14.4
? 5, 11.4
? 999999, 999999
ARE THE DATA VALUES ENTERED CORRECT?
? YES
VALUE OF CHI-SQUARED = 11.8516
YOU HAVE PROVIDED 5 CLASSES (DATA PAIRS)
TYPE NUMBER OF DEGREES OF FREEDOM, OR 0 FOR AN EXPLANATION
? 3

PROBABILITY THAT EXPERIMENTAL DATA WHICH AGREES AS BADLY
OR WORSE THAN THIS COULD HAVE ARISEN BY CHANCE IS 0.8 %

THE HYPOTHESIS IS MOST PROBABLY UNTRUE

WOULD YOU LIKE ANOTHER RUN?
? NO
END OF JOB
```

Program 5.1

```
10  DIM Q$(10), I$(3), F(100), T(100)
20  PRINT TAB(8); "CHI-SQUARED TEST"
30  PRINT TAB(8); "=== ======= ===="
40  REM F ARRAY CONTAINS OBSERVED & T CONTAINS THEORETICAL FREQUENCIES
50  PRINT "WOULD YOU LIKE FULL INSTRUCTIONS?";
60  GOSUB 2580
70  LET I$ = Q$
80  IF I$ = "NO" THEN 180
90  PRINT "THE PURPOSE OF THE CHI-SQUARED TEST IS TO COMPARE A SET OF"
100 PRINT "OBSERVED FREQUENCY READINGS WITH THOSE PREDICTED BY THE"
110 PRINT "HYPOTHESIS UNDER TEST.  A HIGH VALUE SUGGESTS THAT THE"
120 PRINT "HYPOTHESIS IS INCORRECT, WHILST A LOW VALUE INDICATES GOOD"
130 PRINT "AGREEMENT WITH THE HYPOTHESIS."
140 PRINT "TYPE IN A PAIR OF OBSERVED AND EXPECTED DATA FREQUENCIES"
150 PRINT "SEPARATED BY A COMMA."
160 PRINT "PRESS RETURN AFTER EACH PAIR OF VALUES"
170 PRINT "YOU WILL BE GIVEN THE CHANCE TO CORRECT TYPING ERRORS LATER"
180 PRINT
190 REM INPUT & CHECK VALIDITY OF INPUT FREQUENCIES
200 PRINT "INPUT DATA"
210 PRINT "TERMINATE THE DATA WITH 999999, 999999"
220 PRINT "OBS FREQ, THEOR FREQ"
230 LET N = 0
240 FOR I = 1 TO 100
250    REM ENTER SUBROUTINE TO INPUT DATA PAIR & CHECK THAT IT IS VALID
260    GOSUB 1620
270    IF ABS(F(I) - 999999) + ABS(T(I) - 999999) = 0 THEN 310
280    LET N = N + 1
290 NEXT I
300 PRINT "THIS PROGRAM CAN ONLY HANDLE 100 VALUES"
310 IF N > 0 THEN 350
320 PRINT "YOU MUST ENTER SOME VALID DATA"
330 GOTO 240
340 REM ENTER SUBROUTINE TO CHECK & EDIT DATA IF NECESSARY
350 GOSUB 1810
360 IF N > 1 THEN 390
370 PRINT "THERE MUST BE AT LEAST 2 DATA VALUES"
380 GOTO 870
390 LET S = 0
400 FOR I = 1 TO N
410    LET S = S + (F(I) - T(I)) * (F(I) - T(I)) / T(I)
420 NEXT I
430 PRINT "VALUE OF CHI-SQUARED ="; S
440 PRINT "YOU HAVE PROVIDED"; N; "CLASSES (DATA PAIRS)"
450 PRINT "TYPE NUMBER OF DEGREES OF FREEDOM, OR 0 FOR AN EXPLANATION"
460 INPUT D
470 IF D < N THEN 500
480 PRINT "THE NUMBER OF DEGREES OF FREEDOM MUST BE LESS THAN"; N
490 GOTO 530
500 IF D <= 0 THEN 530
510 IF D = INT(D) THEN 710
520 PRINT "THE NUMBER OF DEGREES OF FREEDOM MUST BE A WHOLE NUMBER"
530 PRINT "THE NUMBER OF DEGREES OF FREEDOM IS THE NUMBER OF CLASSES IN"
540 PRINT "THE DATA PROVIDED MINUS THE NUMBER OF RESTRICTIONS IMPOSED"
550 PRINT "IN THE CALCULATION OF THE THEORETICAL FREQUENCY.  EG. IF A"
560 PRINT "NORMAL DISTRIBUTION IS BEING TESTED FOR THEN THE THEORETICAL"
570 PRINT "FREQUENCIES WILL HAVE (1) THE SAME TOTAL, (2) SAME MEAN &"
580 PRINT "(3) THE SAME STANDARD DEVIATION AS THE OBSERVED FREQUENCIES."
590 PRINT "THUS THERE ARE THREE RESTRICTIONS. SIMILARLY IF THE POISSON"
600 PRINT "OR BINOMIAL DISTRIBUTIONS ARE BEING TESTED THERE WILL BE"
610 PRINT "ONE OR TWO RESTRICTIONS DEPENDING ON WHETHER JUST THE TOTALS"
```

```
620 PRINT "OR BOTH THE TOTALS & THE MEANS HAVE BEEN FORCED TO AGREE."
630 PRINT
640 PRINT "IF YOU ARE USING A CONTINGENCY TABLE OF H ROWS AND K COLUMNS"
650 PRINT "NUMBER OF DEGREES OF FREEDOM IS CALCULATED AS (H-1)*(K-1)"
660 PRINT "WHICH MAKES ALLOWANCE FOR AGREEMENT OF ROW AND COLUMN TOTALS"
670 PRINT "BETWEEN OBSERVED AND CALCULATED TABLES."
680 PRINT
690 GOTO 440
700 REM APPLY YATES CORRECTION TO CHI-SQUARED IF DEGREES FREEDOM = 1
710 IF D > 1 THEN 840
720 PRINT
730 PRINT "SINCE THERE IS ONLY ONE DEGREE OF FREEDOM"
740 PRINT "CHI-SQUARED WILL BE RECALCULATED USING YATES CORRECTION"
750 IF I$ = "NO" THEN 780
760 PRINT "ABSOLUTE DIFFERENCE OF EACH OBSERVED FREQ - EXPECTED FREQ"
770 PRINT "TERM HAS BEEN REDUCED BY 0.5"
780 LET S = 0
790 FOR I = 1 TO N
800    IF ABS(F(I) - T(I)) < 0.5 THEN 820
810    LET S = S + (ABS(F(I)-T(I))-0.5) * (ABS(F(I)-T(I))-0.5) / T(I)
820 NEXT I
830 PRINT "RECALCULATED VALUE FOR CHI-SQUARED ="; S
840 IF S <> 0 THEN 890
850 PRINT "THE OBSERVED FREQUENCY VALUES ALL AGREE"
860 PRINT "EXACTLY WITH THE THEORETICAL FREQUENCY"
870 PRINT "RUN ABANDONED ON THIS DATA."
880 GOTO 1410
890 LET E2 = EXP(-10)
900 LET G2 = S / 2
910 LET S1 = 0
920 LET W = 1
930 LET J = 2
940 LET M1 = D + 1
950 LET E1 = 0
960 IF D = 2 * INT(D / 2) THEN 1070
970 LET E1 = 1
980 LET J = 3
990 GOTO 1070
1000 LET S1 = S1 + W
1010 LET W = W * S / J
1020 LET J = J + 2
1030 IF W < 100000 THEN 1070
1040 LET S1 = S1 * E2
1050 LET W = W * E2
1060 LET G2 = G2 - 10
1070 IF J < M1 THEN 1000
1080 LET P2 = EXP(-G2) * S1
1090 LET P = P2
1100 IF E1 = 0 THEN 1160
1110 LET Z = SQR(S)
1120 REM EMPIRICAL POLYNOMIAL TO EVALUATE PROBABILITY
1130 LET P1 = (0.005711 * Z - 0.006523) * Z + 0.038704
1140 LET P1 = ((P1 * Z + 0.094513) * Z + 0.200039) * Z + 1
1150 LET P = 1 / (P1 * P1 * P1 * P1) + SQR(2 * S / 3.14159) * P2
1160 PRINT
1170 LET P = P * 100
1180 PRINT "PROBABILITY THAT EXPERIMENTAL DATA WHICH AGREES AS BADLY"
1190 PRINT "OR WORSE THAN THIS COULD HAVE ARISEN BY CHANCE IS";
1200 LET K = 1
1210 IF (P - 10) * (P - 95) < 0 THEN 1230
1220 LET K = 10
```

```
1230 PRINT INT(K * P + 0.5) / K; " %"
1240 PRINT
1250 REM PRINT COMMENTS ON THE VALIDITY OF THE HYPOTHESIS
1260 IF P > 0.1 THEN 1280
1270 PRINT "THE HYPOTHESIS IS ALMOST CERTAINLY UNTRUE"
1280 IF (P - 0.1) * (P - 1) > 0 THEN 1300
1290 PRINT "THE HYPOTHESIS IS MOST PROBABLY UNTRUE"
1300 IF (P - 1) * (P - 5) > 0 THEN 1320
1310 PRINT "THE HYPOTHESIS IS PROBABLY UNTRUE"
1320 IF (P - 5) * (P - 10) > 0 THEN 1340
1330 PRINT "EVIDENCE TO REJECT THE HYPOTHESIS IS VERY SLIGHT"
1340 IF (P - 10) * (P - 95) > 0 THEN 1360
1350 PRINT "THERE IS NO EVIDENCE TO REJECT THE HYPOTHESIS"
1360 IF (P - 95) * (P - 99) > 0 THEN 1380
1370 PRINT "THE AGREEMENT IS UNBELIEVABLY GOOD & THE DATA ARE SUSPECT"
1380 IF P < 99 THEN 1410
1390 PRINT "THE AGREEMENT IS TOO GOOD TO BE TRUE & THE DATA"
1400 PRINT "ARE HIGHLY SUSPECT"
1410 PRINT
1420 PRINT "WOULD YOU LIKE ANOTHER RUN?";
1430 GOSUB 2570
1440 IF Q$ = "NO" THEN 1580
1450 LET I$ = "NO"
1460 PRINT "TYPE NEW FOR A RUN WITH COMPLETELY NEW DATA"
1470 PRINT "  OR OLD TO EDIT & RERUN THE EXISTING DATA"
1480 INPUT Q$
1490 IF Q$ = "NEW" THEN 1550
1500 IF Q$ = "OLD" THEN 1530
1510 PRINT "REPLY '"; Q$; "' NOT UNDERSTOOD"
1520 GOTO 1460
1530 GOSUB 1860
1540 GOTO 360
1550 PRINT
1560 PRINT "NEW SET OF DATA"
1570 GOTO 180
1580 PRINT "END OF JOB"
1590 STOP
1600 REM SUBROUTINE TO INPUT & CHECK VALIDITY OF INPUT FREQUENCIES
1610 PRINT "OBS FREQ, THEOR FREQ"
1620 INPUT F(I), T(I)
1630 IF F(I) = INT(F(I)) THEN 1660
1640 PRINT "OBSERVED FREQUENCY MUST BE A WHOLE NUMBER"
1650 GOTO 1770
1660 IF F(I) >= 0 THEN 1690
1670 PRINT "OBSERVED FREQUENCY MUST BE POSITIVE"
1680 GOTO 1770
1690 IF T(I) >= 0 THEN 1720
1700 PRINT "A NEGATIVE THEORETICAL FREQUENCY IS IMPOSSIBLE"
1710 GOTO 1770
1720 IF T(I) >= 5 THEN 1790
1730 PRINT "THE CHI-SQUARED DERIVATION REQUIRES THAT ALL EXPECTED"
1740 PRINT "FREQUENCIES ARE SUFFICIENTLY LARGE.  THIS IS USUALLY TAKEN"
1750 PRINT "TO MEAN A MINIMUM OF FIVE."
1760 PRINT "EITHER OMIT THE DATA PAIR, OR MERGE IT WITH ANOTHER GROUP."
1770 PRINT "DATA PAIR REJECTED - RETYPE CORRECTLY"
1780 GOTO 1620
1790 RETURN
1800 REM SUBROUTINE TO CHECK THAT DATA ARE CORRECT & ALTER IF NECESSARY
1810 PRINT "ARE THE DATA VALUES ENTERED CORRECT?";
1820 REM A4 SHOULD BE SET TO THE NUMBER OF LINES ON THE VDU
1830 LET A4 = 20
```

```
1840 GOSUB 2570
1850 IF Q$ = "YES" THEN 2550
1860 PRINT "HERE IS A LIST OF THE CURRENT DATA"
1870 PRINT "LINE NUMBER","OBS FREQ","THEOR FREQ"
1880 FOR I = 1 TO N
1890    PRINT I, F(I), T(I)
1900    IF INT(I / (A4 - 1)) * (A4 - 1) <> I THEN 1940
1910    PRINT "WOULD YOU LIKE TO CONTINUE LISTING";
1920    GOSUB 2570
1930    IF Q$ = "NO" THEN 1950
1940 NEXT I
1950 PRINT "TYPE R TO REPLACE";
1960 IF I$ = "NO" THEN 1980
1970 PRINT " AN EXISTING LINE OF DATA"
1980 IF N = 100 THEN 2030
1990 PRINT TAB(5); " A TO ADD";
2000 IF I$ = "NO" THEN 2020
2010 PRINT " AN EXTRA LINE"
2020 IF N = 1 THEN 2060
2030 PRINT TAB(5); " D TO DELETE";
2040 IF I$ = "NO" THEN 2060
2050 PRINT " AN EXISTING LINE"
2060 PRINT TAB(5); " L TO LIST";
2070 IF I$ = "NO" THEN 2090
2080 PRINT " THE DATA"
2090 PRINT "  OR C TO CONTINUE";
2100 IF I$ = "NO" THEN 2120
2110 PRINT " THE CALCULATION"
2120 INPUT Q$
2130 IF Q$ = "R" THEN 2230
2140 IF N = 100 THEN 2170
2150 IF Q$ = "A" THEN 2330
2160 IF N = 1 THEN 2180
2170 IF Q$ = "D" THEN 2400
2180 IF Q$ = "L" THEN 1860
2190 IF Q$ = "C" THEN 2550
2200 PRINT "REPLY '"; Q$; "' NOT UNDERSTOOD."
2210 GOTO 1950
2220 REM REPLACE LINE
2230 PRINT "TYPE THE LINENUMBER OF THE LINE TO BE REPLACED";
2240 INPUT I
2250 IF I <> INT(I) THEN 2270
2260 IF (I - 1) * (I - N) <= 0 THEN 2300
2270 PRINT "LINENUMBER MUST BE AN INTEGER IN THE RANGE 1 -"; N
2280 PRINT "RE-";
2290 GOTO 2230
2300 PRINT "TYPE THE CORRECT LINE TO REPLACE THE ONE WHICH IS WRONG:"
2310 GOTO 2370
2320 REM ADD A NEW LINE
2330 LET N = N + 1
2340 LET I = N
2350 PRINT "TYPE THE ADDITIONAL LINE OF DATA AS SHOWN:"
2360 REM ENTER SUBROUTINE TO INPUT DATA PAIR & CHECK THAT IT IS VALID
2370 GOSUB 1610
2380 PRINT "OK"
2390 GOTO 1950
2400 REM DELETE A LINE
2410 PRINT "TYPE THE LINENUMBER OF THE LINE TO BE DELETED"
2420 INPUT J
2430 IF (J - 1) * (J - N) > 0 THEN 2450
2440 IF J = INT(J) THEN 2470
```

```
2450 PRINT "LINENUMBER MUST BE AN INTEGER IN THE RANGE 1 -"; N
2460 GOTO 2410
2470 FOR I = J + 1 TO N
2480    LET F(I - 1) = F(I)
2490    LET T(I - 1) = T(I)
2500 NEXT I
2510 LET N = N - 1
2520 PRINT "OK"
2530 IF J > N THEN 1950
2540 GOTO 1860
2550 RETURN
2560 REM SUBROUTINE TO CHECK REPLIES
2570 IF I$ = "NO" THEN 2590
2580 PRINT " TYPE YES OR NO & PRESS RETURN."
2590 PRINT
2600 INPUT Q$
2610 IF Q$ = "YES" THEN 2650
2620 IF Q$ = "NO" THEN 2650
2630 PRINT "REPLY '"; Q$; "' NOT UNDERSTOOD.";
2640 GOTO 2580
2650 RETURN
2660 END
```

5.7 The number of deaths recorded in the obituary column of a newspaper were recorded over a period of 700 publications, and yielded the following results:

Number of deaths recorded on a particular day

0	1	2	3	4	5 or more

Number of days

205	240	165	70	15	5

Test whether the number of deaths differs significantly from a Poisson distribution with the same mean.

5.8 When studying plant genetics, Mendel found 315 round and yellow peas, 108 round and green, 101 wrinkled and yellow, and 32 wrinkled and green. His theory of dominant and recessive characters in heredity predicts that these numbers should be in the ratio 9:3:3:1. Do the observed results agree with the theory, and are the results so good that one must suspect their authenticity?

5.9 The energy intakes for 77 healthy middle-aged men who all had the same sedentary occupation were measured together with their physical activity in leisure time. The results are recorded in Table 5.17 as high, medium and low energy intake and activity. Test whether there is any relationship between energy intake and physical activity in leisure time for this group of men.

Table 5.17

Energy intake	Physical activity in leisure time		
	High	Medium	Low
High	9	10	5
Medium	3	12	9
Low	4	9	16

6

Comparison of Two Samples

Two sets of data have been collected. The first set comprises n_1 readings and has a mean m_1. The second set has n_2 readings and a mean of m_2. The estimated standard deviations s_1 and s_2 of the parent populations from which the two samples were drawn are calculated:

$$s_1 = \sqrt{[\Sigma(x_1 - m_1)^2/(n_1 - 1)]}$$

and

$$s_2 = \sqrt{[\Sigma(x_2 - m_2)^2/(n_2 - 1)]}$$

The problem is to establish if the two sets of data differ significantly, that is whether the difference in means $|m_1 - m_2|$ is significantly large compared with the standard deviations s_1 and s_2.

With large sets of data the method of doing the comparison may be summarised:

(i) The mean m_1 for the first set of data is known. However, if another sample of n_1 readings was drawn from the same parent population, it is probable that a slightly different mean m_1 would be obtained. The first step is to estimate by how much m_1 varies, that is to estimate the standard error of m_1.

(ii) In a similar way an estimate is made of the standard error of m_2 for the second set of data.

(iii) The estimated values for the standard errors for m_1 and m_2 are then used to estimate the standard error of the difference in means $|m_1 - m_2|$.

(iv) The value of $|m_1 - m_2|$ is divided by the standard errors of the difference in means calculated in (iii) to give the number of standard errors variation.

(v) This value is looked up in a *normal distribution table* (Appendix 4) to find the probability of obtaining a difference in means $|m_1 - m_2|$ as large or larger than this.

Derivation of the Test for Comparing Large Samples

'If samples of size n are drawn at random from a parent population of mean μ and standard deviation σ, the sample means constitute a population of mean μ and whose standard deviation tends to $\sigma/\sqrt{n}$.' (See Chapter 3.)

(i) For the first set of data, the standard deviation s_1 is the best estimate for σ the standard deviation of the parent population. The mean value m_1 is a sample mean of n_1 values from the parent population. If further sets of data are taken from the parent population, further values of m_1 will be obtained. From the central limit theorem, the standard error of the m_1 values tends to $\sigma/\sqrt{n_1}$ which equals $s_1/\sqrt{n_1}$.

The best estimate of the standard error of m_1 is $s_1/\sqrt{n_1}$.

(ii) The best estimate of the standard error of m_2 is $s_2/\sqrt{n_2}$.

(iii) It is shown in Appendix 8 that:

The variance of the *sum* of two terms = The *sum* of the variances of the two terms.

69

The variance of the *difference* of two terms = The *sum* of the variances of the two terms.

Thus the variance of $m_1 - m_2$ is the variance of m_1 plus the variance of m_2. Since variance is standard error squared, the variance of $m_1 - m_2$ is:

$$s_1^2/n_1 + s_2^2/n_2$$

and hence the standard error of $m_1 - m_2$ is:

$$\sqrt{\left(\frac{s_1^2}{n_1} + \frac{s_2^2}{n_2} \right)} \quad (1)$$

(iv) The number of standard errors variation of this particular mean difference $|m_1 - m_2|$ is calculated:

$$= \frac{\text{Particular } |m_1 - m_2|}{\text{Standard error for } m_1 - m_2} \quad (2)$$

$$= \frac{|m_1 - m_2|}{\sqrt{\left(\frac{s_1^2}{n_1} + \frac{s_2^2}{n_2} \right)}} \quad (3)$$

(v) The number of standard errors calculated from Equation 3 is looked up in a two-tailed normal distribution table (Appendix 4) to obtain a probability. This probability is the chance that a mean difference smaller than $|m_1 - m_2|$ would occur if the two parent populations have in fact the same mean value.

For example if the value calculated from Equation 3 exceeds 1.96, then there is less than a 5% chance that so large a difference between m_1 and m_2 could have arisen if the means of the two parent populations are the same. (This is because 1.96 standard deviations correspond to a 5% two-tailed normal probability.) Since there is less than a 1 in 20 chance of this occurring, one would suspect that the means of the two parent populations are different.

Example 1

Fifty pupils entered school 1 and sixty pupils

of the same age entered school 2. On leaving school both groups were given a common mathematics examination, which yielded the statistical results shown in Table 6.1. It is required to know if the two groups of pupils

Table 6.1

	School 1	School 2
Number of pupils	50	60
Mean mark obtained	62	58
Estimated standard deviation	12	8

differ significantly in attainment, or whether they are essentially as good and the observed variations attributed to random chance.

Using Equation 3 the difference $(62 - 58 = 4)$ in the mean marks is converted into a number of standard errors:

Number of standard errors

$$= \frac{|62 - 58|}{\sqrt{\left(\frac{12^2}{50} + \frac{8^2}{60} \right)}} = 2.01$$

Reference to the table of areas under a normal curve given in Appendix 4 shows that this corresponds to a one-tailed probability of 97.8%. This indicates that if the pupils from both schools were really equal in attainment then a result such as the one obtained in favour of school 1 would only occur 2.2% of the time. (There is also a 2.2% chance that school 2 would be superior by the same margin.) Since this results will only occur once in about 45 surveys, it lends moderately strong support to the statement that 'pupils at school 1 are better than those at school 2 *in that particular type of mathematics examination*'.

Description of Program to Compare Large Samples (See Program 6.1)

First a heading is printed (lines 20–30). Next a

loop from lines 50–200 is executed twice to input values of mean, standard deviation and number of readings for the two sets. (These values can be obtained using the standard deviation program in Chapter 3.) A number of checks are performed on the input data:

(i) There must be a whole number of readings.
(ii) The number of readings must be large, that is at least 30. (For small samples the t-test should be used—see later in this chapter.)
(iii) The standard deviation must not be negative since this is physically impossible!

Unacceptable data values are rejected, with warning messages, and the user is asked to re-type the correct value.

A check is performed (line 220) and a warning message printed (lines 230–290) if both the standard deviations are zero. Generally this is not the case, and a further check is performed at line 300 to see if both means are different. If the means are the same, a warning

message is printed (line 310). Line 320 calculates the number of standard errors variation using Equation 3.

A subroutine (lines 610–770) is called to calculate the area under the normal curve up to the appropriate number of standard errors. This saves the user the effort of looking the number of standard errors up in a 'normal' table. The subroutine uses an order 11 empirical polynomial, and is based on the constants and method used by the Numerical Algorithms Group. The subroutine differs only slightly from that used in Appendix 4.

The probability is printed (lines 360–410) rounded to a whole number if it is in the range 5%–95%, or rounded to give one decimal figure if it is outside this range. A message which interprets the probability is then printed (lines 420–500). These explain that there is no evidence, some evidence or strong evidence to support the theory that the two parent populations have different means.

Finally, the user is offered another run (lines 520–530). The answer must be YES or NO and is checked in lines 550–580.

Program 6.1   Trial run.

```
SIGNIFICANCE OF THE DIFFERENCE IN MEAN BETWEEN TWO LARGE SETS
============ == === ========== == ==== ======= === ===== ====

TYPE IN THE NUMBER OF READINGS IN SET 1
? 50
TYPE THE MEAN VALUE FOR SET 1
? 62
TYPE IN THE STANDARD DEVIATION OF SET 1
? 12

TYPE IN THE NUMBER OF READINGS IN SET 2
? 60
TYPE THE MEAN VALUE FOR SET 2
? 58
TYPE IN THE STANDARD DEVIATION OF SET 2
? 8

THE PROBABILITY THAT THE TWO MEANS COULD BE SO DIFFERENT
IF THEIR PARENT POPULATIONS HAVE THE SAME MEAN IS 2.2 %

THERE IS SOME  EVIDENCE TO SUPPORT THE THEORY
THAT THE TWO PARENT POPULATIONS HAVE DIFFERENT MEANS

WOULD YOU LIKE ANOTHER RUN  TYPE YES OR NO & PRESS RETURN
? NO
END OF JOB
```

```
10 DIM Q$(10), N(2), M(2), S(2)
20 PRINT "SIGNIFICANCE OF THE DIFFERENCE IN MEAN BETWEEN TWO LARGE SETS"
30 PRINT "============ == === ========== == ==== ======= === ===== ===="
40 PRINT
50 FOR I = 1 TO 2
60    PRINT "TYPE IN THE NUMBER OF READINGS IN SET"; I
70    INPUT N(I)
80    IF N(I) <> INT(N(I)) THEN 100
90    IF N(I) >= 30 THEN 120
100   PRINT "THE NUMBER OF READINGS MUST BE A WHOLE NUMBER AT LEAST 30"
110   GOTO 60
120   PRINT "TYPE THE MEAN VALUE FOR SET"; I
130   INPUT M(I)
140   PRINT "TYPE IN THE STANDARD DEVIATION OF SET"; I
150   INPUT S(I)
160   IF S(I) >= 0 THEN 190
170   PRINT "A STANDARD DEVIATION MUST BE POSITIVE - RETYPE CORRECTLY"
180   GOTO 150
190   PRINT
200 NEXT 1
210 REM CHECK THAT BOTH STANDARD DEVIATIONS ARE NOT ZERO
220 IF S(1) + S(2) > 0 THEN 300
230 PRINT "SINCE BOTH STANDARD DEVIATIONS ARE ZERO, THE CHANCE OF"
240 PRINT "RESULTS SUCH AS THESE OCCURRING BY CHANCE IS";
250 IF M(1) = M(2) THEN 280
260 PRINT "0%"
270 GOTO 510
280 PRINT "100%"
290 GOTO 510
300 IF M(1) <> M(2) THEN 320
310 PRINT "BOTH MEANS ARE THE SAME HENCE"
320 LET S = ABS(M(1)-M(2)) / SQR(S(1)*S(1) / N(1) + S(2)*S(2) / N(2))
330 REM ***CALCULATE AREA UNDER NORMAL CURVE UP TO S STANDARD
340 REM ***ERRORS ABOVE THE MEAN
350 GOSUB 630
360 PRINT "THE PROBABILITY THAT THE TWO MEANS COULD BE SO DIFFERENT"
370 PRINT "IF THEIR PARENT POPULATIONS HAVE THE SAME MEAN IS";
380 LET K = 1
390 IF (P - 0.05) * (P - 0.95) < 0 THEN 410
400 LET K = 10
410 PRINT INT(P * 100 * K + 0.5) / K; "%"
420 PRINT
430 IF P < 0.05 THEN 450
440 PRINT "THERE IS NO";
450 IF (P - 0.05) * (P - 0.01) >= 0 THEN 470
460 PRINT "THERE IS SOME";
470 IF P > 0.01 THEN 490
480 PRINT "THERE IS STRONG";
490 PRINT " EVIDENCE TO SUPPORT THE THEORY"
500 PRINT "THAT THE TWO PARENT POPULATIONS HAVE DIFFERENT MEANS"
510 PRINT
520 PRINT "WOULD YOU LIKE ANOTHER RUN ";
530 PRINT "TYPE YES OR NO & PRESS RETURN"
540 INPUT Q$
550 IF Q$ = "YES" THEN 40
560 IF Q$ = "NO" THEN 590
570 PRINT "REPLY '"; Q$; " NOT UNDERSTOOD.   RE-";
580 GOTO 530
590 PRINT "END OF JOB"
600 STOP
610 REM CALC CUMULATIVE AREA UNDER NORMAL CURVE
```

```
620 REM CONSTANTS SET FOR 8 FIGURE ACCURACY
630 LET X = S * 0.707107
640 LET P = 0
650 IF X >= 9.5 THEN 770
660 LET T = 1 - 7.5 / (ABS(X) + 3.75)
670 LET Y = 0
680 FOR I = 1 TO 12
690    READ C
700    LET Y = Y * T + C
710 NEXT I
720 RESTORE
730 DATA 3.14753E-05, -0.000138746, -6.41279E-06, 0.00178663
740 DATA -0.00823169, 0.0241519, -0.0547992, 0.102602
750 DATA -0.163572, 0.226008, -0.273422, 0.14559
760 LET P = 0.5 * EXP(-X * X) * Y
770 RETURN
780 END
```

t-Test

The t-test should be used for comparing the mean values of two small samples, and the procedure is as follows. *The assumption is made that both samples have been drawn from populations with the same standard deviation (and hence the same variance), even if the means of the two parent populations differ.*

(a) An estimate of the standard deviations of the two parent populations is calculated using n_1, n_2, s_1 and s_2.

(b) The value of s thus obtained is used to estimate the standard error of the difference between the means, that is the standard error of $| m_1 - m_2 |$.

(c) The value of $| m_1 - m_2 |$ is divided by the standard error of the means calculated in (b) to give the number of standard errors variation.

(d) This value is looked up in a t-table (Appendix 6) to find the probability of obtaining a difference in means $| m_1 - m_2 |$ as large or larger than this. The t-distribution must be used because there are only a small number of readings, and a normal approximation is no longer applicable. It should be noted that the t-distribution tends to the normal distribution as the number of readings becomes large.

Derivation of the t-Test for Comparing the Means of Small Samples

(a) The equation to calculate s the estimated standard deviation of the two parent populations is:

$$s = \sqrt{\left[\frac{\Sigma(x_1 - m_1)^2 + \Sigma(x_2 - m_2)^2}{(n_1 - 1) + (n_2 - 1)} \right]}$$
$$= \sqrt{\left[\frac{(n_1 - 1)s_1^2 + (n_2 - 1)s_2^2}{(n_1 - 1) + (n_2 - 1)} \right]} \quad (4)$$

The numerator represents the sum of the squared differences of each of the x_1 values from the mean m_1 plus the sum of the squared differences of each of the x_2 values from the mean m_2. This is similar to the term $\Sigma(x_1 - \bar{x})^2$ used in the usual calculation of standard deviations (Chapter 2, Equation 7). The denominator comprises the number of terms in the first group minus one, and the number of terms in the second group minus one. The minus one corresponds to the loss of one degree of freedom from each group. This is because if one is given $n_1 - 1$ of the differences $(x_1 - m_1)$, the last difference can be calculated since the sum of all the differences is zero (because of the definition of the mean.) The same applies to the second set of data. Further discussion of this point is given in Chapter 2 (explanation of the divisors n and $n - 1$).

(b) Since a single standard deviation s (for

73

both parent populations) is to be used instead of s_1 and s_2 (for the individual sets of data), Equation 1 may be re-written

Standard error of $m_1 - m_2$

$$= \sqrt{\left(\frac{s^2}{n_1} + \frac{s^2}{n_2}\right)} = s\sqrt{\left(\frac{1}{n_1} + \frac{1}{n_2}\right)} \quad (5)$$

(c) Substituting Equation 5 into Equation 2

Number of standard errors variation of $m_1 - m_2$

$$= \frac{|m_1 - m_2|}{s\sqrt{\left(\frac{1}{n_1} + \frac{1}{n_2}\right)}} \quad (6)$$

(d) The number of standard errors calculated in Equation 6 is looked up in a t-table (Appendix 6) for the appropriate number of degrees of freedom v. The value of v may be calculated

$$v = (n_1 - 1) + (n_2 - 1)$$
$$= n_1 + n_2 - 2$$

If for example v is 10 and the number of standard errors is 2.76, the table yields a two-tailed probability of 2%. This means that there is only a 1 in 50 chance that the means of the two parent populations are the same, thus providing strong evidence that the means are different.

It should be noted that in the derivation of the t-test it was assumed that the two parent populations both had the same standard deviation. If there is a large difference in the two standard deviations for the samples (s_1 and s_2), then the assumption of a single standard deviation for the two parent populations is unjustified, and application of the t-test is inappropriate.

Example 2

A dealer has tested motor car tyres from two different manufacturers fitted to the same model of car, to find if one make gives a significantly better mileage than the other.

The mileage covered by each tyre before it reached the legal limit is shown to the nearest 100 miles in Table 6.2. Using sample 1 the estimated standard deviation s_1 for tyres from manufacturer 1 is calculated as

$$s_1 = \sqrt{\left[\frac{\Sigma(x_1 - m_1)^2}{(n_1 - 1)}\right]} = 800 \text{ miles}$$

Similarly $s_2 = 600$ miles.

A superficial examination of these values might suggest that since tyres from manufacturer 2 give on average 1000 miles more wear, they must be better.

The statistical examination of these results requires first the calculation of the estimated standard deviation of both parent populations from Equation 4. This gives a value of $s = 727$. Equation 6 is then used to calculate the number of standard errors variation of $m_1 - m_2$, giving a value of 3.17. Reference to the t-table given in Appendix 6 shows that for $(13 - 1) + (9 - 1) = 20$ degrees of freedom, the value of 3.17 lies in between the values for 1% and 0.1% probability. This means that there is less than a 1% (two-tailed) chance of such a large discrepancy occurring between the two makes of tyres if the two parent populations have the same mean mileage. This provides strong evidence that the two parent populations do not have the same mean, and that tyres from manufacturer 2 last longer.

Table 6.2

Manufacturer 1		Manufacturer 2	
25 800	27 600	28 400	27 500
27 200	25 600	26 900	27 200
26 400	27 000	28 400	
25 800	28 000	27 500	
28 000	26 400	28 400	
27 200	26 800	27 500	
26 600		28 400	
Number of terms		Number of terms	
$n_1 = 13$		$n_2 = 9$	
Mean mileage		Mean mileage	
$m_1 = 26\ 800$		$m_2 = 27\ 800$	

F-Test (Variance Ratio)

The F-test is used to compare the variances (or standard deviations) of two set of data, to determine whether they differ significantly. This test may be applied to any size of set, and is particularly useful to establish whether the assumption made in the t-test can be justified for a particular set of data. (The assumption made in the t-test is that both samples have been drawn from populations with the same variance and hence the same standard deviation.)

The procedure for applying the F-test is as follows:

(i) Adopt the hypothesis that the variances of the two parent populations, from which the two sets of data were drawn, are the same. This is often referred to as the null hypothesis.

(ii) The F-statistic is calculated

$$F = s_1^2/s_2^2 \qquad (7)$$

where s_1 and s_2 are the estimated standard deviations of the two parent populations.

For convenience the values are arranged so that s_1 is greater than s_2 which results in a value of $F > 1$. As a consequence only half as many values need tabulating in significance tables.

(iii) The estimated standard deviations for the two parent populations can be calculated

$$s_1 = \sqrt{[\Sigma(x_1 - m_1)^2/(n_1 - 1)]}$$

and

$$s_2 = \sqrt{[\Sigma(x_2 - m_2)^2/(n_2 - 1)]}$$

(iv) The calculated F value is looked up in the tables given in Appendix 7. First the table showing the appropriate significance level is selected. (Tables are provided for 10%, 5%, 1% and 0.1% significance levels.) The number of degrees of freedom in the two sets v_1 and v_2 are calculated from the number of readings n_1 and n_2

$$v_1 = n_1 - 1 \qquad (8)$$

and

$$v_2 = n_2 - 1 \qquad (9)$$

The calculated F value is then compared with the value from the table for the appropriate values of v_1 and v_2. For example if $n_1 = 21$, $n_2 = 6$, $s_1 = 2$ and $s_2 = 5$. Then s_1 is smaller then s_2, hence the sets 1 and 2 are interchanged giving

$$n_1 = 6, n_2 = 21, s_1 = 5 \text{ and } s_2 = 2$$

$$F = 5^2/2^2 = 6.25$$

$$v_1 = 6 - 1 = 5$$

$$v_2 = 21 - 1 = 20$$

Using the 1% significance table with $v_1 = 5$ and $v_2 = 20$ yields an F value of 4.1. Since the calculated F value of 6.25 is greater than 4.1, there is less than a 1% chance that results as divergent as this could have arisen by chance if the variances of the two parent populations are the same. This provides strong evidence that the two parent populations have different variances. (The null hypothesis is therefore rejected.)

Comparison with the 0.1% significance table yields an F value of 6.46. There is therefore more than 0.1% chance of obtaining the observed result by chance if the parent population variances are the same.

It is interesting to note that the F values with $v_1 = 1$ correspond to the t value (with $v = v_2$) squared.

For example:

the 5% F value for $v_1 = 1$ and $v_2 = 20$ is 4.35

the 5% t value for $v = 20$ is 2.086

and $2.086^2 = 4.351$.

Example 3

The analysis in Example 2 was based on the assumption that the variance and standard deviation of both parent populations were the same. Superficially there appears to be a large difference in the standard deviations s_1 and s_2 which were estimated to be 800 and 600 miles respectively for manufacturers 1 and 2. (The variances show an even bigger difference: 640 000 and 360 000.) The F-test provides a statistical means of determining whether the difference in standard deviations s_1 and s_2 is so large that the assumption must be rejected, and the t-test results invalidated.

Using Equation 7

$$F = 800^2/600^2 = 1.78$$

Using equations 8 and 9

$$v = 13-1 = 12$$

$$v_2 = 9-1 = 8$$

Referring to Appendix 7.1 (10% F-test probability) yields a critical value of 2.50. Since the calculated F of 1.78 is less than 2.50, then there is more than a 10% chance of such results occurring if the two parent populations have the same standard deviation. There is therefore no evidence to reject the hypothesis that the standard deviations of both parent populations are the same. It follows that the assumption made in the t-test was valid.

Had the calculated F value been 5.7, then this is larger than the critical value from the 1% table (Appendix 7.3), then there is less than a 1% chance that such a result could have occurred if the hypothesis (that the standard deviations of both parent populations are the same) is true. This would provide strong evidence that the standard deviations are in fact different, and the t-test result would be invalidated.

Description of Program for Comparing Two Samples by the F and t-Tests

The program (Program 6.2) first prints a heading (lines 20–30), and then executes a loop twice from lines 60 to 260. Inside this loop the data values for each of the two groups are input, the total number of points in each group is counted, the sum of errors squared is calculated, and the mean is evaluated. The data values are not stored in arrays, and hence there is, in principle, no limit on the number of data points. However, in this program the F-test calculation cannot be performed if either group has more than 60 values, for reasons which are discussed later. This restriction does not apply to the t-test, but if the samples are large it is recommended that the program for comparison of large samples be used instead.

Using a loop from lines 280 to 380, the mean, sum of errors squared, estimated standard deviation of parent population, number of data points and the number of degrees of freedom are printed for the two groups in turn.

Next three checks are performed (lines 390–500) to ensure that the data provided are suitable for the F- and t-tests.

(i) There must be some data values in both groups.
(ii) There must be a minimum of three data values in the two groups combined in order to calculate the standard deviation s of the two parent populations. (A value of 2 would cause failure through attempting to divide by zero in Equation 4.)
(iii) All of the values entered must not be identical.

If the data fail any of these checks, a warning message is printed, calculation of both the F- and t-tests is skipped, and a message (lines 1600–1610) asks if another run is required. Provided that the data are acceptable, a message (lines 570–580) asks whether an F-test is to be performed. The reply must be either YES or NO, and is checked in a subroutine (lines 1740–1800). If an F-test is requested, three additional checks are carried out on the data:

(i) Each group must contain at least 2 numbers so that the group has a standard deviation.
(ii) A maximum of 60 values in each group is imposed by the program. This is to prevent underflow and overflow errors, which occurred at 10^{-18} and 10^{18} respectively on one of the test computers. Many computers handle a

Program 6.2   Trial run.

```
    F  AND  T  TESTS
    =  ===  =  =====

INPUT NUMBERS IN GROUP 1
PRESS RETURN AFTER EACH NUMBER  & TERMINATE DATA WITH 999999
? 25800
? 27200
? 26400
? 25800
? 28000
? 27200
? 26600
? 27600
? 25600
? 27000
? 28000
? 26400
? 26800
? 999999
INPUT NUMBERS IN GROUP 2
PRESS RETURN AFTER EACH NUMBER  & TERMINATE DATA WITH 999999
? 28400
? 26900
? 28400
? 27500
? 28400
? 27500
? 28400
? 27500
? 27200
? 999999

GROUP 1
 MEAN = 26800
 SUM OF ERRORS SQUARED = 7.68E+06
 STANDARD DEVIATION = 800
 NUMBER OF DATA POINTS = 13
 NUMBER OF DEGREES OF FREEDOM = 12

GROUP 2
 MEAN = 27800
 SUM OF ERRORS SQUARED = 2.88E+06
 STANDARD DEVIATION = 600
 NUMBER OF DATA POINTS = 9
 NUMBER OF DEGREES OF FREEDOM = 8


WOULD YOU LIKE TO PERFORM AN F TEST? (YES/NO)
? YES
CALCULATED VALUE OF F = 1.77778
PROBABILITY THAT SUCH A DIFFERENCE IN VARIANCES COULD OCCUR
BY CHANCE IF THE PARENT POPULATIONS HAVE THE SAME VARIANCE
IS 21.1 %


WOULD YOU LIKE TO PERFORM A T TEST?
TYPE YES OR NO AND PRESS RETURN
? YES
THIS TEST IS ONLY MEANINGFUL WHEN THE STANDARD DEVIATIONS
OF THE TWO GROUPS DO NOT DIFFER SIGNIFICANTLY
```

```
CALCULATED VALUE OF T = 3.17369
PROBABILITY THAT SUCH A DIFFERENCE IN MEANS COULD OCCUR BY
CHANCE IF THE PARENT POPULATIONS HAVE THE SAME MEAN AND
VARIANCE IS 0.478 %

WOULD YOU LIKE ANOTHER RUN? (YES/NO)
? NO
END OF JOB



10 DIM K(2), A(2), Y(2), B(4), Q$(10)
20 PRINT TAB(6); "F AND T TESTS"
30 PRINT TAB(6); "= === = ====="
40 PRINT
50 REM START LOOP TO INPUT DATA
60 FOR J = 1 TO 2
70    LET A(J) = 0
80    LET Y(J) = 0
90    LET I = 1
100    LET Z = 0
110    PRINT "INPUT NUMBERS IN GROUP"; J
120    PRINT "PRESS RETURN AFTER EACH NUMBER ";
130    PRINT "& TERMINATE DATA WITH 999999"
140    INPUT X
150    IF X = 999999 THEN 250
160    LET Z = Z + X
170    LET X1 = X - Y(J)
180    REM CALCULATE SUM OF ERRORS SQUARED
190    LET A(J) = A(J) + X1 * X1 * (I - 1) / I
200    REM CALCULATE MEAN
210    LET Y(J) = Z / I
220    LET I = I + 1
230    GOTO  140
240    REM K(J) = NUMBER OF POINTS IN CURRENT GROUP
250    LET K(J) = I - 1
260 NEXT J
270 REM LOOP TO PRINT STATISTICS FOR EACH GROUP
280 FOR J = 1 TO 2
290    PRINT
300    PRINT "GROUP"; J
310    IF K(J) = 0 THEN 360
320    PRINT TAB(2); "MEAN ="; Y(J)
330    PRINT TAB(2); "SUM OF ERRORS SQUARED ="; A(J)
340    IF K(J) = 1 THEN 360
350    PRINT TAB(2); "STANDARD DEVIATION ="; SQR(A(J) / (K(J) - 1))
360    PRINT TAB(2); "NUMBER OF DATA POINTS ="; K(J)
370    PRINT TAB(2); "NUMBER OF DEGREES OF FREEDOM ="; K(J) - 1
380 NEXT J
390 LET V1 = K(1) - 1
400 IF V1 > -1 THEN 430
410 PRINT "NO NUMBERS IN FIRST GROUP"
420 GOTO  1590
430 LET V2 = K(2) - 1
440 IF V2 > -1 THEN 470
450 PRINT "NO NUMBERS IN SECOND GROUP"
460 GOTO  1590
470 LET V = V1 + V2
480 IF V > 0 THEN 510
490 PRINT "THERE MUST BE AT LEAST 3 NUMBERS IN BOTH GROUPS COMBINED"
500 GOTO  1590
```

```
510 IF A(1) + A(2) > 0 THEN 560
520 IF Y(1) <> Y(2) THEN 560
530 PRINT "F AND T TEST RESULTS INDETERMINATE"
540 PRINT "BECAUSE ALL THE DATA VALUES ARE EQUAL."
550 GOTO 1590
560 PRINT
570 PRINT "WOULD YOU LIKE TO PERFORM AN F TEST? (YES/NO)"
580 GOSUB 1750
590 IF Q$ = "NO" THEN 1370
600 IF V1 > 0 THEN 630
610 PRINT "THERE MUST BE AT LEAST 2 NUMBERS IN FIRST GROUP FOR F TEST"
620 GOTO 1370
630 IF V2 > 0 THEN 660
640 PRINT "THERE MUST BE AT LEAST 2 NUMBERS IN SECOND GROUP FOR F TEST"
650 GOTO 1370
660 IF V1 < 61 THEN 690
670 PRINT "TOO MANY NUMBERS IN FIRST GROUP FOR F TEST"
680 GOTO 1370
690 IF V2 < 61 THEN 720
700 PRINT "TOO MANY NUMBERS IN SECOND GROUP FOR F TEST"
710 GOTO 1370
720 IF A(1) * A(2) > 0 THEN 750
730 PRINT "F TEST CANNOT BE PERFORMED WITH A STANDARD DEVIATION OF ZERO"
740 GOTO 1370
750 LET F = (A(1) / V1) / (A(2) / V2)
760 PRINT "CALCULATED VALUE OF F ="; F
770 LET E = 0
780 IF V1 = 2 * INT(V1 / 2 + 0.1) THEN 820
790 IF V2 = 2 * INT(V2 / 2 + 0.1) THEN 870
800 GOTO 1050
810 REM CALCULATE PROBABILITY IF V1 EVEN
820 LET U = 1 / (1 + V2 / (F * V1))
830 LET P1 = V1 + 1
840 LET Q = V2 - 2
850 GOTO 910
860 REM CALCULATE PROBABILITY IF V2 EVEN
870 LET E = 1
880 LET U = 1 / (1 + F * V1 / V2)
890 LET P1 = V2 + 1
900 LET Q = V1 - 2
910 LET S = 0
920 LET W = 1
930 LET J = 2
940 LET S = S + W
950 LET W = W * U * (J + Q) / J
960 LET J = J + 2
970 IF J < P1 THEN 940
980 LET Z = SQR(1 - U)
990 IF E = 0 THEN 1020
1000 LET P = 100 * S * (Z ^ V1)
1010 GOTO 1280
1020 LET P = 100 * S * (Z ^ V2)
1030 GOTO 1280
1040 REM CALCULATE PROBABILITY IF V1 & V2 BOTH ODD
1050 LET U = 1 / (1 + F * V1 / V2)
1060 LET X = 1 - U
1070 LET S = 0
1080 LET W = 1
1090 LET J = 2
1100 LET P1 = V2
1110 GOTO 1150
```

```
1120 LET S = S + W
1130 LET W = W * U * J / (J + 1)
1140 LET J = J + 2
1150 IF J < P1 THEN 1120
1160 LET W = W * V2
1170 LET J = 3
1180 LET P1 = V1 + 1
1190 LET Q = V2 - 2
1200 GOTO 1240
1210 LET S = S - W
1220 LET W = W * X * (J + Q) / J
1230 LET J = J + 2
1240 IF J < P1 THEN 1210
1250 LET T1 = ATN(SQR(F * V1 / V2))
1260 LET S1 = S * SQR(X * U)
1270 LET P = 100 * (1 - 2 * (T1 + S1) / 3.14159)
1280 IF P <= 50 THEN 1300
1290 LET P = 100 - P
1300 GOSUB 1670
1310 PRINT "PROBABILITY THAT SUCH A DIFFERENCE IN VARIANCES COULD OCCUR"
1320 PRINT "BY CHANCE IF THE PARENT POPULATIONS HAVE THE SAME VARIANCE"
1330 IF P < 0.01 THEN 1360
1340 PRINT "IS"; P; "%"
1350 GOTO 1370
1360 PRINT "IS LESS THAN 0.01 %"
1370 PRINT
1380 PRINT "WOULD YOU LIKE TO PERFORM A T TEST?"
1390 PRINT "TYPE YES OR NO AND PRESS RETURN"
1400 GOSUB 1750
1410 IF Q$ = "NO" THEN 1590
1420 PRINT "THIS TEST IS ONLY MEANINGFUL WHEN THE STANDARD DEVIATIONS"
1430 PRINT "OF THE TWO GROUPS DO NOT DIFFER SIGNIFICANTLY."
1440 PRINT
1450 IF A(1) + A(2) > 0 THEN 1480
1460 LET P = 0
1470 GOTO 1560
1480 LET S = SQR((A(1) + A(2)) / (V1 + V2))
1490 LET T = ABS(Y(1) - Y(2)) / (S * SQR(1 / K(1) + 1 / K(2)))
1500 PRINT "CALCULATED VALUE OF T ="; T
1510 LET N2 = K(1) - 1 + K(2) - 1
1520 REM CALCULATE T PROBABILITY
1530 GOSUB 1820
1540 REM ROUND ANSWER
1550 GOSUB 1670
1560 PRINT "PROBABILITY THAT SUCH A DIFFERENCE IN MEANS COULD OCCUR BY"
1570 PRINT "CHANCE IF THE PARENT POPULATIONS HAVE THE SAME MEAN AND"
1580 PRINT "VARIANCE IS"; P; "%"
1590 PRINT
1600 PRINT "WOULD YOU LIKE ANOTHER RUN? (YES/NO)"
1610 GOSUB 1750
1620 RESTORE
1630 IF Q$ = "YES" THEN 60
1640 PRINT "END OF JOB"
1650 STOP
1660 REM ***** SUBROUTINE TO ROUND OFF ANSWERS *****
1670 IF P < 5 THEN 1690
1680 LET P = INT(P * 10 + 0.5) * 0.1
1690 IF (P - 5) * (P - 0.5) > 0 THEN 1710
1700 LET P = INT(P * 100 + 0.5) * 0.01
1710 IF P > 0.5 THEN 1730
1720 LET P = INT(P * 1000 + 0.5) * 0.001
```

```
1730 RETURN
1740 REM ***** SUBROUTINE TO CHECK REPLIES *****
1750 INPUT Q$
1760 IF Q$ = "YES" THEN 1800
1770 IF Q$ = "NO" THEN 1800
1780 PRINT "REPLY '"; Q$; "' NOT UNDERSTOOD. RETYPE"
1790 GOTO  1750
1800 RETURN
1810 REM ***** SUBROUTINE TO CALCUATE T PROBABILITY *****
1820 READ B(1), B(2), B(3), B(4), P
1830 DATA 1E6, 10000, 1000, 100, 0
1840 IF N2 > 4 THEN 1870
1850 IF T > B(N2) THEN 2080
1860 GOTO  1880
1870 IF T > 50 THEN 2080
1880 LET A = T / SQR(N2)
1890 LET B = N2 / (N2 + T * T)
1900 LET J = N2 - 2
1910 LET K = N2 - INT(N2 / 2 + 0.1) * 2 + 2
1920 LET S = 1
1930 IF J < 2 THEN 2010
1940 LET C = 1
1950 LET F2 = K
1960 FOR I = K TO J STEP 2
1970    LET C = C * B * (F2 - 1) / F2
1980    LET S = S + C
1990    LET F2 = F2 + 2
2000 NEXT I
2010 IF K > 2 THEN 2040
2020 LET P = 0.5 - 0.5 * A * SQR(B) * S
2030 GOTO  2080
2040 IF N2 > 1 THEN 2060
2050 LET S = 0
2060 LET P = 0.5 - (A * B * S + ATN(A)) / 3.14159
2070 REM CONVERT TO TWO TAILED PERCENTAGE
2080 LET P = 100 * 2 * P
2090 RETURN
2100 END
```

wider range of numbers than this, and the limit of 60 values may be increased with care.

(iii) Neither group may have a standard deviation of zero because attempted division by zero would cause failure with Equation 7.

If any of these tests fail, the F-test is abandoned, and the user is asked if the t-test is required.

Provided that the data pass the extra checks, the variance ratio F is calculated and printed (lines 750–760). This value could be looked up in the F-tables (Appendix 7), but for convenience the program calculates the probability (lines 770–1300). Three different methods are employed depending on whether the number of degrees of freedom v_1 is even, or v_2 is even or both v_1 and v_2 are odd. The derivation of the equations used is not given here, but may be found in a paper by J. D. Lee and D. G. Hayes (*Computers and Education*, 1978, **2**, 165–176). It should be noted that the program does not exchange the two sets of data to guarantee that $s_1 > s_2$ as recommended in the discussion and in Equation 7. This guarantee ensures that the F probability is in the range 0–50%. The program calculates the probability in the range 0%–100%, and values above 50% are appropriately converted. (A probability of 90% is equivalent to a probability of 10%, since a variance ratio of 0·4 is equally significant to a variance ratio of 2.5, since $2.5 = 1/0.4$.) From this it is apparent that the maximum F-test probability will be 50% corresponding to the standard deviations of the two groups being identical.

The calculated F-probability is rounded to give 1, 2 or 3 decimal figures as appropriate in a subroutine (lines 1660–1730), and is printed (lines 1310–1360).

Next a message (lines 1380–1390) asks whether a t-test is required. The answer which must be YES or NO is checked in a subroutine (lines 1740–1800).

If the t-test is requested then the estimated standard deviation s of the two parent populations is calculated (line 1480) using Equation 4. The number of standard errors variation t of $| m_1 - m_2 |$ is calculated in line 1490 using Equation 6, and is printed in line 1500. Next a subroutine is entered (lines 1810–2090) to calculate the probability of such a t-value occurring by chance if the parent populations have the same mean. (This subroutine is the same as that used in the Pearson's correlation coefficient program.) The probability is rounded in a subroutine (lines 1660–1730) to give 1, 2 or 3 decimal figures as appropriate. The probability is then printed out (lines 1560–1580).

Finally a message (line 1600) asks whether the user would like another run. The answer must be YES or NO and is checked by a subroutine (lines 1740–1800).

It should be noted that it is always valid to perform an F-test, but that the result from a t-test is meaningful only if the standard deviations of the two groups do not differ significantly.

Sign Test

The sign test may be used to test if the claimed *median* value for a population is in accord with a random sample of measurements.

The previously described t-test for comparing mean values, and the F test for comparing variances are both based on the assumption that samples are taken from a normal distribution. In contrast, the sign test may be used on distributions which are highly skewed, or differ markedly from normal. Data from many biological experiments are skewed and not normal—for example analysis of a particular drug, hormone or protein in blood may vary in different individuals, but the amount cannot be negative, so the distribution is not symmetrical.

The sign test is also useful for cases where there is no scale for measuring the character being studied, yet it is possible to distinguish the rankings of two products by several judges. For example, a group of animal husbandmen might choose the better animal from a pair of animals, or housewives might choose which of two food preparations they considered to be the most palatable.

Example 4

Suppose the median yield of tomatoes from a new variety is claimed to be 8 lb per plant. Seven plants were chosen at random and their yields recorded:

7.4	7.0	7.7	9.6
7.8	6.8	7.6	

Do these figures suggest that the claimed median is wrong? If the median value is subtracted from each of the measurements, the following values are obtained:

-0.6	-1.0	-0.3	$+1.6$
-0.2	-1.2	-0.4	

In the sign test, only the signs of the differences are considered, and not their numerical values, so the data become:

$-$	$-$	$-$	$+$
$-$	$-$	$-$	

By the definition of the median, there is a 50% chance of obtaining a value larger than the median, and a 50% chance of obtaining a value less than the median. One might expect the results to yield three or four $+$ signs out of seven, and the question is whether the observed result of only one $+$ is so improbable that the value claimed for the median is in doubt.

82

The probability $P(x)$ that an event will occur x times out of n trials may be calculated exactly from the binomial distribution:

$$P(x) = \binom{n}{x} p^x (1-p)^{n-x}$$

where p is the probability of the event occurring in one trial.

The term $\binom{n}{x}$ is the number of different ways of choosing x items out of n values where $x \leqslant n$, and is called the binominal coefficient. This may be calculated:

$$\text{Binominal coefficient} = \frac{n!}{x!(n-x)!}$$

Since the probability p is 0.5, it follows that p is equal to $(1-p)$ and so $p^x(1-p)^{n-x}$ becomes p^n. The total number of permutations of n objects is 2^n, hence p^n is the reciprocal of the number of permutations.

In this case of seven measurements, the number of permutations is $2^7 = 128$. The probability of obtaining no $+$ signs from seven readings $P(0)$ is:

$$P(0) = \frac{7!}{0!(7-0)!} \times \frac{1}{128}$$

$$= \frac{1}{128}$$

(Remember that $0! = 1$.)

Similarly, the probability of obtaining one $+$ sign $P(1)$ is:

$$P(1) = \frac{7!}{1!(7-1)!} \times \frac{1}{128}$$

$$= \frac{7}{128}$$

The chance of obtaining one or less $+$ signs is $P(0) + P(1) = 1/128 + 7/128 = 8/128$.

The test whether the data differ significantly from the claimed median is a two-sided test, and the other extreme probability of obtaining one or less $-$ signs must also be taken into account. The chance of this is also $8/128$. Thus, the chance of obtaining six or more signs the same is:

$$8/128 + 8/128 = 8/64 = 12.5\%$$

This is considerably larger than the usually accepted 5% probability level, hence the data do not provide statistical evidence that the median is wrong.

Three points should be noted:

1. When calculating the differences (value $-$ median) it is possible to obtain a difference of zero. Any zero difference is omitted from the calculation and the number of terms (plants in this case) reduced by one.

2. If pairs of data are available, the signs of the differences may be calculated and used in the same way as the median example. However, the efficiency of the sign test is only about 65% relative to the t-test. Thus, if numeric data are available, and the distribution is normal, it is better to perform a t-test than a sign test. For example, a t-test based on eight data values has about the same probability of detecting a significant difference as 12 values with the sign test. The reason for the lower efficiency of the sign test is that only the signs of the differences are used, but the numerical values are ignored.

3. The evaluation of the probabilities should be performed with care on a computer, since factorial expressions give rise to very large numbers—and computers store a limited number of significant figures. Consider the calculation of the probability of obtaining two plus signs from seven readings:

$$P(2) = \frac{7!}{2!(7-2)!} \times \frac{1}{128}$$

$$= \frac{7.6.5.4.3.2.1}{1.2.(5.4.3.2.1)} \times \frac{1}{128}$$

Clearly, the 5.4.3.2.1 part of the expression will cancel, and it is less work and will avoid producing a very large number with the possibility of numerical rounding errors if the expression is evaluated as:

$$\frac{7.6}{1.2} \times \frac{1}{128}$$

The computer program uses this method of calculation.

Before performing an experiment where it is hoped that the sign test will reveal a significant difference (should such a difference exist), it is essential to consider the design of the experiment—in particular the number of readings that must be taken. If an insufficient number of readings is taken it may be impossible to detect a difference which really exists. The minimum number of readings varies with the significance level chosen, and values are shown in Table 6.3.

Table 6.3 Numbers of Data Values Required for Different Numbers of Like Signs to be Significant in a Two-Tailed Sign Test

Number of data values	Number of like signs		
	1% significance	5% significance	10% significance
5	—	—	0
6	—	0	0
7	—	0	0
8	0	0	1
9	0	1	1
10	0	1	1
11	0	1	2
12	1	2	2
13	1	2	3
14	1	2	3
15	2	3	3
16	2	3	4
17	2	4	4
18	3	4	5
19	3	4	5
20	3	5	5

For example, if five readings all gave the same sign, this would not be significant at the 1% or 5% levels, but would be significant at the 10% level. Similarly, if 13 readings were taken and two signs were different, this would not indicate a significant difference at the 1% level, but would at 5% and 10%.

Example 5

The sign test is often used when there is no numeric scale for measuring, yet it is possible to express a preference between two products. Two turkeys were cooked, and a sample of 9 judges were asked to taste both of them and record which they like best (+ signs indicate best, − signs worst). One of the turkeys had been freshly killed, but the other had been stored for 6 months in a deep freezer. The results are shown in Table 6.4.

The question is whether this result provides evidence of a significant difference in flavour between the two turkeys. If the flavour is the same in both, one would expect half the judges to choose one, and half the other, or some of the judges to be unwilling to rank them. The observed result was one different in nine. The total number of possible permutations is $2^9 = 512$. The probability of obtaining no minus signs $P(0)$ is:

$$P(0) = \frac{9!}{0!(9-0)!} \times \frac{1}{512}$$

$$= \frac{1}{512}$$

The probability of obtaining one minus sign $P(1)$ is:

$$P(1) = \frac{9!}{1!(9-1)!} \times \frac{1}{512}$$

$$= \frac{9}{512}$$

The probability of obtaining one or less minus signs is:

$$P(0) + P(1) = 1/512 + 9/512$$

$$= 10/512$$

There is the same probability of obtaining one or less plus signs. Thus, the chance of obtaining eight or more signs the same is:

$$10/512 + 10/512 = 10/256$$

$$= 3.91\%$$

Since this is less than 5%, there is evidence for

Table 6.4

Judge	'Fresh' turkey	'Frozen' turkey
1	+	−
2	+	−
3	+	−
4	−	+
5	+	−
6	+	−
7	+	−
8	+	−
9	+	−

a difference in flavour of fresh and frozen turkeys at the 5% significance level.

Description of Sign Test Program

The program (Program 6.3) first prints a heading (lines 20–30) and then asks (line 50) if full instructions are required. The answer is obtained by a subroutine (lines 1710–1800), where it is checked to make sure that the reply is either YES or NO. Full or abbreviated instructions are printed as appropriate in the first run, but shortened instructions are always given on a second or subsequent run.

A message (line 140) requests that the median value be typed, and then another message (line 160) asks how many data values are included in the sample. The number of data values is checked (lines 180–210) to make sure that it is an integer and in the range 5–100. An explanatory message is printed if the number is unacceptable, and the user is asked to type the correct value. There is no point in doing the test on less than five values, since Table 6.3 shows that five values which all have the same sign are only just significant at the 10% level. The upper limit of 100 is set by the DIMensions of the X and Y arrays in line 10. This limit may be changed if required.

A message explaining how to input the data is printed (lines 230–240) if full instructions

were requested, otherwise only a short message (line 250) is printed. A loop which extends from lines 260 to 280 is used to input the data values.

Next, a subroutine (lines 960–1700) is entered to check that the input data are correct, and to alter it if necessary. The data may be listed, or lines may be replaced, deleted or added. The subroutine is similar to the one used and described in more detail in the write up for Pearson's correlation coefficient.

A check is then performed (line 330) to check that after editing the data a minimum of five data values remain. Should there be less than this, a message is printed (line 340) that the run has been abandoned, and the user is asked if another run is required (lines 810–820).

The next step in the calculation (lines 360–500) involves first setting the number of positive and negative signs to zero (lines 380–390) and the number of comparisons to the number of data values (line 400). Inside the loop from lines 410 to 500 the difference between each data value and the median is calculated in turn (line 420). The difference is tested and if it is greater than zero or less than zero, the number of positive signs $P1$ or the number of negative signs $N1$ is increased by one as appropriate. Should the difference be zero, the number of comparisons $N2$ is decreased by one, and this data value is excluded from the test.

The data values and the differences from the median are printed (lines 510–550). If full instructions were requested, the number of positive and negative differences, and the number of comparisons (lines 560–590) are also printed.

The smaller value from the number of positive and negative signs is chosen (lines 600–650) since this simplifies the calculation of probabilities.

The number of permutations is evaluated as 2^n (line 680) and the binominal coefficient is calculated using a loop (lines 720–770). If full instructions were requested, the probabilities of 0, 1, 2 . . . negative or positive signs are printed to explain the steps in the calculation.

The sum of these probabilities is multiplied by two since a two-tailed test is required, and the probability is printed both as a fraction, and also as a percentage (rounded to give two decimal figures), in line 800. This is the final result.

The remainder of the program asks (lines 820–830) if another run is required. The answer must be YES or NO, and is checked by a subroutine (lines 1710–1800). If the answer is NO, then a finishing message is printed (line 1810) and the run finished. If the answer is YES, the user is asked whether to start again typing in completely NEW data, or whether to edit the OLD existing data before re-running (lines 860–950).

Program 6.3   Trial run.

```
            SIGN TEST
            ==== ====

WOULD YOU LIKE FULL INSTRUCTIONS?
'TYPE YES OR NO & PRESS RETURN.

? YES

THIS TEST IS USED TO ESTABLISH IF THE MEDIAN VALUE
OF THE SAMPLE OF DATA VALUES DIFFERS SIGNIFICANTLY
FROM THE TRUE MEDIAN VALUE

TYPE THE TRUE MEDIAN VALUE
? 8
TYPE THE NUMBER OF DATA VALUES IN THE SAMPLE ? 7
TYPE A DATA VALUE, PRESS RETURN, NEXT VALUE, RETURN, ...
YOU WILL HAVE CHANCE TO CORRECT WRONGLY TYPED DATA LATER
INPUT DATA X
? 7.4
? 7.0
? 7.7
? 9.6
? 7.8
? 6.8
? 7.6
END OF DATA INPUT
ARE THE DATA VALUES ENTERED CORRECT?   TYPE YES OR NO & PRESS RETURN.

? YES
DATA            DATA - MEDIAN
7.4             -0.6
7               -1
7.7             -0.3
9.6             1.6
7.8             -0.2
6.8             -1.2
7.6             -0.4
NUMBER OF POSITIVE DIFFERENCES = 1
NUMBER OF NEGATIVE DIFFERENCES = 6
NUMBER OF COMPARISONS = 7

PROBABILITY OF 0 PLUS  SIGNS IS  1 / 128
PROBABILITY OF 1 PLUS  SIGNS IS  7 / 128

PROBABILITY OF GETTING UP TO & INCLUDING 1 SIGNS DIFFERENT
IS 2 * 8 / 128 THAT IS 12.5 %
```

WOULD YOU LIKE ANOTHER RUN?
 TYPE YES OR NO & PRESS RETURN.

? NO
END OF JOB


```
10 DIM X(100), Y(100), Q$(10), I$(3)
20 PRINT TAB(12); "SIGN TEST"
30 PRINT TAB(12); "==== ===="
40 PRINT
50 PRINT "WOULD YOU LIKE FULL INSTRUCTIONS?"
60 GOSUB 1730
70 LET I$ = Q$
80 IF I$ = "NO" THEN 130
90 PRINT
100 PRINT "THIS TEST IS USED TO ESTABLISH IF THE MEDIAN VALUE"
110 PRINT "OF THE SAMPLE OF DATA VALUES DIFFERS SIGNIFICANTLY"
120 PRINT "FROM THE TRUE MEDIAN VALUE"
130 PRINT
140 PRINT "TYPE THE TRUE MEDIAN VALUE"
150 INPUT M
160 PRINT "TYPE THE NUMBER OF DATA VALUES IN THE SAMPLE";
170 INPUT N
180 IF N <> INT(N) THEN 200
190 IF (N - 5) * (N - 100) <= 0 THEN 220
200 PRINT "VALUE TYPED MUST BE A WHOLE NUMBER BETWEEN 5 AND 100"
210 GOTO 160
220 IF I$ = "NO" THEN 250
230 PRINT "TYPE A DATA VALUE, PRESS RETURN, NEXT VALUE, RETURN, ..."
240 PRINT "YOU WILL HAVE CHANCE TO CORRECT WRONGLY TYPED DATA LATER"
250 PRINT "INPUT DATA X"
260 FOR I = 1 TO N
270    INPUT X(I)
280 NEXT I
290 PRINT "END OF DATA INPUT"
300 REM CALL SUBROUTINE TO CHECK THAT DATA VALUES ARE CORRECT
310 GOSUB 970
320 REM CHECK THAT THE NUMBER OF DATA VALUES IS SENSIBLE
330 IF N >= 5 THEN 380
340 PRINT "NOT ENOUGH DATA VALUES - RUN ABANDONED"
350 GOTO 810
360 REM CALCULATE THE DIFFERENCES (DATA VALUES - MEDIAN)
370 REM ADD UP THE NUMBER OF POSITIVE & NEGATIVE DIFFERENCES P1 & N1
380 LET P1 = 0
390 LET N1 = 0
400 LET N2 = N
410 FOR I = 1 TO N
420    LET Y(I) = X(I) - M
430    IF Y(I) > 0 THEN 470
440    IF Y(I) < 0 THEN 490
450    LET N2 = N2 - 1
460    GOTO 500
470    LET P1 = P1 + 1
480    GOTO 500
490    LET N1 = N1 + 1
500 NEXT I
510 REM PRINT CONTENTS OF ARRAYS
520 PRINT "DATA", "DATA - MEDIAN"
530 FOR I = 1 TO N
540    PRINT X(I), Y(I)
```

```
550 NEXT I
560 IF I$ = "NO" THEN 610
570 PRINT "NUMBER OF POSITIVE DIFFERENCES ="; P1
580 PRINT "NUMBER OF NEGATIVE DIFFERENCES ="; N1
590 PRINT "NUMBER OF COMPARISONS ="; N2
600 REM CHOOSE SMALLER OF P1 & N1
610 LET K = P1
620 LET Q$ = "PLUS"
630 IF P1 < N1 THEN 670
640 LET K = N1
650 LET Q$ = "MINUS"
660 REM CALCULATE TOTAL NUMBER OF COMBINATIONS
670 PRINT
680 LET S4 = 2 ^ N2
690 LET S3 = 0
700 REM USE BINOMIAL THEOREM TO CALCULATE THE PROBABILITY
710 LET C = 1
720 FOR X = 0 TO K
730    IF I$ = "NO" THEN 750
740    PRINT "PROBABILITY OF"; X; Q$; " SIGNS IS "; C; "/"; S4
750    LET S3 = S3 + C
760    LET C = C * (N2 - X) / (X + 1)
770 NEXT X
780 PRINT
790 PRINT "PROBABILITY OF GETTING UP TO & INCLUDING";K;"SIGNS DIFFERENT"
800 PRINT "IS 2 *"; S3; "/"; S4; "THAT IS"; INT(S3/S4*20000+0.5)/100;"
810 PRINT
820 PRINT "WOULD YOU LIKE ANOTHER RUN?"
830 GOSUB 1720
840 IF Q$ = "NO" THEN 1810
850 REM SEE WHETHER TO INPUT NEW DATA OR EDIT & RERUN THE OLD DATA
860 PRINT "TYPE NEW TO START AGAIN WITH NEW DATA"
870 PRINT "  OR OLD TO EDIT EXISTING DATA"
880 INPUT Q$
890 LET I$ = "NO"
900 IF Q$ = "NEW" THEN 130
910 IF Q$ = "OLD" THEN 940
920 PRINT "REPLY '"; Q$; "' NOT UNDERSTOOD."
930 GOTO 860
940 GOSUB 1020
950 GOTO 330
960 REM *****SUBROUTINE TO CHECK DATA ARE CORRECT & ALTER IF NECESSARY
970 PRINT "ARE THE DATA VALUES ENTERED CORRECT?";
980 REM A4 SHOULD BE SET TO THE NUMBER OF LINES ON THE VDU
990 LET A4 = 20
1000 GOSUB 1720
1010 IF Q$ = "YES" THEN 1700
1020 PRINT "HERE IS A LIST OF THE CURRENT DATA"
1030 PRINT "LINE NUMBER", "X"
1040 FOR I = 1 TO N
1050    PRINT I, X(I)
1060    IF INT(I / (A4 - 1)) * (A4 - 1) <> I THEN 1100
1070    PRINT "WOULD YOU LIKE TO CONTINUE LISTING";
1080    GOSUB 1720
1090    IF Q$ = "NO" THEN 1110
1100 NEXT I
1110 PRINT "TYPE R TO REPLACE";
1120 IF I$ = "NO" THEN 1140
1130 PRINT " AN EXISTING LINE OF DATA"
1140 IF N = 100 THEN 1190
1150 PRINT TAB(5); " A TO ADD";
```

```
1160 IF I$ = "NO" THEN 1180
1170 PRINT " AN EXTRA LINE"
1180 IF N = 1 THEN 1220
1190 PRINT TAB(5); " D TO DELETE";
1200 IF I$ = "NO" THEN 1220
1210 PRINT " AN EXISTING LINE"
1220 PRINT TAB(5); " L TO LIST";
1230 IF I$ = "NO" THEN 1250
1240 PRINT " THE DATA"
1250 PRINT "  OR C TO CONTINUE";
1260 IF I$ = "NO" THEN 1280
1270 PRINT " THE CALCULATION"
1280 INPUT Q$
1290 IF Q$ = "R" THEN 1390
1300 IF N = 100 THEN 1330
1310 IF Q$ = "A" THEN 1510
1320 IF N = 1 THEN 1340
1330 IF Q$ = "D" THEN 1560
1340 IF Q$ = "L" THEN 1020
1350 IF Q$ = "C" THEN 1700
1360 PRINT "REPLY '";Q$;"' NOT UNDERSTOOD."
1370 GOTO 1110
1330 REM REPLACE LINE
1390 PRINT "TYPE THE LINENUMBER OF THE LINE TO BE REPLACED";
1400 INPUT I
1410 IF I <> INT(I) THEN 1430
1420 IF (I - 1) * (I - N) <= 0 THEN 1460
1430 PRINT "LINENUMBER MUST BE AN INTEGER IN THE RANGE 1 -"; N
1440 PRINT "RE-";
1450 GOTO 1390
1460 PRINT "TYPE THE CORRECT LINE TO REPLACE THE ONE WHICH IS WRONG;"
1470 PRINT "X"
1480 INPUT X(I)
1490 GOTO 1540
1500 REM ADD A NEW LINE
1510 LET N = N + 1
1520 PRINT "TYPE THE ADDITIONAL LINE OF DATA AS SHOWN;   X"
1530 INPUT X(N)
1540 PRINT "OK"
1550 GOTO 1110
1560 REM DELETE A LINE
1570 PRINT "TYPE THE LINENUMBER OF THE LINE TO BE DELETED"
1580 INPUT J
1590 IF (J - 1) * (J - N) > 0 THEN 1610
1600 IF J = INT(J) THEN 1630
1610 PRINT "LINENUMBER MUST BE AN INTEGER IN THE RANGE 1 -"; N
1620 GOTO 1570
1630 FOR I = J + 1 TO N
1640    LET X(I - 1) = X(I)
1650 NEXT I
1660 LET N = N - 1
1670 PRINT "OK"
1680 IF J > N THEN 1110
1690 GOTO 1020
1700 RETURN
1710 REM ***** SUBROUTINE TO CHECK REPLIES *****
1720 IF I$ = "NO" THEN 1740
1730 PRINT " TYPE YES OR NO & PRESS RETURN."
1740 PRINT
1750 INPUT Q$
1760 IF Q$ = "YES" THEN 1800
```

```
1770 IF Q$ = "NO" THEN 1800
1780 PRINT "REPLY '";Q$;"' NOT UNDERSTOOD.";
1790 GOTO 1730
1800 RETURN
1810 PRINT "END OF JOB"
1820 END
```

Wilcoxon's Signed Rank Test

The Wilcoxon signed rank test is another substitute for the *t*-test to compare two random samples of measurements that are matched. The matching may be achieved by pairing each member of one group with a member of the other group, or alternatively by using one group and comparing the effects of two different treatments, or different judges or on different occasions.

The test uses the signs of the differences between pairs of readings (as in the sign test), and also takes account of the relative magnitude of the differences by assigning ranks to them. The use of ranks makes the test more sensitive than the sign test, simplifies the arithmetic compared with the *t*-test, and has the advantage that it may be applied to any distribution of measurements (whereas the *t*-test requires a normal distribution). The Wilcoxon test may be used with numeric data measured on a coarse scale such as 1–4 or 1–10. The test is slightly less powerful than the *t*-test, which uses the exact numerical values rather than just ranks, but the simplicity and wide applicability of the Wilcoxon test far outweigh its shortcomings.

Method

First the absolute values (ignoring signs) of the differences between pairs of data are calculated. These values are then ranked in ascending order; that is, the smallest value first. Then rank 1 is assigned to the smallest value, 2 to the next smallest, and so on. Note that because the absolute values are ranked, a difference of +5 would be given a lower rank than a difference of − 10. If two or more of the differences are equal, the average of the ranks

is assigned to each. The signs of the differences are restored to the rankings giving a signed rank table. Next two rank sums, $T(+)$ and $T(-)$, are obtained by adding together all of the ranks with positive signs to give $T(+)$ and adding all of the ranks with negative signs to give $T(-)$. It is more convenient to work with the smaller of the two rank sums, and the question to be answered is could a T value as small or smaller than this arise if there is no real difference between the paired samples? The way in which this is done is best illustrated by an example.

Example 6

Ten pairs of corn seeds were selected at random, and paired at random. One seed out of each pair was treated with a mercury compound (to inhibit attack by fungi), and the seeds were grown for a fixed time under the same conditions. The lengths of the treated and untreated seedlings are given in Table 6.5. The test is to see if the treatment has

Table 6.5

Pair	Length (untreated) (cm)	Length (treated) (cm)	Difference (cm)
1	7.5	6.2	1.3
2	6.0	4.0	2.0
3	8.6	5.2	3.4
4	8.1	7.6	0.5
5	11.1	6.1	5.0
6	9.7	4.8	4.9
7	9.6	6.2	3.4
8	6.2	7.5	− 1.3
9	8·1	5·8	2.3
10	6.3	6.7	− 0.4

significantly changed the growth rate of the seedlings.

Next the absolute values of the differences are ranked. Ignoring the signs, -0.4 is the smallest value and is ranked 1, and the value 0.5 is ranked 2. The values 1.3 and -1.3 are next and should be ranked 3 and 4, but since the differences are equal (ignoring the signs), both are ranked $(3+4)/2 = 3.5$. The differences 2.0 and 2.3 are ranked 5 and 6, and the two values of 3.4 are both ranked 7.5. The differences 4.9 and 5.0 are ranked 9 and 10 respectively. The results are shown in Table 6.6.

Table 6.6

Pair	Difference	Rank	Signed rank
1	1.3	3.5	3.5
2	2.0	5	5
3	3.4	7.5	7.5
4	0.5	2	2
5	5.0	10	10
6	4.9	9	9
7	3.4	7.5	7.5
8	-1.3	3.5	-3.5
9	2.3	6	6
10	-0.4	1	-1

The rank sum $T(-)$ is calculated as the modulus of the sum of -1 and $-3.5 = 4.5$, and similarly $T(+)$ is

$$3.5 + 5 + 7.5 + 2 + 10 + 9 + 7.5 + 6 = 50.5.$$

Taking T as the smaller rank sum of 4.5, it is possible to list the combinations of signs which could produce a value of $T \leqslant 4.5$ (Table 6.7). There are in total $2^{10} = 1024$ permutations of signs, but only the 7 combinations listed give $T \leqslant 4.5$. If there is no difference between the treated and untreated seeds then all of these permutations of sign are equally likely. The probability of obtaining so small a T value from the $-$ signs is therefore 7/1024. The test is two sided, so there is also a probability of 7/1024 getting so small a T value from $+$ signs. The total probabi-

Table 6.7

Rank										T
1	2	3	4	5	6	7	8	9	10	
+	+	+	+	+	+	+	+	+	+	0
−	+	+	+	+	+	+	+	+	+	1
+	−	+	+	+	+	+	+	+	+	2
−	−	+	+	+	+	+	+	+	+	3
+	+	−	+	+	+	+	+	+	+	3
−	+	−	+	+	+	+	+	+	+	4
+	+	+	−	+	+	+	+	+	+	4

lity $= 7/1024 + 7/1024 = 7/512 = 1.37\%$. Since this value is less than 5%, it indicates that the hypothesis that the treated and untreated seeds grow equally well must be rejected at the 5% level. The calculated value exceeds 1%, so the hypothesis cannot be rejected at the 1% level.

Handling Ties

Suppose the data for treated and untreated seedlings given in Table 6.5 had been measured less precisely, and the lengths estimated only to the nearest centimetre. The data become those shown in Table 6.8.

Since the fourth pair of seedlings show no difference in length, they are not ranked, and the number of pairs used in the comparison is reduced from 10 to 9. The value of $T+$ is 40.5 and $T-$ is 4.5. Table 6.7 shows that seven combinations of minus signs give $T \leqslant 4.5$ (and, in addition, seven combinations of plus signs would also give $T \leqslant 4.5$). For 9 pairs, there are a total of $2^9 = 512$ permutations, so for a two-sided test the total probability of obtaining a T value as small or smaller than 4.5 is $7/512 + 7/512 = 7/256 = 2.73\%$. This is less than 5%, so the hypothesis that treated and untreated seeds grow equally well must be rejected at this level.

Table 6.8

Pair	Length (cm)			Rank	Signed Rank
	Un-treated	Treated	Differ-ence		
1	8	6	2	3.5	3.5
2	6	4	2	3.5	3.5
3	9	5	4	6.5	6.5
4	8	8	0		
5	11	6	5	8.5	8.5
6	10	5	5	8.5	8.5
7	10	6	4	6.5	6.5
8	6	8	−2	3.5	−3.5
9	8	6	2	3.5	3.5
10	6	7	−1	1	−1

In this example the result is the same as for the more accurate data, but it should be noted that the probability of obtaining results so different by chance has increased from 1.37% to 2.73%. The less accurate data have resulted in a poorer discrimination. In general, the use of imprecise data makes it more difficult to reach a conclusion that two samples are significantly different, and for this reason ties between pairs of data should be resolved wherever possible.

It is laborious to calculate the probability of obtaining a T value less than or equal to a given value by the method shown in Table 6.7 if the number of pairs tested is not very small. Two ways of overcoming this are to calculate the frequency distribution by a simple itera- tive method, or alternatively if the number of pairs is sufficiently large to use a normal approximation.

Calculation of Frequency Distribution

Suppose that there are two pairs; the possible permutations are:

Rank	1	2	
	+	+	$T=0$ (rank sum for −)
	−	+	$T=1$
	+	−	$T=2$
	−	−	$T=3$

The frequency distribution may be written:

Rank sum	0	1	2	3
frequency	1	1	1	1

The frequency distribution for three pairs can easily be calculated from the previous case of two pairs:

Rank 1	2	3
	(same as 2 pairs)	+
	(same as 2 pairs)	−

The frequency distribution from the first of these terms is identical to that for two pairs, and for the second term is the same as the two pair case with 3 added to each rank:

Rank sum	0	1	2	3	4	5	6
Frequency	1	1	1	1			
				2	1	1	1

In an analogous way, the frequency distri- bution for four pairs can be derived by combining the three pair case, and the three pair case + 4 . . . and so on. The number of combinations less than any given T value is easily obtained by adding the appropriate frequencies from the probability distribution, e.g. with three pairs, the total number of combinations giving $T \leqslant 3$ is

$$P(0) + P(1) + P(2) + P(3) = 1 + 1 + 1 + 2 = 5.$$

This approach is useful on a computer, but the time and memory requirements limit the application above about 80 pairs.

Use of Normal Approximation

It can be shown that for n data pairs the distribution of T has a mean of $n(n+1)/4$, and a standard deviation of $\sqrt{[n(n+1)(2n+1)/24]}$.

When n is sufficiently large, the distribution of T is approximately normal, even though the original experimental data need not be nor- mal. Thus, with more than about 25 usable pairs, the significance of the smaller rank sum T is given by the mean minus T divided by the standard deviation. This term is called the standardised normal deviate z, and is calculated:

$$z = \frac{\frac{n(n+1)}{4} - T}{\sqrt{\left[\frac{n(n+1)(2n+1)}{24}\right]}}$$

The values of z corresponding to some important probability levels are shown in Table 6.9.

If z is less than 1.96, there is more than a 5% probability that results as different as these

Table 6.9

z	1.64	1.96	2.58	3.09
P	10%	5%	1%	0.2%

could have arisen by chance if they were really the same, and a significant difference is not proven. Similarly, if z is between 1.96 and 2.58 there is some evidence for a difference, and if z is more than 2.58 there is strong evidence for a difference between the two samples.

Description of the Wilcoxon Signed Rank Program (Program 6.4)

First, a heading is printed (lines 20–30), and then the user is asked (line 50) whether full instructions are required. The answer, which must be YES or NO, is obtained and checked by a subroutine (lines 1930–2020). Either full or abbreviated instructions are printed as requested for the first run, but shortened instructions are given on a second or subsequent run. An explanation of the test (lines 90–130) is printed if full instructions were requested.

A message (line 150) asks the user to type the number of data pairs. The value typed in (line 160) is checked (lines 170–200) to ensure that it is an integer, and that it is in the range 5–200. The upper limit of 200 is imposed by the DIMensions of the X, Y, A and B arrays in line 10, and details of how to change this are given later.

Either a long or short message asking the user to type the data is printed (lines 210–250), and the specified number of data pairs are input in a loop (lines 260–280). Then a message (line 290) is printed to indicate the end of data input.

Next (line 310) a subroutine is called to check that the input data are correct and to allow the user to add, replace or delete complete lines of data. The subroutine extends from lines 1170 to 1920, and is similar to the one used and described more fully in the write up for Pearson's correlation coefficient. The user may change the value of $A4$ (line 1200) to alter the number of lines listed on a vdu screen from 20 to any other value if required.

On returning from the checking subroutine, a further check is carried out (lines 320–350) to make sure that at least five data pairs remain, and then the difference between the two sets of data values are calculated in a loop (lines 370–390).

The absolute values of the differences are ranked (lines 400–540). Data pairs that are numerically equal (that is, have zero difference) are omitted from the ranking, and where the absolute values of two or more differences are equal then an average rank is assigned to each. Then the sign of each difference is put on the corresponding rank.

Next, the sums of ranks $T(+)$ and $T(-)$ are set to zero (lines 550–560) and $N2$ is set to the number of data pairs (line 570). A loop is used (lines 580–660) to evaluate the sum of positive ranks, $T(+)$, the sum of negative ranks, $T(-)$, and the number of valid pairs of data used in the test. The latter is the total number of data pairs minus any which have zero difference.

A table is then printed (lines 670–710) showing the input data pairs, together with their differences and their signed ranks. The rank sums for both positive and negative ranks are printed (lines 730–740), and the smaller of their absolute values is selected as the test criterion T which is printed (line 770).

A test is performed (line 780) to find if the value calculated for T is less than or equal to

93

Program 6.4   Trial run.

```
        WILCOXSON SIGNED RANK TEST
        ========= ====== ==== ====

WOULD YOU LIKE FULL INSTRUCTIONS?
  TYPE YES OR NO & PRESS RETURN.

? YES

TEST OF SIGNIFICANCE OF THE DIFFERENCE IN MEDIANS
OF TWO RELATED POPULATIONS WHICH HAVE THE SAME DISTRIBUTION
MEASUREMENTS MAY BE ON A ORDINAL SCALE , ON A COARSE NUMERIC
SCALE SUCH AS 1-10, OR 1-100, OR BE ACCURATE DATA READINGS.

TYPE THE NUMBER OF DATA PAIRS ? 10
TYPE A PAIR OF DATA VALUES SEPARATED BY A COMMA
THEN PRESS RETURN, TYPE NEXT PAIR OF DATA VALUES ETC
YOU WILL HAVE CHANCE TO CORRECT WRONGLY TYPED DATA LATER
INPUT DATA X,Y
? 7.5,6.2
? 6.0,4.0
? 8.6,5.2
? 8.1,7.6
? 11.1,6.1
? 9.7,4.8
? 9.6,6.2
? 6.2,7.5
? 8.1,5.8
? 6.3,6.7
END OF DATA INPUT
ARE THE DATA VALUES ENTERED CORRECT?   TYPE YES OR NO & PRESS RETURN.

? YES
DATA 1          DATA 2          DIFF 1-2        SIGNED RANK
7.5             6.2             1.3             3.5
6               4               2               5
8.6             5.2             3.4             7.5
8.1             7.6             0.5             2
11.1            6.1             5               10
9.7             4.8             4.9             9
9.6             6.2             3.4             7.5
6.2             7.5             -1.3            -3.5
8.1             5.8             2.3             6
6.3             6.7             -0.4            -1

RANK SUM FOR POSITIVE TERMS = 50.5
RANK SUM FOR NEGATIVE TERMS = -4.5
TEST CRITERION T = 4.5
NUMBER OF COMPARISONS (EXCLUDING TIES) = 10
NUMBER OF COMBINATIONS GIVING T <= 4.5 IS 7
TOTAL NUMBER OF PERMUTATIONS IS 2 ^ 10  = 1024
TWO TAILED PROBABILITY OF SUCH A T VALUE IS 1.37 %

STANDARDISED NORMAL DEVIATE Z = 2.34438

WOULD YOU LIKE ANOTHER RUN?
  TYPE YES OR NO & PRESS RETURN.

? NO
END OF JOB
```

```
10 DIM X(200), Y(200), A(200), B(200), Q$(10), I$(3)
20 PRINT TAB(5); "WILCOXSON SIGNED RANK TEST"
30 PRINT TAB(5); "========= ====== ==== ===="
40 PRINT
50 PRINT "WOULD YOU LIKE FULL INSTRUCTIONS?"
60 GOSUB 1950
70 LET I$ = Q$
80 IF I$ = "NO" THEN 140
90 PRINT
100 PRINT "TEST OF SIGNIFICANCE OF THE DIFFERENCE IN MEDIANS"
110 PRINT "OF TWO RELATED POPULATIONS WHICH HAVE THE SAME DISTRIBUTION"
120 PRINT "MEASUREMENTS MAY BE ON A ORDINAL SCALE, ON A COARSE NUMERIC"
130 PRINT "SCALE SUCH AS 1-10, OR 1-100, OR BE ACCURATE DATA READINGS."
140 PRINT
150 PRINT "TYPE THE NUMBER OF DATA PAIRS";
160 INPUT N
170 IF N <> INT(N) THEN 190
180 IF (N - 5) * (N - 200) <= 0 THEN 210
190 PRINT "VALUE TYPED MUST BE A WHOLE NUMBER BETWEEN 5 AND 200"
200 GOTO 150
210 IF I$ = "NO" THEN 250
220 PRINT "TYPE A PAIR OF DATA VALUES SEPARATED BY A COMMA"
230 PRINT "THEN PRESS RETURN, TYPE NEXT PAIR OF DATA VALUES ETC"
240 PRINT "YOU WILL HAVE CHANCE TO CORRECT WRONGLY TYPED DATA LATER"
250 PRINT "INPUT DATA X,Y"
260 FOR I = 1 TO N
270    INPUT X(I), Y(I)
280 NEXT I
290 PRINT "END OF DATA INPUT"
300 REM CALL SUBROUTINE TO CHECK THAT DATA VALUES ARE CORRECT
310 GOSUB 1180
320 REM CHECK THAT THE NUMBER OF DATA PAIRS IS SENSIBLE
330 IF N >= 5 THEN 370
340 PRINT "NOT ENOUGH DATA PAIRS - RUN ABANDONED"
350 GOTO 1020
360 REM CALCULATE DIFFERENCES BETWEEN THE TWO JUDGES
370 FOR I = 1 TO N
380    LET A(I) = X(I) - Y(I)
390 NEXT I
400 REM CALCULATE RANKS OF DIFFERENCES (A ARRAY) & STORE IN B ARRAY
410 FOR I = 1 TO N
420    LET B(I) = 0
430    FOR J = 1 TO N
440       IF A(J) = 0 THEN 500
450       IF ABS(A(J)) > ABS(A(I)) THEN 500
460       IF ABS(A(J)) = ABS(A(I)) THEN 490
470       LET B(I) = B(I) + 1
480       GOTO 500
490       LET B(I) = B(I) + 0.5
500    NEXT J
510    REM PUT SIGN FROM A VALUE (DIFF) ON CORRESPONDING B VALUE (RANK)
520    IF A(I) = 0 THEN 540
530    LET B(I) = (B(I) + 0.5) * A(I) / ABS(A(I))
540 NEXT I
550 LET P1 = 0
560 LET N1 = 0
570 LET N2 = N
580 FOR I = 1 TO N
590    IF A(I) <> 0 THEN 610
600    LET N2 = N2 - 1
610    IF B(I) < 0 THEN 640
```

```
620    LET P1 = P1 + B(I)
630    GOTO 650
640    LET N1 = N1 + B(I)
650    LET T = P1
660 NEXT I
670 REM PRINT CONTENTS OF ARRAYS
680 PRINT "DATA 1", "DATA 2", "DIFF 1-2", "SIGNED RANK"
690 FOR I = 1 TO N
700    PRINT X(I), Y(I), A(I), B(I)
710 NEXT I
720 PRINT
730 PRINT "RANK SUM FOR POSITIVE TERMS ="; P1
740 PRINT "RANK SUM FOR NEGATIVE TERMS ="; N1
750 IF P1 < ABS(N1) THEN 770
760 LET T = ABS(N1)
770 PRINT "TEST CRITERION T ="; T
780 IF T <= 80 THEN 830
790 PRINT "T IS SO LARGE THAT PROBABILITY IS NOT WORKED OUT EXACTLY"
800 PRINT "USE THE NORMAL DEVIATE INSTEAD."
810 GOTO 980
820 REM CALCULATE PROBABILITY OF OBTAINING T<= VALUE OBTAINED
830 FOR I = 0 TO INT(T)
840    READ P2
850 NEXT I
860 DATA 1,2,3,5,7,10,14,19,25,33,43,55,70,88,110,137,169,207,253,307
870 DATA 371,447,536,640,762,904,1069,1261,1483,1739,2035,2375,2765,3213
880 DATA 3725,4310,4978,5738,6602,7584,8697,9957,11383,12993,14809,16857
890 DATA 19161,21751,24661,27925,31583,35680,40262,45382,51100,57478
900 DATA 64586,72503,81311,91103,101983,114059,127453,142301,158745
910 DATA 176945,197077,219327,243903,271033,300960,333952,370304,410330
920 DATA 454376,502822,556072,614571,678805,749293,826605
930 PRINT "NUMBER OF COMPARISONS (EXCLUDING TIES) ="; N2
940 PRINT "NUMBER OF COMBINATIONS GIVING T <="; T; "IS"; P2
950 PRINT "TOTAL NUMBER OF PERMUTATIONS IS 2 ^ "; N2; " ="; 2 ^ N2
960 PRINT "TWO TAILED PROBABILITY OF SUCH A T VALUE IS";
970 PRINT INT((2 * P2) / (2 ^ N2) * 10000 + 0.5) / 100    ; "
980 PRINT
990 REM CALCULATE NORMAL DEVIATE
1000 LET Z = ABS((T-(N2*(N2+1))/4) / SQR((N2*(N2+1)*(2*N2+1))/24))
1010 PRINT "STANDARDISED NORMAL DEVIATE Z ="; Z
1020 PRINT
1030 PRINT "WOULD YOU LIKE ANOTHER RUN?"
1040 GOSUB 1940
1050 IF Q$ = "NO" THEN 2030
1060 RESTORE
1070 PRINT "TYPE NEW TO START AGAIN WITH NEW DATA"
1080 PRINT "  OR OLD TO EDIT EXISTING DATA"
1090 INPUT Q$
1100 LET I$ = "NO"
1110 IF Q$ = "NEW" THEN 140
1120 IF Q$ = "OLD" THEN 1150
1130 PRINT "REPLY '";Q$;"' NOT UNDERSTOOD."
1140 GOTO 1070
1150 GOSUB 1230
1160 GOTO 330
1170 REM ***SUBROUTINE TO CHECK DATA ARE CORRECT & ALTER IF NECESSARY
1180 PRINT "ARE THE DATA VALUES ENTERED CORRECT?";
1190 REM A4 SHOULD BE SET TO THE NUMBER OF LINES ON THE VDU
1200 LET A4 = 20
1210 GOSUB 1940
1220 IF Q$ = "YES" THEN 1920
```

```
1230 PRINT "HERE IS A LIST OF THE CURRENT DATA"
1240 PRINT "LINE NUMBER", "X", "Y"
1250 FOR I = 1 TO N
1260    PRINT I, X(I), Y(I)
1270    IF INT(I / (A4 - 1)) * (A4 - 1) <> I THEN 1310
1280    PRINT "WOULD YOU LIKE TO CONTINUE LISTING";
1290    GOSUB 1940
1300    IF Q$ = "NO" THEN 1320
1310 NEXT I
1320 PRINT "TYPE R TO REPLACE";
1330 IF I$ = "NO" THEN 1350
1340 PRINT " AN EXISTING LINE OF DATA"
1350 IF N = 200 THEN 1400
1360 PRINT TAB(5); " A TO ADD";
1370 IF I$ = "NO" THEN 1390
1380 PRINT " AN EXTRA LINE"
1390 IF N = 1 THEN 1430
1400 PRINT TAB(5); " D TO DELETE";
1410 IF I$ = "NO" THEN 1430
1420 PRINT " AN EXISTING LINE"
1430 PRINT TAB(5); " L TO LIST";
1440 IF I$ = "NO" THEN 1460
1450 PRINT " THE DATA"
1460 PRINT "   OR C TO CONTINUE";
1470 IF I$ = "NO" THEN 1490
1480 PRINT " THE CALCULATION"
1490 INPUT Q$
1500 IF Q$ = "R" THEN 1600
1510 IF N = 200 THEN 1540
1520 IF Q$ = "A" THEN 1720
1530 IF N = 1 THEN 1550
1540 IF Q$ = "D" THEN 1770
1550 IF Q$ = "L" THEN 1230
1560 IF Q$ = "C" THEN 1920
1570 PRINT "REPLY '";Q$;"' NOT UNDERSTOOD."
1580 GOTO 1320
1590 REM REPLACE LINE
1600 PRINT "TYPE THE LINENUMBER OF THE LINE TO BE REPLACED";
1610 INPUT I
1620 IF I <> INT(I) THEN 1640
1630 IF (I - 1) * (I - N) <= 0 THEN 1670
1640 PRINT "LINENUMBER MUST BE AN INTEGER IN THE RANGE 1 -"; N
1650 PRINT "RE-";
1660 GOTO 1600
1670 PRINT "TYPE THE CORRECT LINE TO REPLACE THE ONE WHICH IS WRONG:"
1680 PRINT "X, Y"
1690 INPUT X(I), Y(I)
1700 GOTO 1750
1710 REM ADD A NEW LINE
1720 LET N = N + 1
1730 PRINT "TYPE THE ADDITIONAL LINE OF DATA AS SHOWN:   X,Y"
1740 INPUT X(N), Y(N)
1750 PRINT "OK"
1760 GOTO 1320
1770 REM DELETE A LINE
1780 PRINT "TYPE THE LINENUMBER OF THE LINE TO BE DELETED"
1790 INPUT J
1800 IF (J - 1) * (J - N) > 0 THEN 1820
1810 IF J = INT(J) THEN 1840
1820 PRINT "LINENUMBER MUST BE AN INTEGER IN THE RANGE 1 -"; N
1830 GOTO 1780
```

```
1840 FOR I = J + 1 TO N
1850    LET X(I - 1) = X(I)
1860    LET Y(I - 1) = Y(I)
1870 NEXT I
1880 LET N = N - 1
1890 PRINT "OK"
1900 IF J > N THEN 1320
1910 GOTO 1230
1920 RETURN
1930 REM ****SUBROUTINE TO CHECK REPLIES****
1940 IF I$ = "NO" THEN 1960
1950 PRINT " TYPE YES OR NO & PRESS RETURN."
1960 PRINT
1970 INPUT Q$
1980 IF Q$ = "YES" THEN 2020
1990 IF Q$ = "NO" THEN 2020
2000 PRINT "REPLY '";Q$;"' NOT UNDERSTOOD.";
2010 GOTO 1950
2020 RETURN
2030 PRINT "END OF JOB"
2040 END
```

80. In the unlikely circumstances that T exceeds 80, then the calculation to obtain the exact probability of obtaining so small a value is skipped. If T is 80 or less, the program (lines 820–920) obtains the number of combinations which would give a sum of ranks $\leqslant$ the calculated T value from a stored table. Then the number of comparisons (excluding ties), the number of combinations giving $T \leqslant$ the calculated value, and the two-tailed probability of such a T value (rounded to two decimal places), are printed (lines 930–970).

Regardless of whether T is greater than or less than 80, the standardised normal deviate is calculated and printed (lines 990–1010).

A message (line 1030) asks if another run is required. The answer, which must be YES or NO, is checked by a subroutine (lines 1930–2020). If another run is required, the user is given the option (lines 1070–1160) of typing in completely NEW data, or editing the OLD existing data before re-running the calculation. If another run is not required, a finishing message is printed (line 2030).

How to Alter the Size of the Program

If the program is too large to fit into the memory available on the computer, the size of the arrays may be reduced to say 50, which limits the maximum number of data pairs to 50. The number 200 should be replaced by 50 in the DIMensions of the X, Y, A and B arrays in line 10, and also in lines 180, 190, 1350 and 1510. If necessary, the number of terms stored in the DATA statements (lines 860–920) may also be reduced, and the value of 80 in line 780 changed to one less than the number of values stored in the DATA statements.

Exercises

6.1 Firm A manufactures electric light bulbs and claims that its product has a longer life than that of firm B. One hundred bulbs from firm A had an average life of 1234 hours, with a standard deviation of 84 hours, whereas 75 bulbs from firm B had an average life of 1210 hours with a standard deviation of 72 hours. Use the large sample test to see if bulbs from A are significantly better than bulbs from B.

6.2 Compare the electric bulbs made by firm A in Question 6.1 with those of firm C. In a test of 150 bulbs from C, the mean life was 1210 hours and the standard deviation was 110 hours.

6.3 A survey was conducted to establish whether the average height of eleven-year-old boys has increased from 1950 to 1980. The following data were collected

1950: 600 boys, mean height 149 cm with standard deviation 6.1 cm.

1980: 500 boys, mean height 150 cm with standard deviation 5.9 cm.

6.4 The Army has tested two types of anti-tank missiles. Type A missed the target by an average of 5.3 metres with a standard deviation of 1.2 metres in a test of 11 missiles. Eight type B missiles were tested and missed by an average of 4.2 metres with a standard deviation of 1.4 metres. Use the t-test to determine if missile B is significantly better than missile A.

6.5 The speeds of nine cars travelling along road A were found to be 29.6, 31.1, 30.4, 29.0, 29.9, 30.3, 29.4, 30.1 and 30.2 m.p.h. Estimate the mean speed and the standard deviation. At the same time the speeds of seven cars travelling along road B were measured, and were found to be 31.8, 30.0, 31.9, 29.7, 32.1, 30.3 and 31.2 m.p.h. Estimate the mean and standard deviation. Use a t-test to find if the means of A and B differ significantly. Perform an F-test to find if the standard deviations of A and B differ significantly. Are the results from the t-test still valid?

6.6 15 people were put on a special weight-reducing diet for a fortnight. Their 'before' and 'after' weights were recorded, and it was found that the weight of three people had increased, two people weighed the same, and ten people had lost weight. Use the sign test to test if the special diet gives significantly different results from a normal diet.

6.7 14 people were tested with each of two tranquillisers A and B. One person found no difference in effect, eleven preferred A and two preferred B. Use the sign test to see if there is a significant difference between A and B.

6.8 It is suspected that lack of sleep affects the ability of people to perform simple tests. In an experiment to study this, subjects were asked to press a button as quickly as possible after a red light came on. The response time was recorded automatically to the nearest 0.01 s. Each subject was tested five times after having a good nights sleep, and five times after being awake for 24 h. The mean response times are given in Table 6.10. Use (a) the sign test and (b) Wilcoxon's signed rank test to see if there is a significant difference in response times between the tired and rested values.

6.9 In 1876 Charles Darwin reported the results of a series of experiments in which he compared the growth of matched pairs of self-fertilized and cross-fertilized plants grown under the same conditions. The results for 15 pairs of maize seeds are given in Table 6.11.

Table 6.10 Mean Response Times in Seconds

Subject	Rested	Tired	Subject	Rested	Tired	Subject	Rested	Tired
1	0.52	0.56	5	0.67	0.74	9	0.63	0.68
2	0.61	0.60	6	0.58	0.64	10	0.58	0.55
3	0.55	0.59	7	0.59	0.59	11	0.53	0.63
4	0.54	0.82	8	0.55	0.62	12	0.61	0.72

Table 6.11

| Pair | Height of plants (inches) | |
	Self-fertilized	Cross-fertilized
1	17.375	23.5
2	20.375	12
3	20	21
4	20	22
5	18.375	19.125
6	18.625	21.5
7	18.625	22.125
8	15.25	20.375
9	16.5	18.25
10	18	21.625
11	16.25	23.25
12	18	21
13	12.75	21.125
14	15.5	23
15	18	12

Test if there is a significant difference between the growth of self-fertilized and cross-fertilized plants (a) using Wilcoxon's signed rank test, and (b) using the t-test.

6.10 A trial was conducted to find if there was any significant difference between the effect of a new sleep inducing drug and the old accepted phenobarbitone. Eight chronic insomniacs were selected, and randomly split into two groups of four. One group was given the new drug and the other group was given phenobarbitone, and the hours of sleep were recorded. Several days were allowed to elapse to make sure that the effect of the first sedative had worn off, then the patients were given the other drug (i.e. the one they did not have in the first test). The hours of sleep were again noted. The results are shown in Table 6.12. Use Wilcoxon's signed pair test to find if

Table 6.12

| Patient | Hours of sleep | |
	With phenobarb	With new drug
1	6.0	7.0
2	5.5	5.0
3	5.5	5.5
4	8.0	7.5
5	4.25	4.5
6	6.25	7.0
7	3.5	3.75
8	5.0	6.5

there is a significant difference between these drugs.

6.11 Ten athletes ran a 1500 m race at sea level, and at a later date they ran another 1500 m race at a high altitude. Their times are given in Table 6.13. Use Wilcoxon's signed rank test and the t-test to test the hypothesis that the athletes' performance is unaffected by altitude, and state the assumptions made in these tests.

Table 6.13

| Athlete | Time | |
	Sea level	High altitude
1	3 min 43 s	3 min 57 s
2	3 min 47 s	3 min 59 s
3	3 min 52 s	4 min 01 s
4	3 min 59 s	3 min 57 s
5	3 min 44 s	3 min 42 s
6	4 min 02 s	4 min 14 s
7	3 min 55 s	3 min 53 s
8	3 min 40 s	3 min 48 s
9	3 min 57 s	4 min 12 s
10	3 min 45 s	3 min 47 s

7

Comparison of More than Two Samples—Analysis of Variance

The previous chapter considered methods for comparing two samples to determine whether they differ significantly. Whenever there are more than two samples these methods are inapplicable, and a different technique called 'analysis of variance' is used instead. Two different procedures are described, namely one-way analysis of variance for single classifications and two-way analysis of variance for double classifications.

If an experiment yields two sets of replicate readings, for example the amount of photosynthesis which occurs in eight tubes of algae in red light and blue light respectively, then the data from red and blue light may be compared using the F and t-tests as previously described. If an experiment yields three or more sets of data, for example using red, yellow, tungsten (white) and white fluorescent light in the photosynthesis experiment above, then these pairwise tests are both laborious and unreliable as explained below.

Since the F- and t-tests compare two sets of data, with four sets of data it would be necessary to perform the test six times: red and yellow, red and tungsten, red and fluorescent, yellow and tungsten, yellow and fluorescent, tungsten and fluorescent. This is plainly laborious. Furthermore, if the 5% level of significance is used in the test, there is a 1:20 chance of concluding that the mean values of the two parent populations are different when in reality they are equal. Thus for each test there is only a 95% chance of making the correct conclusion. With six tests there is only a

$0.95^6 = 73.5\%$ chance of drawing the correct conclusion and hence a 26.5% chance of an incorrect conclusion. This is plainly unacceptable. Suppose that the 1% significance level was used rather than 5%. There will then only be a $0.99^6 = 94.1\%$ chance of concluding that the means of the parent populations are different when they are equal, but it would greatly increase the chance of ignoring real differences. Table 7.1 shows how the chance of error from the first of the reasons above increases with the number of samples.

A better technique is analysis of variance.

One-Way (Single Classification) Analysis of Variance

This method should be used whenever three or more sets of replicated readings are to be compared. First the mean value for each set of replicated readings is calculated. Next the deviations (errors) for each reading are calculated as (reading − sample mean). Then the errors squared are summed, and the variance (of the errors) is evaluated by dividing the sum of errors squared by the number of degrees of freedom. (The number of degrees of freedom is equal to the total number of readings minus the number of sets.)

The grand mean of all of the values is calculated from the sum of the values divided by the number of values. The residuals between the sample means and the grand mean are evaluated, and the sum of their

Table 7.1

Number of samples	Number of combinations	Chance of error at 5% significance (%)
2	1	5
3	3	14.3
4	6	26.5
5	10	40.1
6	15	53.7
7	21	65.9

squares is collected and divided by the number of degrees of freedom to give the variance of the sample means. (In this case the number of degrees of freedom is the number of sets minus one.)

Consider the case where there is no systematic difference betweeen the readings in the different sets. All of the readings are subject to random experimental errors whose magnitude is the variance of error (calculated in the first paragraph). The sample means are also subject to random experimental error, but since they have been produced by averaging, which tends to cancel errors, the variance of the sample means tends to the variance of error divided by number of replicates in a sample. Thus the variance of sample means multiplied by the number of replicates in a sample provides and estimate for the variance of error. The ratio

$$\frac{\text{Variance of sample means} \times \times \text{Number of replicates in a sample}}{\text{variance of error}} \quad (1)$$

is close to one.

Consider the case where there is a systematic difference between the readings in different sets. This will make the sample means differ by a systematic amount in addition to the random experimental error. This will result in a larger value for the variance of sample means, and hence the ratio calculated in Equation 1 will be greater than one.

To determine whether the ratio so obtained is significantly greater than one (which corresponds to testing whether or not there is a systematic difference between the sets), the ratio is looked up in an F table (Appendix 7). If the calculated F value exceeds the table F value, the observed difference is too large to be explained by random experimental error, and one concludes that the samples have not been drawn from the same parent population, and hence there is a significant difference between the sets. Conversely if the calculated F value is smaller than the table F value, the observed differences may be explained in terms of random experimental error, and there is no significant evidence for a systematic difference. The above steps are illustrated in the example below.

Example 1

An investigation was performed into the effect of four different colours of light on the rate of photosynthesis of a blue-green alga *Anacystis*. 32 tubes containing algae were used, providing eight replicates for each of the colours red, yellow, tungsten white and fluorescent white with equal energy light. The amount of photosynthesis was measured by the amount of oxygen evolved, which was determined by a Winkler titration, and the results are given in Table 7.2 as ml of $M/100$ $Na_2S_2O_3$ solution used.

(i) Calculate the mean value for each set.

$$\text{red} = (5.66 + 5.70 + 6.11 + 5.83 + 6.25 + 5.97 + 5.92 + 5.84)/8 = 5.91$$

Similarly,

Yellow	= 5.50
Tungsten	= 6.21
Fluorescent	= 5·98

(ii) Calculate the deviations (reading − sample mean), deviations squared and sum of deviations squared.

Table 7.2

Colour of light	Replicate measurements							
	1	2	3	4	5	6	7	8
Red	5.66	5.70	6.11	5.83	6.25	5.97	5.92	5.84
Yellow	5.24	5.45	5.34	5.79	5.33	5.76	5.40	5.69
Tungsten white	5.89	6.20	6.47	6.46	5.73	6.60	6.11	6.22
Fluorescent white	5.96	6.22	5.67	5.87	6.21	6.01	5.55	6.35

See Tables 7.3 and 7.4 for deviations and deviations squared.

Sum of deviations squared

$$= 1.7666$$

Number of degrees of freedom
$$= \text{Number of readings}$$
$$- \text{Number of samples}$$
$$v_2 = 32 - 4 = 28$$

Variance of error

$$= \frac{\text{Sum of deviations squared}}{\text{Number of degrees of freedom}}$$

$$= \frac{1.7666}{28} = 0.0631$$

(iii) Calculate grand mean

Grand mean

$$= \frac{\begin{array}{c}(\text{red mean} \times 8) + (\text{yellow mean} \times 8)\\ + (\text{tungsten mean} \times 8) + (\text{fluorescent mean} \times 8)\end{array}}{8+8+8+8}$$

$$= \frac{(5.91 \times 8) + (5.50 \times 8) + (6.21 \times 8) + 5.98 \times 8}{32}$$

$$= 5.90$$

The equation used to calculate the grand mean may appear to be unnecessarily long, but is written in this form so that the calculation may be carried out using the sample means regardless of whether each of the samples contains the same number of replicates.

(iv) Calculate the residuals between the sample means and the grand mean.

Residual for
Red	$5.91 - 5.90 =$	0.01
Yellow	$5.50 - 5.90 =$	-0.40
Tungsten	$6.21 - 5.90 =$	0.31
Fluorescent	$5.98 - 5.90 =$	0.08

(v) Calculate the residuals squared, then multiply for the number of replicates for each sample (see Table 7.5).

(vi) The sum of residuals squared as calculated in (v) is divided by the number of degrees of freedom to give the mean square for the variance between sets.

number of degrees of freedom
$$= \text{number of sets minus one}$$
$$v_1 = 4 - 1$$
$$= 3$$

mean square for variance between sets
$$= \frac{2.1008}{3}$$
$$= 0.7003$$

(vii) The ratio of the mean square for the between sets variance to the mean square for the within sets variance is calculated as

$$F = \frac{0.7003}{0.0631} = 11.1$$

This value should be compared with the F value for $v_1 = 3$ and $v_2 = 28$ given in Appendix 7.

Since the calculated F of 11.1 exceeds the 5% F value of 2.95 there is less than a 5% chance that such results could occur by chance if there was in reality no systematic difference

Table 7.3 Deviations

Red	-0.25	-0.21	0.20	-0.08	0.34	0.06	0.01	-0.07
Yellow	-0.26	-0.05	-0.16	0.29	-0.17	0.26	-0.10	0.19
Tungsten	-0.32	-0.01	0.26	0.25	-0.48	0.39	-0.10	0.01
Fluorescent	-0.02	0.24	-0.31	-0.11	0.23	0.03	-0.43	0.37

Table 7.4 Deviations Squared

Red	0.0625	0.0441	0.0400	0.0064	0.1156	0.0036	0.0001	0.0049
Yellow	0.0676	0.0025	0.0256	0.0841	0.0289	0.0676	0.0100	0.0361
Tungsten	0.1024	0.0001	0.0676	0.0625	0.2304	0.1521	0.0100	0.0001
Fluorescent	0.0004	0.0576	0.0961	0.0121	0.0529	0.0009	0.1849	0.1369

Table 7.5

	Residual	(Residual)^2	Number of replicates	(Residual)^2 $\times$ Number of replicates
Red	0.01	0.0001	8	0.0008
Yellow	-0.40	0.1600	8	1.2800
Tungsten	0.31	0.0961	8	0.7688
Fluorescent	0.08	0.0064	8	0.0512
				$\Sigma 2.1008$

in the rate of photosynthesis with different colours of light. Furthermore, the 0.1% F value is 7.19, so there is less than a 0.1% chance that the difference could have arisen by chance. This provides very strong evidence for a relation between the amount of photosynthesis and the wavelength of light by the alga *Anacystis*.

Description of Program to Compare More than Two Samples

First the program (Program 7.1) prints a heading (lines 20–30) and then asks if full instructions are required. The answer, which must be YES or NO, is checked in a subroutine (lines 1790–1870), and full or shortened instructions are printed as appropriate.

Next the user is asked to type the number of samples to be compared, and the number typed is checked to ensure that it is an integer, and in the range 3–10 inclusive (lines 140–220). The lower limit is set at three since comparison of two samples should be carried out using the F and t-tests described in Chapter 6. The upper limit of ten is empirical (and may be changed by the user if required).

A loop (lines 240–530) is used to input each of the several sets of data. The X values are input (line 360) and a dummy number of 999999 is used to indicate the end of each data set. The sum of differences squared for each sample is calculated in line 410 and the mean is calculated (line 430). After the terminator has been typed, a check is performed (lines 480–520) to ensure that there are at least two numbers in each sample.

After all the samples of data have been input, the mean, sum of differences squared, number of points and number of degrees of freedom are printed for each sample in a loop (lines 580–670). Next, the total sum of differences squared, total number of degrees of freedom and the mean square for the within sets variance are calculated and printed (lines 680–720).

A test is performed (line 740) to check that the mean square for the within sets variance is not zero. If the value is zero, it is impossible to calculate F since the denominator is zero, and an explanatory message (lines 750–770) is printed, the run terminated on these data, and the user asked if another run is required. Generally the value is not zero, and the calculation proceeds.

Program 7.1 Trial run.

```
              COMPARISON OF MORE THAN TWO SAMPLES
              ========== == ==== ==== === =======

WOULD YOU LIKE FULL INSTRUCTIONS
TYPE YES OR NO AND PRESS RETURN
? YES

THIS PROGRAM COMPARES THREE OR MORE SAMPLES OF DATA TO
DETERMINE IF THEY DIFFER SIGNIFICANTLY, OR IF THEY ARE
SUFFICIENTLY SIMILAR TO BE POOLED INTO A SINGLE SAMPLE

TYPE THE NUMBER OF SAMPLES TO BE COMPARED
THEN PRESS RETURN.
? 4

TYPE IN THE NUMBERS FOR SAMPLE 1  ONE AT A TIME.
PRESS RETURN AFTER EACH NUMBER.
TERMINATE DATA WITH 999999
? 5.66
? 5.70
? 6.11
? 5.83
? 6.25
? 5.97
? 5.92
? 5.84
? 999999

TYPE IN THE NUMBERS FOR SAMPLE 2  ONE AT A TIME.
PRESS RETURN AFTER EACH NUMBER.
TERMINATE DATA WITH 999999
? 5.24
? 5.45
? 5.34
? 5.79
? 5.33
? 5.76
? 5.40
? 5.69
? 999999

TYPE IN THE NUMBERS FOR SAMPLE 3  ONE AT A TIME.
PRESS RETURN AFTER EACH NUMBER.
TERMINATE DATA WITH 999999
? 5.89
? 6.20
```

```
? 6.47
? 6.46
? 5.73
? 6.60
? 6.11
? 6.22
? 999999

TYPE IN THE NUMBERS FOR SAMPLE 4  ONE AT A TIME.
PRESS RETURN AFTER EACH NUMBER.
TERMINATE DATA WITH 999999
? 5.96
? 6.22
? 5.67
? 5.87
? 6.21
? 6.01
? 5.55
? 6.35
? 999999

SAMPLE 1
  MEAN = 5.91
  SUM OF DIFFERENCES SQUARED = 0.277201
  NUMBER OF DATA POINTS = 8
  NUMBER OF DEGREES OF FREEDOM = 7

SAMPLE 2
  MEAN = 5.5
  SUM OF DIFFERENCES SQUARED = 0.322402
  NUMBER OF DATA POINTS = 8
  NUMBER OF DEGREES OF FREEDOM = 7

SAMPLE 3
  MEAN = 6.21
  SUM OF DIFFERENCES SQUARED = 0.6252
  NUMBER OF DATA POINTS = 8
  NUMBER OF DEGREES OF FREEDOM = 7

SAMPLE 4
  MEAN = 5.98
  SUM OF DIFFERENCES SQUARED = 0.541801
  NUMBER OF DATA POINTS = 8
  NUMBER OF DEGREES OF FREEDOM = 7

TOTAL SUM OF DIFFERENCES SQUARED = 1.7666
TOTAL NUMBER OF DEGREES OF FREEDOM = 28
MEAN SQUARE FOR WITHIN SAMPLES VARIANCE = 0.063093

GRAND MEAN = 5.9
SUM OF RESIDUALS SQUARED = 2.1008
NUMBER OF DEGREES OF FREEDOM = 3
MEAN SQUARE FOR THE BETWEEN SAMPLES VARIANCE = 0.700267


CALCULATED VALUE OF F = 11.099
PROBABILITY THAT SUCH A DIFFERENCE IN THE WITHIN SAMPLES
MEAN SQUARE AND THE BETWEEN SAMPLES MEAN SQUARE COULD OCCUR
BY CHANCE IF THE PARENT POPULATIONS HAVE THE SAME VARIANCE
IS LESS THAN 0.01 %
```

```
WOULD YOU LIKE ANOTHER RUN?
TYPE YES OR NO AND PRESS RETURN
? NO
END OF JOB

10 DIM K(10), A(10), Y(10), Q$(10), I$(3)
20 PRINT TAB(15); "COMPARISON OF MORE THAN TWO SAMPLES"
30 PRINT TAB(15); "========== == ==== ==== === ======="
40 PRINT
50 PRINT "WOULD YOU LIKE FULL INSTRUCTIONS"
60 GOSUB  1810
70 LET I$ = Q$
80 IF I$ = "NO" THEN 130
90 PRINT
100 PRINT "THIS PROGRAM COMPARES THREE OR MORE SAMPLES OF DATA TO"
110 PRINT "DETERMINE IF THEY DIFFER SIGNIFICANTLY, OR IF THEY ARE"
120 PRINT "SUFFICIENTLY SIMILAR TO BE POOLED INTO A SINGLE SAMPLE"
130 PRINT
140 PRINT "TYPE THE NUMBER OF SAMPLES TO BE COMPARED"
150 IF I$ = "NO" THEN 170
160 PRINT "THEN PRESS RETURN."
170 INPUT M
180 IF (M - 3) * (M - 10) > 0 THEN 200
190 IF M = INT(M) THEN 230
200 PRINT "NUMBER OF SAMPLES MUST BE AN INTEGER BETWEEN 3 & 10."
210 PRINT "RE-";
220 GOTO  140
230 REM START LOOP TO INPUT DATA
240 FOR J = 1 TO M
250    LET A(J) = 0
260    LET Y(J) = 0
270    LET I = 1
280    LET Z = 0
290    PRINT
300    IF I$ = "YES" THEN 330
310    PRINT "INPUT NUMBERS IN SAMPLE"; J
320    GOTO  350
330    PRINT "TYPE IN THE NUMBERS FOR SAMPLE"; J; " ONE AT A TIME."
340    PRINT "PRESS RETURN AFTER EACH NUMBER."
350    PRINT "TERMINATE DATA WITH 999999"
360    INPUT X
370    IF X = 999999 THEN 470
380    LET Z = Z + X
390    LET X1 = X - Y(J)
400    REM CALCULATE SUM OF DIFFERENCES SQUARED
410    LET A(J) = A(J) + X1 * X1 * (I - 1) / I
420    REM CALCULATE MEAN OF SAMPLE
430    LET Y(J) = Z / I
440    LET I = I + 1
450    GOTO  360
460    REM K(J) = NUMBER OF POINTS IN CURRENT SAMPLE
470    LET K(J) = I - 1
480    IF K(J) >= 2 THEN 530
490    PRINT
500    PRINT "THERE MUST BE AT LEAST TWO NUMBERS IN EACH SAMPLE."
510    PRINT "RUN TERMINATED ON THIS DATA."
520    GOTO  1700
530 NEXT J
540 LET E2 = 0
550 LET V2 = 0
560 LET R2 = 0
```

```
570 REM LOOP TO PRINT STATISTICS FOR EACH SAMPLE
580 FOR J = 1 TO M
590    PRINT
600    PRINT "SAMPLE"; J
610    PRINT "  MEAN ="; Y(J)
620    PRINT "  SUM OF DIFFERENCES SQUARED ="; A(J)
630    LET E2 = E2 + A(J)
640    LET V2 = V2 + K(J) - 1
650    PRINT "  NUMBER OF DATA POINTS ="; K(J)
660    PRINT "  NUMBER OF DEGREES OF FREEDOM ="; K(J) - 1
670 NEXT J
680 PRINT
690 PRINT "TOTAL SUM OF DIFFERENCES SQUARED ="; E2
700 PRINT "TOTAL NUMBER OF DEGREES OF FREEDOM ="; V2
710 REM CALCULATE MEAN SQUARE FOR THE WITHIN SAMPLES VARIANCE
720 LET M2 = E2 / V2
730 PRINT "MEAN SQUARE FOR WITHIN SAMPLES VARIANCE ="; M2
740 IF M2 > 0 THEN 800
750 PRINT
760 PRINT "BECAUSE THE VARIANCE WITHIN EACH SAMPLE IS ZERO"
770 PRINT "IT IS IMPOSSIBLE TO CALCULATE F.  RUN ABANDONED"
780 GOTO  1710
790 REM CALCULATE GRAND MEAN G
800 LET G = 0
810 FOR J = 1 TO M
820    LET G = G + Y(J) * K(J)
330 NEXT J
840 LET G = G / (V2 + M)
850 PRINT
860 PRINT "GRAND MEAN ="; G
870 REM CALCULATE SUM OF RESIDUALS (SAMPLE MEANS - GRAND MEAN) SQUARED
880 FOR J = 1 TO M
890    LET R1 = Y(J) - G
900    LET R2 = R2 + K(J) * R1 * R1
910 NEXT J
920 PRINT "SUM OF RESIDUALS SQUARED ="; R2
930 REM CALCULATE MEAN SQUARE FOR THE BETWEEN SAMPLES VARIANCE
940 LET V1 = M - 1
950 PRINT "NUMBER OF DEGREES OF FREEDOM ="; V1
960 LET M1 = R2 / V1
970 PRINT "MEAN SQUARE FOR THE BETWEEN SAMPLES VARIANCE ="; M1
980 REM CALCULATE F
990 LET F = M1 / M2
1000 PRINT
1010 PRINT
1020 PRINT "CALCULATED VALUE OF F ="; F
1030 LET E = 0
1040 IF V1 = 2 * INT(V1 / 2 + 0.1) THEN 1080
1050 IF V2 = 2 * INT(V2 / 2 + 0.1) THEN 1130
1060 GOTO  1310
1070 REM CALCULATE PROBABILITY IF V1 EVEN
1080 LET U = 1 / (1 + V2 / (F * V1))
1090 LET P1 = V1 + 1
1100 LET Q = V2 - 2
1110 GOTO  1170
1120 REM CALCULATE PROBABILITY IF V2 EVEN
1130 LET E = 1
1140 LET U = 1 / (1 + F * V1 / V2)
1150 LET P1 = V2 + 1
1160 LET Q = V1 - 2
1170 LET S = 0
```

```
1180 LET W = 1
1190 LET J = 2
1200 LET S = S + W
1210 LET W = W * U * (J + Q) / J
1220 LET J = J + 2
1230 IF J < P1 THEN 1200
1240 LET Z = SQR(1 - U)
1250 IF E = 0 THEN 1280
1260 LET P = 100 * S * (Z ^ V1)
1270 GOTO 1540
1280 LET P = 100 * S * (Z ^ V2)
1290 GOTO 1540
1300 REM CALCULATE PROBABILITY IF V1 & V2 BOTH ODD
1310 LET U = 1 / (1 + F * V1 / V2)
1320 LET X = 1 - U
1330 LET S = 0
1340 LET W = 1
1350 LET J = 2
1360 LET P1 = V2
1370 GOTO 1410
1380 LET S = S + W
1390 LET W = W * U * J / (J + 1)
1400 LET J = J + 2
1410 IF J < P1 THEN 1380
1420 LET W = W * V2
1430 LET J = 3
1440 LET P1 = V1 + 1
1450 LET Q = V2 - 2
1460 GOTO 1500
1470 LET S = S - W
1480 LET W = W * X * (J + Q) / J
1490 LET J = J + 2
1500 IF J < P1 THEN 1470
1510 LET T1 = ATN(SQR(F * V1 / V2))
1520 LET S1 = S * SQR(X * U)
1530 LET P = 100 * (1 - 2 * (T1 + S1) / 3.14159)
1540 IF P <= 50 THEN 1570
1550 LET P = 100 - P
1560 REM ROUND OFF ANSWER
1570 IF P < 5 THEN 1590
1580 LET P = INT(P * 10 + 0.5) * 0.1
1590 IF (P - 5) * (P - 0.5) > 0 THEN 1610
1600 LET P = INT(P * 100 + 0.5) * 0.01
1610 IF P > 0.05 THEN 1630
1620 LET P = INT(P * 1000 + 0.5) * 0.001
1630 PRINT "PROBABILITY THAT SUCH A DIFFERENCE IN THE WITHIN SAMPLES"
1640 PRINT "MEAN SQUARE AND THE BETWEEN SAMPLES MEAN SQUARE COULD OCCUR"
1650 PRINT "BY CHANCE IF THE PARENT POPULATIONS HAVE THE SAME VARIANCE"
1660 IF P < 0.01 THEN 1690
1670 PRINT "IS"; P; "%"
1680 GOTO 1700
1690 PRINT "IS LESS THAN 0.01 %"
1700 PRINT
1710 PRINT
1720 PRINT "WOULD YOU LIKE ANOTHER RUN?"
1730 GOSUB 1800
1740 LET I$ = "NO"
1750 RESTORE
1760 IF Q$ = "YES" THEN 130
1770 PRINT "END OF JOB"
1780 STOP
```

```
1790 REM ***** SUBROUTINE TO CHECK REPLIES   *****
1800 IF I$ = "NO" THEN 1820
1810 PRINT "TYPE YES OR NO AND PRESS RETURN"
1820 INPUT Q$
1830 IF Q$ = "YES" THEN 1870
1840 IF Q$ = "NO" THEN 1870
1850 PRINT "REPLY '"; Q$; "' NOT UNDERSTOOD. RE-";
1860 GOTO  1810
1870 RETURN
1880 END
```

The grand mean of all the data values is calculated and printed (lines 790–860) and the sum of residuals squared is calculated (lines 870–920). The mean square for the between sets variance is calculated and printed (lines 930–970) and F is calculated and printed (lines 980–1020).

The significance of the calculated F value is evaluated (lines 1030–1550). This routine is the same as that used in the F and t-test program where it is described. The calculated probability is rounded as appropriate and printed (lines 1560–1690).

Finally the user is asked whether another run is required (lines 1720–1730). The answer must be either YES or NO, and is checked by a subroutine (lines 1790–1870).

Elementary Points in Experimental Design

In many physics and chemistry experiments it is possible to keep all of the factors constant except the one to be varied in the experiment. However, in biological experiments variation in the animals or plants, the sample of material, the handling of the material, or of the environment inevitably lead to additional variation in the results. Consequently biological experiments frequently have a wider range of results than are usually found in physics or chemistry. Three essential points are:

1. to minimise the experimental errors,
2. to randomise the experimental material to avoid biasing the result, and
3. to distinguish between variations arising from the treatments and variations

arising from errors: this is dealt with by using analysis of variance.

Individual errors may be reduced by using uniform experimental material, and making their treatments exactly the same. Sample errors can be minimised by increasing the number of replicates and the size of counts.

Several methods of analysing numerical data and testing a hypothesis have already been described using data collected from samples or surveys. The design of the experiment or survey should be considered at the stage where the experiment is being planned— before the experiment is carried out, and before the results are collected, since the design can affect the results. No amount of fancy mathematics or statistical tests can counteract the effect of badly planned experiments or surveys.

Randomisation

Experimental animals or plants should be a random selection. If, for example, only the largest and healthiest individuals are used for testing, then the final results cannot be claimed to be representative of the whole population. An experiment to see if supplementary vitamins improve the health of schoolchildren would have little meaning either if the children tested came only from deprived areas, or if the teachers were allowed to select which children received the treatment, since the teachers would probably select those children who they thought most needed it. Clearly such an experiment would show if needy children benefitted, but would not show if children in general would benefit.

Extraneous factors, which are frequently unknown, must also be randomised over the sample of plants or animals used. This is necessary for two reasons:

1. An extraneous factor may have a direct effect on the effect being measured. Unless the effect of the extraneous factor is randomised, it will be wrongly attributed to the treatment.
2. A further reason for a randomisation is the necessity of obtaining an unbiassed estimate of the error inherent in the experiment. To determine whether a particular treatment has a significant effect, or whether one treatment is significantly better than another, the change produced must be compared with the experimental error. If the experimental error is incorrectly estimated, wrong conclusions will be drawn about the significance of the effects. The experimental error will be incorrectly estimated unless the extraneous factors affect all treatments equally.

Point 2 is best illustrated by means of an example.

A botanist is growing geranium plants in pots in a greenhouse. There is a strong temptation to have all the pots receiving one treatment together at one end of the greenhouse, and another treatment elsewhere. If any of the extraneous effects (e.g. temperature, humidity, ventilation or the amount of light) is different in the two locations then the difference will be ascribed to the treatments. Furthermore, the error introduced by this factor is not measured in the estimate of the error in the experiment. Not only is the effect due to the treatment incorrectly measured, but the estimated experimental error is also too small. Consequently the significance associated with the treatment will be incorrect. The correct way of overcoming these difficulties is to mix the treatments randomly between the locations. Extraneous factors then affect both treatments and are also included in the estimated experimental error.

Replication

More reliable results will be obtained if an experiment (or treatment) is replicated a large number of times. This is because a better estimate of the experimental error is made and the sample (or treatment) means become more reliable; that is, the standard error of the sample means is decreased. Unfortunately, replication alone is a laborious way of achieving accuracy. To double the accuracy it is necessary to make four times as many readings, since the standard error tends to the standard deviation of the parent population divided by the square root of the number of replicates. There is a conflict in experimental design between the two factors of (1) having uniform samples of material with identical treatments and conditions and (2) having a large number of samples and adequate replication.

Two-Way (Double Classification) Analysis of Variance

In many biological experiments a number of different treatments are applied, and the results of the treatments are judged by measuring a particular attribute. For example, the treatments might be dosages of fertilizer and the attribute measured the height of the plants or the yield of the crop. Clearly there are other factors affecting the attribute measured. In the one-way analysis of variance, these factors are considered as errors, and for the analysis to be valid these extraneous factors must be random. In practice this may not be so; for example more fertile land, better drainage or more shade at one end of a field, and consequently experiments utilising one-way classification of variance are not usual.

Problems such as this require a different form of experimental design, based on a two-way classification. The main purpose of introducing blocks is to make allowance for unwanted but unavoidable heterogeneity. A simple block design is described using plant growth in relation to fertilizer for the experiment above. Agricultural research was the area in which such experiments were originally developed, but they may be applied to most branches of biology. In agriculture a block consists of plots of land, but the terminology is now applied to any group of experiments expected to possess some degree of homogeneity—for example, a rack of test tubes in an incubator, or several animals from the same litter.

In the fertilizer example above, the field should be divided into a number of blocks, in such a way that each block is as uniform as possible—that is, the variation within blocks is minimised but variation between blocks is permitted. Each block is then divided into a series of plots, and each plot is given a particular treatment. Each block must contain at least one of each of the treatments, and the allocation of plots with different treatments within each block should be made at random. In the simplest case, if there are n treatments, each block is divided into n plots. Such an arrangement is described as an orthogonal two-way classification of blocks and treatments. The word orthogonal comes from mathematics, and its original meaning was at right-angles. In mechanics, orthogonal forces are important because one force does not affect the other, and hence the effect of either may be considered in isolation. In the example above, the blocks do not affect the treatments since if one block is in the shade and another in the sun, all of the treatments in one block are in the shade and all of the treatments in the other block are in the sun. In the same way, the treatments do not affect the blocks.

Interpretation of the results from an orthogonal two-way classification experiment is performed using analysis of variance.

The steps in the analysis are as follows:

(i) Consider an experiment in which there are b blocks and c different treatments.

(ii) The b values for the first treatment are summed to produce a treatment total, and similarly for the other $(c\text{-}1)$ treatments.

(iii) The c values for the first block are summed to give a block total, and similarly for the other $(b\text{-}1)$ blocks.

(iv) The sum of all the individual treatment totals from (i) is collected.

(v) The value from (iv) is squared and then divided by $(b \cdot c)$ to give the 'correction factor'.

(vi) The treatment totals are squared, summed, and then divided by b. The correction factor is then subtracted from this value to produce a sum of squares for the treatments.

(vii) The block totals are squared, summed, and then divided by c. The correction factor is then subtracted from this value to produce a sum of squares for the blocks.

(viii) The individual experimental values are squared and summed. The correction factor is subtracted to produce a total sum of squares.

(ix) The sum of squares for the residuals is calculated as the total sum of squares minus the sum of squares for treatments, minus the sum of squares for blocks. The value so obtained corresponds to all of the factors other than blocks and treatments, which are collectively called experimental error.

(x) The mean square for treatments is calculated as the sum of squares for treatments divided by the number of degrees of freedom $(c\text{-}1)$.

(xi) The mean square for blocks is calculated as the sum of squares for the blocks divided by $(b\text{-}1)$.

(xii) The mean square for residuals is calculated as the sum of squares for the residuals divided by $(b\text{-}1)(c\text{-}1)$.

It is desirable that the experiment is designed with the number of blocks and the number of treatments such that the number of degrees of freedom $(b\text{-}1)(c\text{-}1)$ is at least 10 and preferably more. With a lower number, a poor estimate will be made for the error in the experiment. To overcome this, a large F value is required to show a significant difference, hence it is more difficult to obtain a conclusive answer.

(xiii) An F value is obtained from the ratio of the mean square for treatments to the mean square for residuals. This calculated value is compared with the F probability values tabulated in Appendix 7 to find if the differences observed due to the treatments are significantly larger than those expected from experimental error. If the calculated F value exceeds the 5% table value with degrees of freedom $v_1 = (b\text{-}1)$ and $v_2 = (b\text{-}1)(c\text{-}1)$, there is some evidence that the treatments produce different effects. If the calculated F value exceeds the 1% table value there is strong evidence that the treatments produce different effects.

(xiv) It is instructive, though not essential, to calculate another F value as the ratio of mean square for blocks divided by the mean square of residuals. The purpose of choosing blocks is to eliminate one or more extraneous factors, and a high F value indicates that the choice of blocks has succeeded. If the calculated F value is smaller than the table value for a 5% probability, the observed differences could have arisen by chance, and the choice of blocks has not removed the effect of extraneous factors. In this case, no benefit has been obtained from carrying the experiment out in blocks, and the choice of new blocks should be considered before repeating the experiment. These steps are illustrated in Example 2.

Example 2

The effects of four possible growth-promoting substances on pea plants were compared with an untreated control in a randomised block experiment. Seeds of the same variety but from different sources were germinated and planted in six boxes, corresponding to six blocks. Each box contained 25 plants, which were divided into five plots comprising five plants each. Each plot received a different treatment—either the application of 1.5 μg of one of the growth-promoting substances to the first leaf, or no application in the case of the control. The number of the node which produced the first flower was recorded, and the results in Table 7.6 give the sum total of the nodes for the five plants in each treatment.

Table 7.6

Treatment	Block					
	1	2	3	4	5	6
Indole acetic acid	73	79	80	76	88	86
Gibberellic acid	77	81	85	80	86	83
Kinetin	73	75	77	79	81	78
Adenine	69	74	80	76	85	82
Untreated (control)	69	80	81	73	81	81

Carry out a two-way analysis of variance to find if the treatments cause any significant difference.

(i) Blocks $b = 6$, treatments $c = 5$.

(ii) Treatment totals are calculated:

Indole acetic acid 482
Gibberellic acid 492
Kinetin 463
Adenine 466
Untreated 465

(iii) Block totals are calculated:

361, 389, 403, 384, 421, 410

(iv) The sum of the treatment totals is calculated as 2368.

(v) The correction term is calculated as

$$(2368)^2/(6 \times 5) = 186914.1$$

(vi) The treatment totals are squared and summed to give an answer of 1122138. This is divided by b to give 187023, and the correction factor is

113

subtracted giving a sum of squares for the treatments of 108.9.

(vii) Similarly the sum of squares for blocks is calculated:

$$(936848/5) - 186914.1 = 455.5$$

(viii) The total sum of squares is calculated:

$$187590 - 186914.1 = 675.9$$

(ix) The sum of squares for residuals is calculated:

$$675.9 - 108.9 - 455.5 = 111.5$$

(x) The mean square for treatments is calculated:

$$108.9/(5-1) = 27.23$$

(xi) The mean square for blocks is calculated:

$$455.5/(6-1) = 91.1$$

(xii) The mean square for residuals is calculated:

$$111.5/[(6-1)\cdot(5-1)] = 5.575$$

(xiii) F is calculated:

$$27.23/5.575 = 4.88$$

This F value shows that differences arising from the treatments are almost five times greater than would be expected from the magnitude of the errors inherent in the experiment. The calculated F value of 4.88 is compared with the table of significance values (Appendix 7) with $v_1 = 4$ and $v_2 = 20$. The table value with 1% significance is 4.43. The calculated F is larger than this value, hence there is less than 1% chance—less than 1 in 100 chance—that so wide a difference in variance could have arisen by chance. The difference in treatments is therefore highly significant. Since one of the treatments was a control, it follows that at least one of the other treatments has had a significant effect on growth—either accelerating or retarding it. The result does not show if more than one of the tests has an effect.

(xiv) Another F value is calculated

$$91.1/5.575 = 16.3$$

This F value shows that differences arising between blocks are about 16 times greater than would be expected from experimental errors. The calculated F value greatly exceeds the 0.1% probability table value of 6.46 (with $v_1 = 5$ and $v_2 = 20$). Differences between blocks are exceedingly significant. This shows that seeds from different sources are significantly different, and that the choice of a block design of experiment was essential to remove this source of variation.

Numerical Accuracy

From this example it is apparent that loss of numerical accuracy may become acute if only a small number of significant figures are carried. If, for example, only five significant figures are carried, the F value calculated in (xiii) is 5.00 rather than 4.88; that is, the answer is accurate only to one significant figure. A major reason for this is the loss of accuracy which occurs when two nearly equal numbers are subtracted. This occurs in several places, e.g. step (viii) where numbers accurate to seven figures yield a result accurate to four figures—a loss of three significant figures. The numerical accuracy may be improved by reducing the magnitude of the numbers by subtracting the overall mean value for all of the blocks and all of the treatments from each of the data values before the calculation is commenced. This is done in the computer program described later, and this has the effect of making the correction factor zero.

Least Significant Difference

If the F value for the treatments is significant, the treatments produce significantly different results, and it is reasonable to enquire whether any particular pair of treatments produce results significantly different from each other.

The standard error of the difference between two treatment means is:

$$\sqrt{(2 \cdot \text{Mean square for residuals}/b)}$$

If the difference in treatment means divided by the standard error exceeds the 5% t-value with $c(b-1)$ degrees of freedom, then the treatments produce significantly different results. For hand calculations it is easier to multiply the 5% significance t-value to produce the *least significant difference*. Any pair of means whose difference is larger than the least significant difference may be declared different.

This procedure is very similar to the t-test for comparing two samples. The method used here is more reliable than the t-test because the estimate of the error in the experiment is based on all of the treatments rather than just the two which are being compared.

Latin Squares

The two-way classification previously described provides a method of removing one source of variation from the analysis of an experiment, thus improving the precision compared with the single classification. Latin squares provide a means of removing two sources of variation. Some examples are:

(a) The yields from different varieties of the same crop in a field may vary because of the position in the field. If the extraneous factors are believed to be the slope of the field (which runs for example in a north–south direction) and the prevailing wind (which runs in an east–west direction), then the effects of these two factors may be removed by a Latin square design of experiment.

(b) If six intelligence tests were evaluated by testing six different people on each of six days (one test per person per day), the results will be affected by their true aptitude and also by their learning experience on how to handle such tests, which will improve as the days go by. To obtain a true comparison of the intelligence tests, the effects due to the people's different aptitudes and their learning experience should be eliminated using a Latin square design.

(c) If a number of different drugs, or a number of different doses of the same drug are being compared, then the results will be complicated because different individuals will have differing sensitivity to the drug, and furthermore the response may be affected by drugs administered previously. A Latin square design may be used to eliminate the last two factors.

If there are c different treatments (i.e. different drugs, different doses or different varieties of crops) then a (c by c) Latin square should be produced. In the crop example described in (a) this would correspond to dividing the field into c rows and c columns, to produce c^2 plots. Each variety must occur once in each row and once in each column.

In the intelligence test example described in (b) the six tests are labelled A, B, C, D, E and F. A Latin square arrangement showing the days on which the different people take the different tests is shown in Table 7.7.

Table 7.7

Day	People					
	1	2	3	4	5	6
1	A	B	F	C	E	D
2	E	F	D	A	C	B
3	B	C	A	D	F	E
4	D	E	C	F	B	A
5	C	D	B	E	A	F
6	F	A	E	B	D	C

The important feature of this arrangement is that each test (letters A–F) appears once in every row and once in every column. The analysis of the data is very similar to that described for two-way classification except that sums of squares for both rows and columns are collected rather than just for blocks. The residual sum of squares is calculated as the difference between the total sum of squares and the sum of the sums of squares from treatments, rows and columns. The residual mean square has $(c-1)(c-2)$ degrees of freedom. The ratio of the treatment mean square to the residual mean square is evaluated, and this calculated F value is compared with significance tables for F.

Latin square type tests are usually used only when c is between 4 and 8. Above 8 the number of plots may become difficult to implement. Below 4 the number of degrees of freedom in the residual sum of squares is too small, for the same reasons as were outlined in the description of a two-way test section (xii).

Description of the Program to Perform a Two-Way Analysis of Variance

The program (Program 7.2) prints a heading (lines 20–30) and then the user is asked if full instructions are required (lines 50–70). The answer must be YES or NO, and is checked in a subroutine (lines 1530–1590). Full or abbreviated instructions are printed as appropriate on the first run, but shortened instructions are always given on the second or subsequent runs.

The user is requested to type the number of blocks used in the experiment, and the number typed is checked to make sure that it is an integer in the range 2–10 (lines 100–150).

Program 7.2   Trial run.

```
ANALYSIS OF DOUBLE CLASSIFICATION
======== == ====== ==============

WOULD YOU LIKE INSTRUCTIONS
TYPE YES OR NO AND PRESS RETURN.
? YES

TYPE THE NUMBER OF BLOCKS USED IN THE EXPERIMENT
? 6

TYPE THE NUMBER OF TREATMENTS IN EACH BLOCK.
? 5


TYPE THE VALUES FOR THE FIRST TREATMENT AND
PRESS RETURN AFTER EACH VALUE.
THEN TYPE THE SECOND AND SUBSEQUENT TREATMENTS.

PLEASE TYPE DATA
TREATMENT 1 TYPE THE 6   VALUES CORRESPONDING TO THE BLOCKS
? 73
? 79
? 80
? 76
? 88
? 36
TREATMENT 2 TYPE THE 6   VALUES CORRESPONDING TO THE BLOCKS
? 77
? 81
? 85
? 80
? 86
? 83
TREATMENT 3 TYPE THE 6   VALUES CORRESPONDING TO THE BLOCKS
? 73
? 75
? 77
? 79
? 81
? 78
```

116

```
TREATMENT 4 TYPE THE 6   VALUES CORRESPONDING TO THE BLOCKS
? 69
? 74
? 80
? 76
? 85
? 82

TREATMENT 5 TYPE THE 6   VALUES CORRESPONDING TO THE BLOCKS
? 69
? 80
? 81
? 73
? 81
? 81
```

TREATMENT	TOTAL	MEAN
1	482	80.3333
2	492	82
3	463	77.1667
4	466	77.6667
5	465	77.5

	SUM OF SQUARES	MEAN SQUARE
TOTAL	675.865	
TREATMENT	108.867	27.2167
BLOCK	455.467	91.0933
RESIDUAL	111.532	5.57661

```
F RATIO :  TREATMENTS/RESIDUAL = 4.8805
DEGREES OF FREEDOM   NU(1) = 4   NU(2) = 20
PROBABILITY THAT SUCH A DIFFERENCE IN SUMS OF SQUARES COULD
OCCUR BY CHANCE IF THE PARENT POPULATIONS HAVE THE SAME
SUM OF SQUARES IS 0.65 %

F RATIO :  BLOCKS/RESIDUAL = 16.3349
DEGREES OF FREEDOM  NU(1) = 5   NU(2) = 20
PROBABILITY THAT SUCH A DIFFERENCE IN SUMS OF SQUARES COULD
OCCUR BY CHANCE IF THE PARENT POPULATIONS HAVE THE SAME
SUM OF SQUARES IS LESS THAN 0.01 %


WOULD YOU LIKE TO COMPARE TREATMENTS USING A T TEST

THE T TEST COMPARES EACH TREATMENT IN TURN WITH A PARTICULAR
TREATMENT TO DETERMINE WHETHER THEY DIFFER SIGNIFICANTLY.
THIS IS MOST USEFUL FOR COMPARING WITH PLACEBOS OR CONTROLS.

? YES

WHICH TREATMENT 1 - 5  WAS THE CONTROL
? 5

NUMBER OF DEGREES OF FREEDOM = 25
STANDARD ERROR OF DIFFERENCE IN MEANS = 6.09733

COMPARISON OF TREATMENT 1  AGAINST TREATMENT 5
DIFFERENCE IN MEANS = 3.4
T VALUE = 2.78811
PROBABILITY OF SUCH A LARGE DIFFERENCE IN MEANS OCCURRING
BY CHANCE IF BOTH TREATMENTS HAVE THE SAME EFFECT IS 1.00 %
```

COMPARISON OF TREATMENT 2 AGAINST TREATMENT 5
DIFFERENCE IN MEANS = 5.4
T VALUE = 4.42817
PROBABILITY OF SUCH A LARGE DIFFERENCE IN MEANS OCCURRING
BY CHANCE IF BOTH TREATMENTS HAVE THE SAME EFFECT IS 0.0163773 %

COMPARISON OF TREATMENT 3 AGAINST TREATMENT 5
DIFFERENCE IN MEANS = -0.4
T VALUE = 0.328013
PROBABILITY OF SUCH A LARGE DIFFERENCE IN MEANS OCCURRING
BY CHANCE IF BOTH TREATMENTS HAVE THE SAME EFFECT IS 74.6 %

COMPARISON OF TREATMENT 4 AGAINST TREATMENT 5
DIFFERENCE IN MEANS = 0.2
T VALUE = 0.164006
PROBABILITY OF SUCH A LARGE DIFFERENCE IN MEANS OCCURRING
BY CHANCE IF BOTH TREATMENTS HAVE THE SAME EFFECT IS 87.1 %

WOULD YOU LIKE ANOTHER RUN? (YES/NO)
? NO

END OF JOB


```
10 DIM Q$(10), I$(3), A(10, 10), X(10), B(4)
20 PRINT TAB(10); "ANALYSIS OF DOUBLE CLASSIFICATION"
30 PRINT TAB(10); "======== == ====== =============="
40 PRINT
50 PRINT "WOULD YOU LIKE INSTRUCTIONS"
60 PRINT "TYPE YES OR NO AND PRESS RETURN."
70 GOSUB  1540
80 LET I$ = Q$
90 PRINT
100 PRINT "TYPE THE NUMBER OF BLOCKS USED IN THE EXPERIMENT"
110 INPUT B1
120 IF B1 <> INT(B1) THEN 140
130 IF (B1 - 2) * (B1 - 10) <= 0 THEN 160
140 PRINT "THE NUMBER OF BLOCKS MUST BE AN INTEGER BETWEEN 2 AND 10"
150 GOTO  100
160 PRINT
170 PRINT "TYPE THE NUMBER OF TREATMENTS IN EACH BLOCK."
180 INPUT C1
190 IF C1 <> INT(C1) THEN 210
200 IF (C1 - 2) * (C1 - 10) <= 0 THEN 230
210 PRINT "THE NUMBER OF TREATMENTS MUST BE AN INTEGER BETWEEN 2 AND 10"
220 GOTO  170
230 PRINT
240 IF I$ = "NO" THEN 300
250 PRINT
260 PRINT "TYPE THE VALUES FOR THE FIRST TREATMENT AND"
270 PRINT "PRESS RETURN AFTER EACH VALUE."
280 PRINT "THEN TYPE THE SECOND AND SUBSEQUENT TREATMENTS."
290 PRINT
300 LET T = 0
310 PRINT "PLEASE TYPE DATA"
320 FOR I = 1 TO C1
330   PRINT "TREATMENT"; I;
340   PRINT "TYPE THE"; B1; " VALUES CORRESPONDING TO THE BLOCKS"
350   FOR J = 1 TO B1
```

```
360      INPUT A(I, J)
370      LET T = T + A(I, J)
380    NEXT J
390 NEXT I
400 LET S1 = 0
410 REM CALCULATE MEAN VALUE AND SUBTRACT FROM ALL VALUES
420 LET T = T / (B1 * C1)
430 FOR I = 1 TO C1
440    FOR J = 1 TO B1
450      LET A1 = A(I, J) - T
460      LET A(I, J) = A1
470      LET S1 = S1 + A1 * A1
480    NEXT J
490 NEXT I
500 PRINT
510 PRINT "TREATMENT       TOTAL           MEAN"
520 REM CALCULATE SUM OF SQUARES OF TREATMENT TOTALS
530 LET S2 = 0
540 FOR I = 1 TO C1
550    LET T1 = 0
560    FOR J = 1 TO B1
570      LET T1 = T1 + A(I, J)
580    NEXT J
590    LET S2 = S2 + T1 * T1
600    LET X(I) = T1
610    PRINT I; TAB(15); X(I) + T * B1; TAB(32); X(I) / B1 + T
620 NEXT I
630 LET S2 = S2 / B1
640 REM CALCULATE SUM OF SQUARES OF BLOCK TOTALS
650 LET S3 = 0
660 FOR J = 1 TO B1
670    LET T1 = 0
680    FOR I = 1 TO C1
690      LET T1 = T1 + A(I, J)
700    NEXT I
710    LET S3 = S3 + T1 * T1
720 NEXT J
730 LET S3 = S3 / C1
740 REM CALCULATE VARIANCE DUE TO ERROR
750 LET S4 = S1 - S2 - S3
760 PRINT
770 PRINT TAB(15); "SUM OF SQUARES    MEAN SQUARE"
780 PRINT "TOTAL"; TAB(15); S1
790 PRINT "TREATMENT"; TAB(15); S2; TAB(32); S2 / (C1 - 1)
800 PRINT "BLOCK"; TAB(15); S3; TAB(32); S3 / (B1 - 1)
810 PRINT "RESIDUAL "; TAB(15); S4; TAB(32); S4 / ((B1 - 1) * (C1 - 1))
820 PRINT
830 IF S4 > 0 THEN 870
840 PRINT "NO ERROR IN EXPERIMENT - IMPOSSIBLE TO CALCULATE F"
850 GOTO  1370
860 REM TEST WHETHER THE TREATMENTS DIFFER SIGNIFICANTLY
870 LET V1 = C1 - 1
880 LET V2 = (B1 - 1) * (C1 - 1)
890 LET F = (S2 / V1) / (S4 / V2)
900 PRINT "F RATIO ;   TREATMENTS/RESIDUAL ="; F
910 PRINT "DEGREES OF FREEDOM    NU(1) ="; V1; "   NU(2) ="; V2
920 GOSUB  1900
930 REM TEST WHETHER IT WAS WORTH DOING BLOCKS
940 LET V1 = B1 - 1
950 LET F = (S3 / V1) / (S4 / V2)
960 PRINT "F RATIO ;   BLOCKS/RESIDUAL ="; F
```

```
970 PRINT "DEGREES OF FREEDOM   NU(1) ="; V1; "   NU(2) ="; V2
980 GOSUB   1900
990 PRINT
1000 PRINT "WOULD YOU LIKE TO COMPARE TREATMENTS USING A T TEST"
1010 IF I$ = "NO" THEN 1070
1020 PRINT
1030 PRINT"THE T TEST COMPARES EACH TREATMENT IN TURN WITH A PARTICULAR"
1040 PRINT "TREATMENT TO DETERMINE WHETHER THEY DIFFER SIGNIFICANTLY."
1050 PRINT"THIS IS MOST USEFUL FOR COMPARING WITH PLACEBOS OR CONTROLS."
1060 PRINT
1070 GOSUB   1540
1080 IF Q$ = "NO" THEN 1380
1090 PRINT
1100 PRINT "WHICH TREATMENT 1 -"; C1; " WAS THE CONTROL"
1110 INPUT Y
1120 IF Y <> INT(Y) THEN 1140
1130 IF (Y - 1) * (Y - C1) <= 0 THEN 1160
1140 PRINT "TREATMENT NUMBER MUST BE AN INTEGER BETWEEN 1 AND"; C1
1150 GOTO   1100
1160 LET N2 = (B1 - 1) * C1
1170 PRINT
1180 PRINT "NUMBER OF DEGREES OF FREEDOM ="; N2
1190 PRINT "STANDARD ERROR OF DIFFERENCE IN MEANS ="; SQR(2 * S4 / B1)
1200 PRINT
1210 FOR L = 1 TO C1
1220    IF L = Y THEN 1360
1230    LET T = ABS(X(L) - X(Y)) / SQR(2 * S4 / B1)
1240    PRINT "COMPARISON OF TREATMENT"; L; " AGAINST TREATMENT"; Y
1250    PRINT "DIFFERENCE IN MEANS ="; (X(L) - X(Y)) / C1
1260    PRINT "T VALUE ="; T
1270    REM CALCULATE T PROBABILITY
1280    GOSUB   1610
1290    RESTORE
1300    REM ROUND ANSWER
1310    GOSUB   1460
1320    PRINT "PROBABILITY OF SUCH A LARGE DIFFERENCE IN MEANS OCCURRING"
1330    PRINT "BY CHANCE IF BOTH TREATMENTS HAVE THE SAME EFFECT IS";
1340    PRINT P; "%"
1350    PRINT
1360 NEXT L
1370 PRINT
1380 PRINT "WOULD YOU LIKE ANOTHER RUN? (YES/NO)"
1390 GOSUB   1540
1400 RESTORE
1410 LET I$ = "NO"
1420 IF Q$ = "YES" THEN 90
1430 PRINT "END OF JOB"
1440 STOP
1450 REM ***** SUBROUTINE TO ROUND OFF ANSWERS *****
1460 IF P < 5 THEN 1480
1470 LET P = INT(P * 10 + 0.5) * 0.1
1480 IF (P - 5) * (P - 0.5) > 0 THEN 1500
1490 LET P = INT(P * 100 + 0.5) * 0.01
1500 IF P > 0.5 THEN 1520
1510 LET P = INT(P * 1000 + 0.5) * 0.001
1520 RETURN
1530 REM ***** SUBROUTINE TO CHECK REPLIES *****
1540 INPUT Q$
1550 IF Q$ = "YES" THEN 1590
1560 IF Q$ = "NO" THEN 1590
1570 PRINT "REPLY '"; Q$; "' NOT UNDERSTOOD. RETYPE"
```

```
1580 GOTO  1540
1590 RETURN
1600 REM ***** SUBROUTINE TO CALCUATE T PROBABILITY *****
1610 READ B(1), B(2), B(3), B(4), P
1620 DATA 1E6, 10000, 1000, 100, 0
1630 IF N2 > 4 THEN 1660
1640 IF T > B(N2) THEN 1870
1650 GOTO  1670
1660 IF T > 50 THEN 1870
1670 LET A = T / SQR(N2)
1680 LET B = N2 / (N2 + T * T)
1690 LET J = N2 - 2
1700 LET K = N2 - INT(N2 / 2 + 0.1) * 2 + 2
1710 LET S = 1
1720 IF J < 2 THEN 1800
1730 LET C = 1
1740 LET F2 = K
1750 FOR I = K TO J STEP 2
1760    LET C = C * B * (F2 - 1) / F2
1770    LET S = S + C
1780    LET F2 = F2 + 2
1790 NEXT I
1800 IF K > 2 THEN 1830
1810 LET P = 0.5 - 0.5 * A * SQR(B) * S
1820 GOTO  1870
1830 IF N2 > 1 THEN 1850
1840 LET S = 0
1850 LET P = 0.5 - (A * B * S + ATN(A)) / 3.14159
1860 REM CONVERT TO TWO TAILED PERCENTAGE
1870 LET P = 100 * 2 * P
1880 RETURN
1890 REM ***** SUBROUTINE TO CALCULATE F PROBABILITY *****
1900 LET E = 0
1910 IF V1 = 2 * INT(V1 / 2 + 0.1) THEN 1950
1920 IF V2 = 2 * INT(V2 / 2 + 0.1) THEN 2000
1930 GOTO 2180
1940 REM CALCULATE PROBABILITY IF V1 EVEN
1950 LET U = 1 / (1 + V2 / (F * V1))
1960 LET P1 = V1 + 1
1970 LET Q = V2 - 2
1980 GOTO  2040
1990 REM CALCULATE PROBABILITY IF V2 EVEN
2000 LET E = 1
2010 LET U = 1 / (1 + F * V1 / V2)
2020 LET P1 = V2 + 1
2030 LET Q = V1 - 2
2040 LET S = 0
2050 LET W = 1
2060 LET K = 2
2070 LET S = S + W
2080 LET W = W * U * (K + Q) / K
2090 LET K = K + 2
2100 IF K < P1 THEN 2070
2110 LET Z = SQR(1 - U)
2120 IF E = 0 THEN 2150
2130 LET P = 100 * S * (Z ^ V1)
2140 GOTO 2410
2150 LET P = 100 * S * (Z ^ V2)
2160 GOTO 2410
2170 REM CALCULATE PROBABILITY IF V1 & V2 BOTH ODD
2180 LET U = 1 / (1 + F * V1 / V2)
```

```
2190 LET X = 1 - U
2200 LET S = 0
2210 LET W = 1
2220 LET K = 2
2230 LET P1 = V2
2240 GOTO 2280
2250 LET S = S + W
2260 LET W = W * U * K / (K + 1)
2270 LET K = K + 2
2280 IF K < P1 THEN 2250
2290 LET W = W * V2
2300 LET K = 3
2310 LET P1 = V1 + 1
2320 LET Q = V2 - 2
2330 GOTO 2370
2340 LET S = S - W
2350 LET W = W * X * (K + Q) / K
2360 LET K = K + 2
2370 IF K < P1 THEN 2340
2380 LET T1 = ATN(SQR(F * V1 / V2))
2390 LET R1 = S * SQR(X * U)
2400 LET P = 100 * (1 - 2 * (T1 + R1) / 3.14159)
2410 IF P <= 50 THEN 2440
2420 LET P = 100 - P
2430 REM ROUND OFF ANSWER
2440 GOSUB 1460
2450 PRINT "PROBABILITY THAT SUCH A DIFFERENCE IN SUMS OF SQUARES COULD"
2460 PRINT "OCCUR BY CHANCE IF THE PARENT POPULATIONS HAVE THE SAME"
2470 IF P < 0.01 THEN 2500
2480 PRINT "SUM OF SQUARES IS"; P; "%"
2490 GOTO 2510
2500 PRINT "SUM OF SQUARES IS LESS THAN 0.01 %"
2510 PRINT
2520 RETURN
2530 END
```

The upper limit of 10 is arbitrary, but is imposed by the DIMension of the A array, in line 10. Next the number of treatments (plots) in each block is requested and checked to ensure that it is an integer in the range 2–10 (lines 170–220). The upper limit is arbitrary but is imposed by the DIMension of the A and X arrays in line 10.

The program then asks for the data to be typed in (lines 260–310), and the data are input using two nested loops from lines 320 to 390.

The mean value for all of the readings is calculated and subtracted from each of the data values (lines 410–490). This is done to preserve numerical accuracy in the calculations. The same loops are also used to evaluate the total sum of squares.

The sum of squares for treatments is calculated (lines 520–630) and the sum of squares for blocks is worked out (lines 640–730). The residual sum of squares is evaluated (line 750) and a table showing the sum of squares and the mean square is printed (lines 770–810).

The F value for treatments is worked and printed (lines 870–910) together with the number of degrees of freedom. The program then evaluates the probability of such an F value with this number of degrees of freedom using a subroutine (lines 1890–2520). The F value for blocks and the probability of this value are calculated and printed in an analogous manner (lines 930–970).

The user is then asked if comparison of individual treatments is required using a t-test. This should only be used if the F value for treatments is significant, and is particularly useful for comparing each of the treatments with a control. The estimate of error used in this test is based on all of the treatments rather than just the two treatments being compared. Because of this, the result is more reliable than would be obtained by comparing the two treatments using the *t*-test program. The user specifies which treatment was used as a control, and the program compares all of the other treatments with this one (lines 1000–1360). The probability of obtaining each of these *t*-values is calculated in a subroutine (lines 1600–1880).

Finally the user is asked if another run is required (line 1380), and the answer is checked in a subroutine (lines 1530–1590).

Exercises

7.1 A field test was carried out to assess the amount of growth of six different varieties of blackcurrant. Cuttings from six varieties were planted in square plots in a nursery. Four plots of each variety were planted, and all the plots contained the same number of cuttings. The length of shoots made during the first growing season were measured, and the plot totals were recorded (see Table 7.8). Determine

Table 7.8

Variety	Plot 1	Plot 2	Plot 3	Plot 4
1	46	31	32	41
2	44	43	48	56
3	23	48	38	30
4	30	26	24	37
5	36	42	38	50
6	37	28	48	31

whether there is any significant difference between varieties.

7.2 A test was carried out using 18 plots of onion plants in which three different fertilizer treatments were applied to each of six plots. The total crop yield from each of the plots were recorded (see Table 7.9).

Table 7.9

Fertilizer	Weights of crop (kg/plot)					
1	40,	37,	33,	51,	45,	34
2	56,	48,	46,	55,	56,	38
3	30,	49,	33,	56,	48,	41

Find if there is any significant difference in crop yields between the fertilizer treatments.

7.3 Four different varieties of wheat were planted in equal-sized plots of land, and the yield of corn obtained recorded for each. Test whether there is a significant difference in yield between varieties (see Table 7.10).

Table 7.10

Variety	Plot yields (kg)				
1	45.2,	39.2,	41.1,	42.1,	41.8
2	42.6,	40.4,	36.7,	39.6	
3	36.8,	41.0,	42.0,	39.6,	38.1
4	35.5,	37.1,	33.9,	40.2	

7.4 24 aliquots of a plant extract were analysed for potassium using three different analytical methods. Test whether the analytical methods give significantly different results (see Table 7.11).

Table 7.11

Method	Potassium content (mg/l)
1	1.80, 1.82, 1.79, 1.81, 1.80, 1.84
2	1.84, 1.83, 1.79, 1.81, 1.83, 1.81
3	1.83, 1.84, 1.81, 1.85, 1.81, 1.83

7.5 The yield from four varieties of barley were measured in a block experiment. Five blocks were used, each divided into four plots, and one variety sown in each plot. Each plot measured 0.1 hectare. The yields obtained from each plot (in kg) are shown in Table 7.12. Use a two-way analysis of variance to test if there is a significant difference between the varieties.

7.6 A two-way block design was used to study the weights of lettuce plants when grown at different plant densities. Three different planting densities were tested, and five blocks were used to take out systematic variations in the environment. The same number of plants were harvested from each treatment, and their combined weight in kg is shown in Table 7.13. Perform a two-way analysis of variance to test if the planting density significantly affects the yield.

7.7 Three different formulations of copper-containing sprays were tested as protection for potato plants against potato blight, *Phytophthera infestans*. A total of 48 plants were used, and these were split into 4 blocks of 12 plants each. Four treatments were carried out within each block: three plants were treated with each of the three sprays, and three plants were left untreated as a control. The plants were grown in a closed environment which was infected with *Phytophthera* spores, and after a period of growth the number of lesions on the first leaf on each plant was counted. The total number of lesions for the three plants in each treatment are recorded in Table 7.14. Perform a two-way analysis of variance to find if the four treatments differ significantly. Then carry out a least significant difference or a *t*-test to see if the treatments with each of the three sprays differ significantly from the control. Repeat the analysis of variance omitting the control data, to determine whether the three different sprays differ significantly from one another.

7.8 An experiment was carried out to evaluate three different seed treatments against a control which was untreated. Five blocks were taken to cancel systematic variations, and each block was divided

Table 7.12

Treatment	Block				
	1	2	3	4	5
Variety 1	151.9	159.1	146.2	155.5	171.6
Variety 2	152.2	149.4	158.0	163.8	168.3
Variety 3	173.6	159.3	148.2	157.1	165.6
Variety 4	160.0	169.9	159.6	171.2	185.0

Table 7.13

Treatment	Blocks				
	1	2	3	4	5
Low density	3.95	3.70	4.01	4.32	4.07
Medium density	3.70	3.47	3.83	3.95	3.46
High density	3.70	3.33	3.51	3.70	3.50

Table 7.14

Treatment	Block			
	1	2	3	4
Spray 1	19	10	15	13
Spray 2	24	21	12	18
Spray 3	19	12	10	15
Control	34	26	23	27

Table 7.15

Treatment	Block				
	1	2	3	4	5
1	7	4	2	8	5
2	8	3	6	7	8
3	7	4	6	4	3
Control	8	6	9	9	10

into four plots—one for each of the three treatments and one for the control. 100 soya bean seeds were planted in each plot, and the number which failed to germinate are recorded in Table 7.15. Test if the treatments differ significantly by using two-way analysis of variance, and perform a least significant difference or a t-test to see if the individual treatments differ significantly from the control.

8

Correlation Coefficients

Pearson's Correlation Coefficient r

Data from many surveys or experiments are collected in the form of pairs of readings (called x and y). The purpose of this correlation coefficient is to establish if the x and y values are related in a linear manner. A simple way of doing this is to plot a scatter diagram. Fig. 8.1 shows that there is an approximately linear relationship between x and y (i.e. the

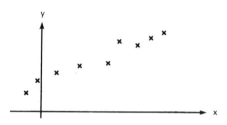

Fig. 8.1   Scatter diagram

points lie close to a straight line). In many cases the fit is less good than in the above example, and a non-subjective value is required to indicate how closely related or correlated are the x and y values. Pearson's correlation coefficient r provides such a non-subjective indicator, and is defined:

Pearson's correlation coefficient

$$r = \frac{s_{xy}}{s_x \cdot s_y} \tag{1}$$

where s_{xy} is the covariance of x and y

$$s_{xy} = \frac{\Sigma(x_i - \bar{x})(y_i - \bar{y})}{n} \tag{2}$$

(In the above expression $\bar{x}$ and $\bar{y}$ are the mean

(average) values of x and y, and n is the number of (x, y) points. The terms x_i and y_i are the values of point i and Σ indicates the summation of $(x_i - \bar{x})(y_i - \bar{y})$ over all of the (x, y) points —that is from $i = 1, 2, 3, \ldots, n$.)

s_x is the standard deviation of the x_i values (see Chapter 2 on standard deviations).

$$s_x = \sqrt{\left[\frac{\Sigma(x_i - \bar{x})^2}{n}\right]} \tag{3}$$

and s_y is the standard deviation of the y_i values

$$s_y = \sqrt{\left[\frac{\Sigma(y_i - \bar{y})^2}{n}\right]} \tag{4}$$

Substituting Equations 2, 3 and 4 into 1 gives:

$$r = \frac{\Sigma(x_i - \bar{x})(y_i - \bar{y})/n}{\sqrt{\left[\frac{\Sigma(x_i - \bar{x})^2}{n}\right]} \cdot \sqrt{\left[\frac{\Sigma(y_i - \bar{y})^2}{n}\right]}}$$

hence

$$r = \frac{\Sigma(x_i - \bar{x})(y_i - \bar{y})}{\sqrt{[\Sigma(x_i - \bar{x})^2 \Sigma(y_i - \bar{y})^2]}} \tag{5}$$

Equation 5 is used in the computer program.

It should be noted that if the n pairs of (x, y) values constitute a sample from a larger population, then to estimate the standard deviations of the x and y values of the population the divisor in Equations 3 and 4 should be $(n-1)$ rather than n. In the same way Equation 2 would require the divisor $(n-1)$ to give the estimated covariance from a sample of readings. It can be seen that the denominator, whether it is $(n-1)$ or n, cancels out in Equation 5.

An alternative to Equation 5 for evaluating Pearson's r is:

$$r = \frac{n\,\Sigma x_i y_i - \Sigma x_i \Sigma y_i}{\sqrt{[(n\Sigma x_i^2 - (\Sigma x_i)^2)\,(n\Sigma y_i^2 - (\Sigma y_i)^2)]}}$$

This equation is mathematically equivalent to Equation 5, and is commonly used on electronic calculators because it is not necessary to store the (x_i, y_i) values. However, Equation 5 gives a more accurate result if only a limited number of significant figures are carried.

The numerical value of r is zero if there is absolutely no linear correlation between the x and y values. This does not preclude the existence of a non-linear relationship. If there is some linear correlation, then a positive value for r indicates direct correlation (y increases as x increases) while a negative value for r indicates inverse correlation (y decreases as x increases). Should all of the points lie exactly on a straight line then this will result in a value for r of either $+1$ or -1. It can be seen that the numerical value of Pearson's correlation coefficient must lie in the range $+1$ to -1. Generally the calculated value of r is not exactly ± 1 or 0.

It cannot be emphasised too strongly that correlation does NOT imply causality. For example the number of colour television sets has increased over the last 10 years, and the number of whales has decreased. Though there is a negative correlation, no reasonable person would suggest that one of these factors had caused the other!

It is worth noting that Pearson's r does not depend on the scale of the x or y values. Because of this, calculations may sometimes be simplified by adding a suitable constant to the x or y values (or both), or by multiplying the x and y values by *positive* constants.

Significance of r

Table 8.1 gives the 5% significance values for varying numbers of points. Its use is best illustrated by considering an example using 10 (x, y) points. If 10 random points are taken then there is a 95% chance that the calculated

Table 8.1

Number of (x, y) points from which r is calculated n	5% significance value for Pearson's r (two-tailed)
3	0.997
4	0.950
5	0.878
6	0.811
7	0.754
8	0.707
9	0.666
→10	0.632←
11	0.602
12	0.576

value of r lies in the range -0.632 to $+0.632$. If the value of r calculated from 10 experimental points lies outside this range (that is larger than 0.632 or smaller than -0.632) then there is less than a 5% chance that this degree of correlation could have occurred from random points. Hence one concludes that it is likely that x and y are linearly related (correlated). A fuller table of 5% significance values is given in Appendix 1. Similar significance tables exist for the 1% value, and if the calculated value for r lies outside the 1% limits then there is less than a 1% chance that such a degree of correlation could have occurred from random points, hence it is highly likely that x and y are linearly related.

Finally, it should be noted that Pearson's r only indicates a linear (straight line) relationship.

Care should be exercised when using significance tables, since some tables are 'one-tailed' and others are 'two-tailed'. In the example above there is a 95% chance that the value of r lies between -0.632 and $+0.632$ (Fig. 8.2). There is therefore a 5% chance that the value of r lies outside this range, that is a $2\frac{1}{2}\%$ chance that the value of r lies between -1 and -0.632, and a $2\frac{1}{2}\%$ chance that r lies between 0.632 and 1. The value of 0.632 is thus the 5%

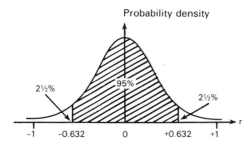

Probability density

2½%   95%   2½%

-1   -0.632   0   +0.632   +1

Fig. 8.2

two-tailed significance value *and* the $2\frac{1}{2}\%$ one-tailed significance value.

In some cases it is appropriate to use a two-tailed test, while in other cases a one-tailed test is appropriate. A one-tailed test should be used when only a direct relationship *or* only an inverse relationship between the x and y values is of importance, while a two-tailed test should be used whenever *both* direct *and* inverse relationships between the x and y values are equally important.

Example 1

An experiment was performed to investigate whether the rate of growth of a mould *(mucor)* is linearly related to the concentration of glucose present in the nutrient medium. Five cultures were grown for three days on nutrient solutions containing 10, 20, 30, 40 and 50 mg per litre of glucose, and the diameter of the colonies was measured.

x: Concentration of glucose mg l^{-1}

   10   20   30   40   50

y: Diameter of colony mm

   12   23   22   25   30

Mean value of x

$$= (10+20+30+40+50)/5 = 30$$

Mean value of y

$$= (12+23+22+25+30)/5 = 22.4$$

$$\Sigma(x_i-\bar{x})^2 = (-20)^2+(-10)^2+(0)^2+$$
$$+(10)^2+(20)^2 = 1000$$

128

$$\Sigma(y_i-\bar{y})^2 = (-10.4)^2+(0.6)^2+$$
$$+(-0.4)^2+(2.6)^2+(7.6)^2$$
$$= 173.2$$

$$\Sigma(x_i-\bar{x})(y_i-\bar{y}) = (-20\times-10.4)+$$
$$+(-10\times0.6)+$$
$$+(0\times-0.4)+$$
$$+(10\times2.6)+$$
$$+(20\times7.6) = 380$$

using Equation 5

$$r = \frac{380}{\sqrt{(1000\times173.2)}} = 0.913$$

If this r value is looked up in a significance table (Appendix 1) for three degrees of freedom (the number of degrees of freedom is the number of x, y pairs minus two), it is found to lie between the two-tailed 5% value of 0.8783 and the 2% value of 0.93433. This means that there is between a 2% and 5% probability of obtaining a value of r greater than 0.913 or less than -0.913 from random values. It is thus somewhat unlikely that these results have occurred by chance, and consequently some support is given to a linear relationship between rate of growth and glucose concentration. In statistical terms the hypothesis that there is no linear relationship is rejected at the 5% level, but cannot be rejected at the 2% level.

Description of Program to Calculate Pearson's Correlation Coefficient, r (see Program 8.1)

After printing a heading the program asks if full instructions are required. The answer which must be YES or NO is checked in a subroutine (lines 1830–1920). Any other reply is rejected by the program which prints a message requesting an answer of YES or NO. Depending on the reply, either long or short instructions are printed out during the first run, but on a second or subsequent run only short instructions are provided.

Next (lines 100–200) the user is asked to type in the number of data values, that is the number of (x, y) pairs. The value typed is checked first to ensure that it is an integer, and secondly to ensure that it is in the range 2–100 inclusive. The upper limit of 100 is imposed by the program for two reasons:

(i) The X and Y arrays are dimensioned at 100 in line 10.
(ii) The accuracy of the method used to calculate the significance is only guaranteed up to this limit.

If a number outside the range 2–100 is typed, the number is rejected, and a message invites the user to re-type an acceptable value.

The user is then invited to type in the data values, one (x, y) pair at a time, with a comma between the terms (lines 210–320). Next a subroutine is entered (lines 1070–1820) to permit the verification, correction or alteration of the data which have just been input.

Subroutine to Check Data

This subroutine is common to many of the programs in this book. A full explanation of its operation is given here, but an abbreviated description is given in several other sections.

A message asks if the data values entered are correct. The reply which must be YES or NO is checked by a subroutine (lines 1830–1920). If the answer is YES then the rest of the subroutine is skipped. Otherwise a list of the current data is printed in the form:

line number x y

The line numbers run sequentially 1, 2, 3 . . . and are used by the program to identify a particular line (X and Y in this case). Listing is formatted to fit on a 20 line VDU screen. When 19 lines of data have been displayed, the last line of the screen is used to ask whether to continue listing. A reply of YES or NO is checked in a subroutine. Stopping the listing is essential to allow time for the data to be read and checked before the next 19 lines are displayed. Some VDUs display a different number of lines. To make the program run satisfactorily on different sizes (number of lines) of VDU the variable A4 (line 1100) should be set to the number of lines which can be displayed, for example with a 16 line screen

LET A4 = 16

If a printing terminal is used, then it is not necessary to stop the listing of data, and A4 should be set to 999. Either when listing is complete, or when the user replies that he does not wish to continue listing then a list of options for altering the data is displayed.

The list of options is as follows:

Type R to replace an existing line of data
    A to add an extra line
    D to delete an existing line
    L to list the current data
or   C to continue the calculation.

Any reply other than R, A, D, L or C is rejected and the user is instructed to re-type a valid command letter.

Replace:
    The user is asked to type the line number of the line to be replaced. This is checked (lines 1490–1560) to ensure that it is an integer in the range 1 to the number of points. Provided the line number is valid, the user is prompted to type in two numbers (lines 1570–1600) separated by a comma, to replace the old (incorrect) X, Y values.

Add:
    This option is only available if the arrays hold less than 100 values. Provided that there is space in the arrays for an additional value, the user is invited to type in a pair of values for X and Y separated by a comma (lines 1610–1660). This line of data is added at the end of the existing data.

Delete:
    This option is only available if more than one line of data exists, that is one is not allowed to delete all the data. Deletion of a line is accomplished by specifying the line number of the line of data to be removed. The line number is checked (lines 1670–1730) to ensure that it is an integer, and that it actually exists. If the line to be deleted is the last in the data list, it is just removed. Otherwise following the deletion of a line, all subsequent lines

of data are moved one place (lines 1740–1790) to fill the gap in the table. Since this changes the line numbers of lines which have been moved, the data are automatically re-listed showing the new line numbers.

List:

After performing any editing function, the current data may be re-listed to permit inspection for other errors.

Continue:

This is typed when the user is satisfied that the data are correct. It finishes the editing, and allows calculations in the main program to continue.

Calculation of r

Before attempting to calculate r, the program verifies that at least two data points remain (lines 350–380). If only one point remains the run is terminated, and the user is asked if another run is required. A warning message is printed (lines 390–400) if there are only two data points since the correlation coefficient must be either $+1$ or -1. Next the mean of the X and the mean of the Y values are calculated (lines 440–500). A loop (lines 550–590) calculates the three terms $\Sigma(x_i - \bar{x}) \cdot (y_i - \bar{y})$, $\Sigma(x_i - \bar{x})^2$ and $\Sigma(y_i - \bar{y})^2$ used to calculate r from Equation 5. Checks are performed (lines 600–650) to ensure that all the x values are not equal (corresponding to a vertical straight line) and also that all the y values are not equal (which corresponds to a horizontal straight line). These checks are essential to avoid division by zero when evaluating Equation 5. The correlation coefficient is then calculated and printed (lines 660–670).

Significance of the r *Value Obtained*

A subroutine (lines 1930–2260) is used to calculate the significance of the r value just evaluated. This converts the Pearson's r value to a Student's t value using:

$$t = \frac{r \cdot \sqrt{\text{Number of degrees of freedom}}}{\sqrt{(1 - r^2)}}$$

where the number of degrees of freedom is the number of x, y pairs minus two. If one was performing the calculation manually, this t value would be looked up in significance tables with the appropriate number of degrees of freedom to determine if it was significant at the 5% or 1% levels. The computer does not look up a table of values, but evaluates the probability from the t-value. The method used was derived from that used by the Numerical Algorithms Group (NAG) Library.

The one-tailed probability is printed (lines 720–770), rounded to the nearest 1% if the probability is not significant, and rounded to the nearest $\frac{1}{2}$% if the probability is significant. One of the five messages shown in Table 8.2 is printed:

Finally the user is asked (lines 900–910) if another run is required, and if so whether (lines 960–1060) to use completely new data, or to edit and re-run the data already typed in.

Table 8.2

Probability range	Message
0%–1%	There is strong evidence for a direct relation between x and y
1%–5%	There is some evidence for a direct relation between x and y
5%–95%	There is no evidence for a relation between x and y
95%–99%	There is some evidence for an inverse relation between x and y
99%–100%	There is strong evidence for an inverse relation between x and y

Correlation by Ranks

Two different methods of rank correlation are developed. Both use the relative positions (i.e. orders or ranks) of the x and y values, rather than the actual x and y values. This idea was originally introduced by C. Spearman in 1906, and later developed by M. G. Kendall in 1948. Spearman's method is simply Pearson's correlation applied to ranks.

Numeric data which could be tested with Pearson's correlation coefficient could be arranged in rank order and either Spearman's or Kendall's test applied. The last two tests have the advantage that they may be used in situations where numerical data are not available, for example

   (i) judging the contestants in a beauty competition;

  (ii) judging the whiteness of laundry using different detergents;
 (iii) judging the entries in a painting competition;
  (iv) judging the fragrance of roses;
   (v) judging the flavour of different brews of beer.

It is because the judging of these attributes is subjective rather than quantitative that numeric data are not available. Nevertheless each judge can arrange the contestants, laundry, paintings, roses or beers in order of preference, though a different judge may well choose a different order of preference. The reason for performing correlation by ranks is to establish how closely the ranked order from one judge resembles that from a second judge.

Consider an example where a brewer has

Program 8.1   Trial run.

```
PEARSON'S CORRELATION COEFFICIENT R
======= = =========== =========== =

WOULD YOU LIKE FULL INSTRUCTIONS?
  TYPE YES OR NO & PRESS RETURN.

? YES
TEST TO MEASURE HOW CLOSELY TWO SETS OF DATA ARE RELATED

TYPE IN THE NUMBER OF DATA VALUES ? 5

TYPE IN A PAIR OF X & Y VALUES SEPARATED BY A COMMA
THEN PRESS RETURN, TYPE THE NEXT PAIR OF VALUES, RETURN ETC
YOU WILL HAVE THE CHANCE TO CORRECT TYPING ERRORS LATER
? 10, 12
? 20, 23
? 30, 22
? 40, 25
? 50, 30
END OF DATA INPUT
ARE THE DATA VALUES ENTERED CORRECT?   TYPE YES OR NO & PRESS RETURN.

? YES
CORRELATION COEFFICIENT = 0.913032

THE ONE-TAILED PROBABILITY OF OBTAINING A CORRELATION AS
GOOD OR BETTER THAN THIS BY CHANCE FROM RANDOM RANKS IS 1.5 %
THERE IS SOME  EVIDENCE FOR A DIRECT  RELATION BETWEEN X & Y

WOULD YOU LIKE ANOTHER RUN?
  TYPE YES OR NO & PRESS RETURN.

? NO
JOB FINISHED
```

```
10 DIM X(100), Y(100), Q$(10), I$(3), B(4)
20 PRINT "PEARSON'S CORRELATION COEFFICIENT R"
30 PRINT "======= = =========== ===========  ="
40 PRINT
50 PRINT "WOULD YOU LIKE FULL INSTRUCTIONS?"
60 GOSUB 1850
70 LET I$ = Q$
80 IF I$ = "NO" THEN 110
90 PRINT "TEST TO MEASURE HOW CLOSELY TWO SETS OF DATA ARE RELATED"
100 PRINT
110 PRINT "TYPE IN THE NUMBER OF DATA VALUES";
120 INPUT N
130 IF N = INT(N) THEN 180
140 PRINT "VALUE MUST BE A WHOLE NUMBER"
150 PRINT "RE-";
160 GOTO 110
170 REM CHECK THAT THE NUMBER OF POINTS IS BETWEEN 2 & 100
180 IF (N - 2) * (N - 100) <= 0 THEN 220
190 PRINT "NUMBER OF DATA PAIRS MUST BE IN THE RANGE 2 - 100"
200 GOTO 100
210 REM INPUT THE DATA
220 PRINT
230 IF I$ = "YES" THEN 260
240 PRINT "TYPE IN THE DATA"
250 GOTO 290
260 PRINT "TYPE IN A PAIR OF X & Y VALUES SEPARATED BY A COMMA"
270 PRINT "THEN PRESS RETURN, TYPE THE NEXT PAIR OF VALUES, RETURN ETC"
280 PRINT "YOU WILL HAVE THE CHANCE TO CORRECT TYPING ERRORS LATER"
290 FOR I = 1 TO N
300    INPUT X(I), Y(I)
310 NEXT I
320 PRINT "END OF DATA INPUT"
330 REM CALL SUBROUTINE TO CHECK & EDIT DATA
340 GOSUB 1080
350 REM CHECK THAT NUMBER OF DATA PAIRS IS SENSIBLE
360 IF N >= 2 THEN 390
370 PRINT "NOT ENOUGH DATA PAIRS - RUN ON THIS DATA TERMINATED"
380 GOTO 910
390 IF N > 2 THEN 420
400 PRINT "BY DEFINITION A STRAIGHT LINE JOINS TWO POINTS, THUS"
410 REM SET INITIAL VALUES
420 READ X1, Y1, X2, Y2, C2, R
430 DATA 0, 0, 0, 0, 0, 1
440 REM CALCULATE THE MEAN VALUES
450 FOR I = 1 TO N
460    LET X1 = X1 + X(I)
470    LET Y1 = Y1 + Y(I)
480 NEXT I
490 LET X1 = X1 / N
500 LET Y1 = Y1 / N
510 REM CALCULATE SUMS
520 REM X2 IS N * VARIANCE OF THE X VALUES
530 REM Y2 IS N * VARIANCE OF THE Y VALUES
540 REM C2 IS N * COVARIANCE OF THE X & Y VALUES
550 FOR I = 1 TO N
560    LET X2 = X2 + (X(I) - X1) * (X(I) - X1)
570    LET Y2 = Y2 + (Y(I) - Y1) * (Y(I) - Y1)
580    LET C2 = C2 + (X(I) - X1) * (Y(I) - Y1)
590 NEXT I
600 IF X2 <> 0 THEN 630
610 PRINT "ALL THE X VALUES ARE THE SAME! ";
```

```
620 GOTO 670
630 IF Y2 <> 0 THEN 660
640 PRINT "ALL THE Y VALUES ARE THE SAME! ";
650 GOTO 670
660 LET R = C2 / SQR(X2 * Y2)
670 PRINT "CORRELATION COEFFICIENT ="; R
680 PRINT
690 IF X2 * Y2 = 0 THEN 910
700 REM CALCULATE PROBABILITY OF SUCH A CORRELATION BY CHANCE
710 GOSUB 1940
720 PRINT "THE ONE-TAILED PROBABILITY OF OBTAINING A CORRELATION AS"
730 PRINT "GOOD OR BETTER THAN THIS BY CHANCE FROM RANDOM RANKS IS";
740 LET F = 1
750 IF (P - 0.95) * (P - 0.05) <= 0 THEN 770
760 LET F = 2
770 PRINT INT(P * F * 100 + 0.5) / F; "%"
780 IF (P - 0.05) * (P - 0.95) > 0 THEN 810
790 PRINT "THERE IS NO EVIDENCE FOR A";
800 GOTO 890
810 IF (P - 0.01) * (P - 0.99) > 0 THEN 840
820 PRINT "THERE IS SOME";
830 GOTO 850
840 PRINT "THERE IS STRONG";
850 IF P > 0.5 THEN 880
860 PRINT " EVIDENCE FOR A DIRECT";
870 GOTO 890
880 PRINT " EVIDENCE FOR AN AN INVERSE";
890 PRINT " RELATION BETWEEN X & Y"
900 PRINT
910 PRINT "WOULD YOU LIKE ANOTHER RUN?"
920 GOSUB 1840
930 IF Q$ = "NO" THEN 1050
940 RESTORE
950 LET I$ = "NO"
960 PRINT "TYPE NEW FOR A RUN WITH COMPLETELY NEW DATA"
970 PRINT "  OR OLD TO EDIT AND RERUN THE EXISTING DATA"
980 INPUT Q$
990 IF Q$ = "NEW" THEN 100
1000 IF Q$ = "OLD" THEN 1030
1010 PRINT "REPLY '"; Q$; " NOT UNDERSTOOD"
1020 GOTO 960
1030 GOSUB 1130
1040 GOTO 360
1050 PRINT "JOB FINISHED"
1060 STOP
1070 REM SUBROUTINE TO CHECK THAT DATA ARE CORRECT & ALTER IF NECESSARY
1080 PRINT "ARE THE DATA VALUES ENTERED CORRECT?";
1090 REM A4 SHOULD BE SET TO THE NUMBER OF LINES ON THE VDU
1100 LET A4 = 20
1110 GOSUB 1840
1120 IF Q$ = "YES" THEN 1820
1130 PRINT "HERE IS A LIST OF THE CURRENT DATA"
1140 PRINT "LINE NUMBER","X","Y"
1150 FOR I = 1 TO N
1160    PRINT I, X(I), Y(I)
1170    IF INT(I / (A4 - 1)) * (A4 - 1) <> I THEN 1210
1180    PRINT "WOULD YOU LIKE TO CONTINUE LISTING";
1190    GOSUB 1840
1200    IF Q$ = "NO" THEN 1220
1210 NEXT I
1220 PRINT "TYPE R TO REPLACE";
```

```
1230 IF I$ = "NO" THEN 1250
1240 PRINT " AN EXISTING LINE OF DATA"
1250 IF N = 100 THEN 1300
1260 PRINT TAB(5); " A TO ADD";
1270 IF I$ = "NO" THEN 1290
1280 PRINT " AN EXTRA LINE"
1290 IF N = 1 THEN 1330
1300 PRINT TAB(5); " D TO DELETE";
1310 IF I$ = "NO" THEN 1330
1320 PRINT " AN EXISTING LINE"
1330 PRINT TAB(5); " L TO LIST";
1340 IF I$ = "NO" THEN 1360
1350 PRINT " THE DATA"
1360 PRINT "  OR C TO CONTINUE";
1370 IF I$ = "NO" THEN 1390
1380 PRINT " THE CALCULATION"
1390 INPUT Q$
1400 IF Q$ = "R" THEN 1500
1410 IF N = 100 THEN 1440
1420 IF Q$ = "A" THEN 1620
1430 IF N = 1 THEN 1450
1440 IF Q$ = "D" THEN 1670
1450 IF Q$ = "L" THEN 1130
1460 IF Q$ = "C" THEN 1820
1470 PRINT "REPLY '"; Q$; "' NOT UNDERSTOOD."
1480 GOTO 1220
1490 REM REPLACE LINE
1500 PRINT "TYPE THE LINENUMBER OF THE LINE TO BE REPLACED";
1510 INPUT I
1520 IF I <> INT(I) THEN 1540
1530 IF (I - 1) * (I - N) <= 0 THEN 1570
1540 PRINT "LINENUMBER MUST BE AN INTEGER IN THE RANGE 1 -"; N
1550 PRINT "RE-";
1560 GOTO 1500
1570 PRINT "TYPE THE CORRECT LINE TO REPLACE THE ONE WHICH IS WRONG:"
1580 PRINT "X, Y"
1590 INPUT X(I), Y(I)
1600 GOTO 1650
1610 REM ADD A NEW LINE
1620 LET N = N + 1
1630 PRINT "TYPE THE ADDITIONAL LINE OF DATA AS SHOWN:   X,Y"
1640 INPUT X(N), Y(N)
1650 PRINT "OK"
1660 GOTO 1220
1670 REM DELETE A LINE
1680 PRINT "TYPE THE LINENUMBER OF THE LINE TO BE DELETED"
1690 INPUT J
1700 IF (J - 1) * (J - N) > 0 THEN 1720
1710 IF J = INT(J) THEN 1740
1720 PRINT "LINENUMBER MUST BE AN INTEGER IN THE RANGE 1 -"; N
1730 GOTO 1680
1740 FOR I = J + 1 TO N
1750    LET X(I - 1) = X(I)
1760    LET Y(I - 1) = Y(I)
1770 NEXT I
1780 LET N = N - 1
1790 PRINT "OK"
1800 IF J > N THEN 1220
1810 GOTO 1130
1820 RETURN
1830 REM SUBROUTINE TO CHECK REPLIES
```

```
1840 IF I$ = "NO" THEN 1860
1850 PRINT " TYPE YES OR NO & PRESS RETURN."
1860 PRINT
1870 INPUT Q$
1880 IF Q$ = "YES" THEN 1920
1890 IF Q$ = "NO" THEN 1920
1900 PRINT "REPLY '"; Q$; "' NOT UNDERSTOOD.";
1910 GOTO 1850
1920 RETURN
1930 REM *** SUBROUTINE TO CALCULATE PROBABILITY
1940 LET N2 = N - 2
1950 READ B(1), B(2), B(3), B(4), P
1960 DATA 1.E+06, 10000, 1000, 100, 0
1970 REM AVOID CALCULATION IF R = 1 OR R = -1
1980 IF ABS(R) = 1 THEN 2240
1990 REM CONVERT PEARSON R TO STUDENT T VALUE
2000 LET T = ABS(R) * SQR(N2) / SQR(1 - R * R)
2010 IF N2 > 4 THEN 2040
2020 IF T > B(N2) THEN 2240
2030 GOTO 2050
2040 IF T > 50 THEN 2240
2050 LET A = T / SQR(N2)
2060 LET B = N2 / (N2 + T * T)
2070 LET J = N2 - 2
2080 LET K = N2 - INT(N2 / 2) * 2 +2
2090 LET S = 1
2100 IF J < 2 THEN 2180
2110 LET C = 1
2120 LET F2 = K
2130 FOR I = K TO J STEP 2
2140    LET C = C * B * (F2 - 1) / F2
2150    LET S = S + C
2160    LET F2 = F2 + 2
2170 NEXT I
2180 IF K > 2 THEN 2210
2190 LET P = 0.5 - 0.5 * A * SQR(B) * S
2200 GOTO 2240
2210 IF N2 > 1 THEN 2230
2220 LET S = 0
2230 LET P = 0.5 - (A * B * S + ATN(A)) / 3.14159
2240 IF R > 0 THEN 2260
2250 LET P = 1 - P
2260 RETURN
2270 END
```

used different ingredients to produce 10 different beers which are labelled A, B, C, D, E, F, G, H, I and J. Two different beer tasters tried all 10 samples, and arranged them in order of preference as shown in Table 8.3. The brewer would like to know how well the two tasters agree. If the tasters are unable to distinguish between different beers then the orders they have chosen will be random, hence no significant correlation will exist. One could make the subjective judgement that since beer D is ranked first and second, and beer E is ranked ninth and tenth, the tasters clearly showed some preference. However, this result could have arisen by chance, and the brewer needs to know probability that such a result could have arisen by chance. To achieve this the brewer must calculate a correlation coefficient and use statistical tables to find the probability of obtaining such a correlation value by chance.

Table 8.3

Rank	1st	2nd	3rd	4th	5th	6th	7th	8th	9th	10th
Beer taster 1	F	D	G	H	C	A	J	B	I	E
Beer taster 2	D	H	G	F	J	I	A	C	E	B

Spearman's Rank Correlation Coefficient ρ (see Program 8.2)

Spearman's rank correlation coefficient (denoted by the Greek letter rho, ρ) is based on comparisons of the ranks given to each item by two different evaluators.

Spearman's rank correlation coefficient

$$\rho = 1 - \frac{6\Sigma D^2}{n(n^2 - 1)} \qquad (6)$$

where D is the difference between the ranks given for one item by the two judges, Σ is the summation of D^2 for all of the items, n is the number of items and ρ ranges from $+1$ for identical ranks to -1 for opposite ranks.

The method of calculating Spearman's correlation coefficient is illustrated in Table 8.4 by using the data on beer tasting mentioned previously.

Table 8.4

Beer	Beer taster 1 rank	Beer taster 2 rank	Rank difference D	D^2
A	6	7	-1	1
B	8	10	-2	4
C	5	8	-3	9
D	2	1	1	1
E	10	9	1	1
F	1	4	-3	9
G	3	3	0	0
H	4	2	2	4
I	9	6	3	9
J	7	5	2	4

$$\Sigma D = 0 \quad \Sigma D^2 = 42$$

Spearman's rank correlation coefficient

$$\rho = 1 - \frac{6 \times 42}{10(10^2 - 1)}$$

$$\rho = 0.745$$

The brewer's question still remains unanswered. How likely is it that a value of $\rho = 0.745$ from 10 readings could have arisen by chance? The question is answered indirectly by comparing the calculated value of ρ with the values from significance tables.

Table 8.5 gives the $2\frac{1}{2}\%$ significance values for varying numbers of items (beer samples in this case).

If 10 random ranks are taken then there is a $2\frac{1}{2}\%$ chance that the calculated value of ρ is greater than 0.6364. In the case with 10 different beer samples the calculated value of $\rho = 0.745$ lies outside this range. There is therefore less than a $2\frac{1}{2}\%$ chance that a correlation as good as this could have arisen by chance from random ranks. Since it is unlikely that this correlation has arisen by

Table 8.5

Number of items from which ρ is calculated n	$2\frac{1}{2}\%$ significance value for Spearman's ρ (one-tailed)
5	0.9000
6	0.8286
7	0.7450
8	0.7143
9	0.6833
10	0.6364
20	0.4451

A fuller significance table for ρ is given in Appendix 2.

chance, the brewer must assume that the tasters can distinguish between the different beers, and also that they have similar tastes.

The brewer is using a one-tailed test since he is only interested in agreement, that is direct correlation, between the two beer tasters. The brewer should not use a two-tailed test because this includes inverse correlation. The latter implies that the tasters can distinguish

between the different beers and that they have opposite preferences, resulting in a negative value for ρ.

Equal ranks

It is possible that the beer taster may rank two or more beers equally, in which case they

Program 8.2 Trial run.

```
SPEARMAN'S RANK CORRELATION COEFFICIENT RHO
========== ==== =========== =========== ===

WOULD YOU LIKE FULL INSTRUCTIONS?
 TYPE YES OR NO & PRESS RETURN.

? YES

TEST TO COMPARE RANKS (IE. RELATIVE POSITIONS)
FOR TWO SETS OF EXPERIMENTAL DATA.  TYPE PAIRS OF RANKS
OR PAIRS OF DATA VALUES & THE PROGRAM SORTS OUT THE RANKS

TYPE THE NUMBER OF DATA PAIRS ? 10
TYPE A PAIR OF DATA VALUES SEPARATED BY A COMMA
THEN PRESS RETURN, TYPE NEXT PAIR OF DATA VALUES ETC
YOU WILL HAVE CHANCE TO CORRECT WRONGLY TYPED DATA LATER
INPUT DATA X,Y
? 6, 7
? 8, 10
? 5, 8
? 2, 1
? 10, 9
? 1, 4
? 3, 3
? 4, 2
? 9, 6
? 7, 5
END OF DATA INPUT
ARE THE DATA VALUES ENTERED CORRECT?  TYPE YES OR NO & PRESS RETURN.

? YES
WOULD YOU LIKE A LIST OF DATA VALUES & RANKS?
 TYPE YES OR NO & PRESS RETURN.

? NO

SPEARMAN'S RANK CORRELATION COEFFICIENT RHO = 0.745455
THE ONE-TAILED PROBABILITY OF OBTAINING A CORRELATION AS
GOOD OR BETTER THAN THIS BY CHANCE FROM RANDOM RANKS IS 0.7 %
THERE IS STRONG  EVIDENCE FOR A DIRECT  RELATION BETWEEN X & Y

WOULD YOU LIKE ANOTHER RUN?
 TYPE YES OR NO & PRESS RETURN.

? NO
END OF JOB
```

```
10 DIM X(100), Y(100), A(100), B(100), Q$(10), I$(3)
20 PRINT "SPEARMAN'S RANK CORRELATION COEFFICIENT RHO"
30 PRINT "========= ==== =========== =========== ==="
40 PRINT
50 PRINT "WOULD YOU LIKE FULL INSTRUCTIONS?"
60 GOSUB 1840
70 LET I$ = Q$
80 IF I$ = "NO" THEN 130
90 PRINT
100 PRINT "TEST TO COMPARE RANKS (IE. RELATIVE POSITIONS)"
110 PRINT "FOR TWO SETS OF EXPERIMENTAL DATA.  TYPE PAIRS OF RANKS"
120 PRINT "OR PAIRS OF DATA VALUES & THE PROGRAM SORTS OUT THE RANKS"
130 PRINT
140 PRINT "TYPE THE NUMBER OF DATA PAIRS";
150 INPUT N
160 IF N <> INT(N) THEN 180
170 IF (N - 2) * (N - 100) <= 0 THEN 200
180 PRINT "VALUE TYPED MUST BE A WHOLE NUMBER BETWEEN 2 AND 100"
190 GOTO 140
200 IF I$ = "NO" THEN 240
210 PRINT "TYPE A PAIR OF DATA VALUES SEPARATED BY A COMMA"
220 PRINT "THEN PRESS RETURN, TYPE NEXT PAIR OF DATA VALUES ETC"
230 PRINT "YOU WILL HAVE CHANCE TO CORRECT WRONGLY TYPED DATA LATER"
240 PRINT "INPUT DATA X,Y"
250 FOR I = 1 TO N
260    INPUT X(I), Y(I)
270 NEXT I
280 PRINT "END OF DATA INPUT"
290 REM CALL SUBROUTINE TO CHECK THAT DATA VALUES ARE CORRECT
300 GOSUB 1070
310 REM CHECK THAT THE NUMBER OF DATA PAIRS IS SENSIBLE
320 IF N >= 2 THEN 360
330 PRINT "NOT ENOUGH DATA PAIRS - RUN ABANDONED"
340 GOTO 910
350 REM CALCULATE RANKED A & B VALUES FROM RAW X & Y VALUES
360 FOR I = 1 TO N
370    LET A1 = 0.5
380    LET B1 = 0.5
390    FOR J = 1 TO N
400      IF X(J) < X(I) THEN 450
410      IF X(J) = X(I) THEN 440
420      LET A1 = A1 + 1
430      GOTO 450
440      LET A1 = A1 + 0.5
450      IF Y(J) < Y(I) THEN 500
460      IF Y(J) = Y(I) THEN 490
470      LET B1 = B1 + 1
480      GOTO 500
490      LET B1 = B1 + 0.5
500    NEXT J
510    LET A(I) = A1
520    LET B(I) = B1
530 NEXT I
540 PRINT "WOULD YOU LIKE A LIST OF DATA VALUES & RANKS?"
550 GOSUB 1830
560 IF Q$ = "NO" THEN 620
570 PRINT "X", "RANK X", "Y", "RANK Y"
580 FOR I = 1 TO N
590    PRINT X(I), A(I), Y(I), B(I)
600 NEXT I
610 REM CALCULATE SUM OF SQUARED RANK DIFFERENCES
```

```
620 LET R2 = 0
630 FOR I = 1 TO N
640    LET D = A(I) - B(I)
650    LET R2 = R2 + D * D
660 NEXT I
670 REM CALCULATE SPEARMANS RANK CORRELATION
680 LET R = 1 - R2 / (N / 6 * (N - 1) * (N + 1))
690 PRINT
700 PRINT "SPEARMAN'S RANK CORRELATION COEFFICIENT RHO ="; R
710 REM CALCULATE PROBABILITY OF SUCH A CORRELATION BY CHANCE
720 GOSUB 1930
730 PRINT "THE ONE-TAILED PROBABILITY OF OBTAINING A CORRELATION AS"
740 PRINT "GOOD OR BETTER THAN THIS BY CHANCE FROM RANDOM RANKS IS";
750 LET F = 1
760 IF (P - 0.95) * (P - 0.05) <=0 THEN 780
770 LET F = 10
780 PRINT INT(P * F * 100 + 0.5) / F; "%"
790 IF (P - 0.05) * (P - 0.95) > 0 THEN 820
800 PRINT "THERE IS NO EVIDENCE FOR A";
810 GOTO 900
820 IF (P - 0.01) * (P - 0.99) > 0 THEN 850
830 PRINT "THERE IS SOME";
840 GOTO 860
850 PRINT "THERE IS STRONG";
860 IF P > 0.5 THEN 890
870 PRINT " EVIDENCE FOR A DIRECT";
880 GOTO 900
890 PRINT " EVIDENCE FOR AN AN INVERSE";
900 PRINT " RELATION BETWEEN X & Y"
910 PRINT
920 PRINT "WOULD YOU LIKE ANOTHER RUN?"
930 GOSUB 1830
940 IF Q$ = "NO" THEN 2260
950 RESTORE
960 PRINT "TYPE NEW TO START AGAIN WITH NEW DATA"
970 PRINT "  OR OLD TO EDIT EXISTING DATA"
980 INPUT Q$
990 LET I$ = "NO"
1000 IF Q$ = "NEW" THEN 130
1010 IF Q$ = "OLD" THEN 1040
1020 PRINT "REPLY '"; Q$; "' NOT UNDERSTOOD."
1030 GOTO 960
1040 GOSUB 1120
1050 GOTO 320
1060 REM SUBROUTINE TO CHECK THAT DATA ARE CORRECT & ALTER IF NECESSARY
1070 PRINT "ARE THE DATA VALUES ENTERED CORRECT?";
1080 REM A4 SHOULD BE SET TO THE NUMBER OF LINES ON THE VDU
1090 LET A4 = 20
1100 GOSUB 1830
1110 IF Q$ = "YES" THEN 1810
1120 PRINT "HERE IS A LIST OF THE CURRENT DATA"
1130 PRINT "LINE NUMBER", "X", "Y"
1140 FOR I = 1 TO N
1150    PRINT I, X(I), Y(I)
1160    IF INT(I / (A4 - 1)) * (A4 - 1) <> I THEN 1200
1170    PRINT "WOULD YOU LIKE TO CONTINUE LISTING";
1180    GOSUB 1830
1190    IF Q$ = "NO" THEN 1210
1200 NEXT I
1210 PRINT "TYPE R TO REPLACE";
1220 IF I$ = "NO" THEN 1240
```

```
1230 PRINT " AN EXISTING LINE OF DATA"
1240 IF N = 100 THEN 1290
1250 PRINT TAB(5); " A TO ADD";
1260 IF I$ = "NO" THEN 1280
1270 PRINT " AN EXTRA LINE"
1280 IF N = 1 THEN 1320
1290 PRINT TAB(5); " D TO DELETE";
1300 IF I$ = "NO" THEN 1320
1310 PRINT " AN EXISTING LINE"
1320 PRINT TAB(5); " L TO LIST";
1330 IF I$ = "NO" THEN 1350
1340 PRINT " THE DATA"
1350 PRINT "  OR C TO CONTINUE";
1360 IF I$ = "NO" THEN 1380
1370 PRINT " THE CALCULATION"
1380 INPUT Q$
1390 IF Q$ = "R" THEN 1490
1400 IF N = 100 THEN 1430
1410 IF Q$ = "A" THEN 1610
1420 IF N = 1 THEN 1440
1430 IF Q$ = "D" THEN 1660
1440 IF Q$ = "L" THEN 1120
1450 IF Q$ = "C" THEN 1810
1460 PRINT "REPLY '"; Q$; "' NOT UNDERSTOOD."
1470 GOTO 1210
1480 REM REPLACE LINE
1490 PRINT "TYPE THE LINENUMBER OF THE LINE TO BE REPLACED";
1500 INPUT I
1510 IF I <> INT(I) THEN 1530
1520 IF (I - 1) * (I - N) <= 0 THEN 1560
1530 PRINT "LINENUMBER MUST BE AN INTEGER IN THE RANGE 1 -"; N
1540 PRINT "RE-";
1550 GOTO 1490
1560 PRINT "TYPE THE CORRECT LINE TO REPLACE THE ONE WHICH IS WRONG:"
1570 PRINT "X, Y"
1580 INPUT X(I), Y(I)
1590 GOTO 1640
1600 REM ADD A NEW LINE
1610 LET N = N + 1
1620 PRINT "TYPE THE ADDITIONAL LINE OF DATA AS SHOWN:   X,Y"
1630 INPUT X(N), Y(N)
1640 PRINT "OK"
1650 GOTO 1210
1660 REM DELETE A LINE
1670 PRINT "TYPE THE LINENUMBER OF THE LINE TO BE DELETED"
1680 INPUT J
1690 IF (J - 1) * (J - N) > 0 THEN 1710
1700 IF J = INT(J) THEN 1730
1710 PRINT "LINENUMBER MUST BE AN INTEGER IN THE RANGE 1 -"; N
1720 GOTO 1670
1730 FOR I = J + 1 TO N
1740    LET X(I - 1) = X(I)
1750    LET Y(I - 1) = Y(I)
1760 NEXT I
1770 LET N = N - 1
1780 PRINT "OK"
1790 IF J > N THEN 1210
1800 GOTO 1120
1810 RETURN
1820 REM SUBROUTINE TO CHECK REPLIES
1830 IF I$ = "NO" THEN 1850
```

```
1840 PRINT " TYPE YES OR NO & PRESS RETURN."
1850 PRINT
1860 INPUT Q$
1870 IF Q$ = "YES" THEN 1910
1880 IF Q$ = "NO" THEN 1910
1890 PRINT "REPLY '"; Q$; "' NOT UNDERSTOOD.";
1900 GOTO 1840
1910 RETURN
1920 REM *** SUBROUTINE TO CALCULATE PROBABILITY
1930 LET N2 = N - 2
1940 READ B(1), B(2), B(3), B(4), P
1950 DATA 1.E+06, 10000, 1000, 100, 0
1960 REM AVOID CALCULATION IF R = 1 OR R = -1
1970 IF ABS(R) = 1 THEN 2230
1980 REM CONVERT SPEARMAN RHO TO STUDENT T VALUE
1990 LET T = ABS(R) * SQR(N2) / SQR(1 - R * R)
2000 IF N2 > 4 THEN 2030
2010 IF T > B(N2) THEN 2230
2020 GOTO 2040
2030 IF T > 50 THEN 2230
2040 LET A = T / SQR(N2)
2050 LET B = N2 / (N2 + T * T)
2060 LET J = N2 - 2
2070 LET K = N2 - INT(N2 / 2) * 2 + 2
2080 LET S = 1
2090 IF J < 2 THEN 2170
2100 LET C = 1
2110 LET F2 = K
2120 FOR I = K TO J STEP 2
2130    LET C = C * B * (F2 - 1) / F2
2140    LET S = S + C
2150    LET F2 = F2 + 2
2160 NEXT I
2170 IF K > 2 THEN 2200
2180 LET P = 0.5 - 0.5 * A * SQR(B) * S
2190 GOTO 2230
2200 IF N2 > 1 THEN 2220
2210 LET S = 0
2220 LET P = 0.5 - (A * B * S + ATN(A)) / 3.14159
2230 IF R > 0 THEN 2250
2240 LET P = 1 - P
2250 RETURN
2260 PRINT "END OF JOB"
2270 END
```

should all be assigned their average rank. For example if beer taster 1 ranked G and H equally, they should both have the rank 3.5.

Description of Program to Calculate Spearman's Correlation Coefficient ρ

The program (Program 8.2) first prints a heading (lines 20–30) and then asks if full instructions are required (line 50). The answer, which must be either YES or NO, is checked in a subroutine (lines 1820–1910). Either full or abbreviated instructions are printed as appropriate in the first run, but abbreviated instructions are always given in subsequent runs.

The program then asks for the number of data pairs (line 140). The value typed is checked (lines 160–190) to ensure that it is an integer, and between 2 and 100 inclusive. The upper limit is imposed by the DIMension of the X, Y, A and B arrays in line 10, and the accuracy of the method used to calculate the significance of ρ is only guaranteed up to this limit.

A message (lines 210–240) then invites the user to type in pairs of data values. The data input loop extends from lines 250 to 270.

A subroutine (lines 1060–1810) is then called which asks if the data values are correct. The answer must be YES or NO, and is checked in a subroutine (lines 1820–1910). If the data are not correct, the values are listed and instructions are given to permit the addition, deletion or replacement of lines of data, re-listing the data, or continuing with the calculation. This subroutine is described in more detail in the write-up for Pearson's correlation coefficient.

Once the data are correct, a check is performed (lines 320–340) to ensure that at least two data pairs remain.

Next the data values which are stored in the X and Y arrays are converted into ranks and stored in the A and B arrays (lines 350–530). A message (line 540) asks if a list of the data values and ranks is required. If requested the data and ranks are printed (lines 570–600).

The sum of the squared rank differences is then calculated (lines 610–660). Using this, Spearman's rank correlation coefficient is evaluated (line 680) using Equation 6, and is printed out (line 700). A subroutine (lines 1920–2250) is then entered to work out the probability that a correlation as good or better than this could arise by chance. The subroutine is the same as that used in Pearson's correlation, where it is described more fully. The probability is then printed out (lines 730–780) together with one of five explanatory messages (lines 790–900).

Finally the user is offered another run (line 920). If another run is required, a choice is given between typing in a completely NEW set of data, or editing and re-running using the OLD (existing) data.

Kendall's Rank Correlation Coefficient τ

Kendall's rank correlation coefficient (denoted by the Greek letter tau τ) is based on whether two items out of the whole set, have been placed in the same relative position by the two ranking processes or judges. The coefficient ranges from $+1$ for identical ranks to -1 for opposite ranks. In the example about beer testing, this correlation coefficient is based on whether two different beers have been placed in the same relative positions by both beer tasters. If the two beers have been placed in the same relative position by both beer tasters, then this counts as 'agreement', whereas if their relative positions are reversed this counts as 'disagreement'.

This may be illustrated by first comparing beers A and B, then comparing beers A and C.

Beer taster 1 prefers beer A to beer B
Beer taster 2 prefers beer A to beer B
(this counts as an agreement)

Beer taster 1 prefers beer C to beer A
Beer taster 2 prefers beer A to beer C
(this counts as a disagreement)

If either taster prefers two beers equally then this counts as a tie. When all combinations of the beers have been compared (in this case after 45 comparisons), Kendall's rank correlation coefficient τ is calculated:

$$\tau = \frac{\substack{\text{number of agreements} \\ -\text{number of disagreements}}}{\text{number of comparisons}} \quad (7)$$

If n different samples are tested then the number of comparisons is $\frac{1}{2}n(n-1)$. Thus in the beer example where $n = 10$, the number of comparisons is 45. Equation 7 may therefore be re-written

$$\tau = \frac{\substack{\text{number of agreements} \\ -\text{number of disagreements}}}{\frac{1}{2}n(n-1)} \quad (8)$$

To calculate Kendall's rank correlation coefficient manually, it is easiest if the original data are sorted into the order chosen by one of the beer tasters as shown in Table 8.6.

Table 8.6

Beer	F	D	G	H	C	A	J	B	I	E
Beer taster 1 ranks	1	2	3	4	5	6	7	8	9	10
Beer taster 2 ranks	4	1	3	2	8	7	5	10	6	9

Comparison of F with D, G, H, C, A, J, B, I, E gives	6 agreements and	3 disagreements
Comparison of D with G, H, C, A, J, B, I, E gives	8 agreements and	0 disagreements
Comparison of G with H, C, A, J, B, I, E gives	6 agreements and	1 disagreements
Comparison of H with C, A, J, B, I, E gives	6 agreements and	0 disagreements
Comparison of C with A, J, B, I, E gives	2 agreements and	3 disagreements
Comparison of A with J, B, I, E gives	2 agreements and	2 disagreements
Comparison of J with B, I, E gives	3 agreements and	0 disagreements
Comparison of B with I, E gives	0 agreements and	2 disagreements
Comparison of I with E gives	1 agreements and	0 disagreements
TOTAL	34 agreements	11 disagreements

Since this manual comparison is laborious and may be error prone, it is worth checking that the number of agreements plus ties plus disagreements is equal to the total number of comparisons which should have been performed, i.e. $34 + 0 + 11 = 45$.

Kendall's rank correlation coefficient is evaluated using Equation 8.

$$\tau = \frac{34 - 11}{45} = 0.511$$

In a similar way to the other correlations, the value of $\tau = 0.511$ is looked up in a significance table (see Table 8.7).

The calculated value for τ is just on the $2\frac{1}{2}\%$ significance value, hence there is a $2\frac{1}{2}\%$ chance that a correlation as good as this could have arisen from random ranks. This provides some evidence of correlation between the ranks assigned by the two beer tasters. Clearly a larger calculated value of τ might lie within the 1% significance level, and there would then be only a 1 in 100 chance of such good correlation occurring from random ranks. This would strengthen the evidence for agreement between the two beer tasters.

A relationship between Spearman's ρ and Kendall's τ was shown by H.E. Daniels in 1950. This is

$$-1 \leqslant 3\tau - 2\rho \leqslant +1$$

In the beer tasting example
$$(3 \times 0.511) - (2 \times 0.745) = 0.043$$

Description of Program to Calculate Kendall's Correlation Coefficient τ

The program (Program 8.3) first prints a heading, and then asks if full instructions are required. The answer which must be YES or NO is checked in a subroutine (lines

Table 8.7

Number of items from which τ is calculated n	$2\frac{1}{2}\%$ significance value for Kendall's τ (one-tailed)
5	1
6	0.87
7	0.71
8	0.64
9	0.56
10	0.51
20	0.33

A fuller significance table for τ is given in Appendix 3.

1380–1470). If the answer is YES then full instructions are given throughout the first run.

Next the program asks for the number of data pairs and performs a number of checks on the value typed in (lines 80–140). The value must be an integer and lie in the range 2–100 inclusive. The test is meaningless on less than two values, and array sizes $X(100)$ and $Y(100)$ impose the upper limit. Any unacceptable value is rejected, and the user is prompted with a message to re-type the value correctly.

Following this the user is invited to type in the pairs of data values (lines 150–210). These data values will generally be the ranks from the two judges, but the program will also work if numerical scores awarded by the judges are entered instead of ranks. Examples of numeri-

cal scores include the judging of gymnastics, ice skating and diving, and typing in the numeric data saves the user the effort of ranking the scores. A consequence of this feature is that the program cannot check that the ranks are sensible values. (Ranks must be either whole numbers, e.g. 1, 2, 3 or halves, e.g. 4.5 as a result of a tie for fourth and fifth places. Furthermore ranks must be in the range one to the number of points.)

A subroutine (lines 620–1370) is then called to permit the verification and alteration of the data pairs. This allows listing, deletion, addition or replacement of data, and is fully described in the description of the program for Pearson's correlation coefficient.

The number of 'agreements' and 'disagree-

Program 8.3   Trial run.

```
KENDALL'S RANK CORRELATION COEFFICIENT
======= = ==== =========== ===========

WOULD YOU LIKE FULL INSTRUCTIONS?
 TYPE YES OR NO & PRESS RETURN.

? YES
TYPE THE NUMBER OF DATA PAIRS
? 10
TYPE DATA PAIRS ONE AT A TIME SEPARATED BY A COMMA
THEN PRESS RETURN, TYPE THE NEXT PAIR OF VALUES, RETURN ETC
YOU WILL HAVE THE CHANCE TO CORRECT TYPING ERRORS LATER
? 1, 4
? 2, 1
? 3, 3
? 4, 2
? 5, 8
? 6, 7
? 7, 5
? 8, 10
? 9, 6
? 10, 9
ARE THE DATA VALUES ENTERED CORRECT?   TYPE YES OR NO & PRESS RETURN.

? YES

KENDALL'S RANK CORRELATION COEFFICIENT = 0.511111

THE ONE TAILED PROBABILITY OF A BETTER CORRELATION ARISING
BY CHANCE FROM RANDOM RANKS = 2 %

WOULD YOU LIKE ANOTHER RUN?
 TYPE YES OR NO & PRESS RETURN.

? NO
END OF JOB
```

```
10 DIM X(100), Y(100), Q$(10), I$(3)
20 PRINT "KENDALL'S RANK CORRELATION COEFFICIENT"
30 PRINT "======= = ==== =========== ==========="
40 PRINT
50 PRINT "WOULD YOU LIKE FULL INSTRUCTIONS?"
60 GOSUB 1400
70 LET I$ = Q$
80 PRINT "TYPE THE NUMBER OF DATA PAIRS"
90 INPUT N
100 IF N <> INT(N) THEN 120
110 IF (N - 2) * (N - 100) <= 0 THEN 150
120 PRINT "THERE MUST BE A WHOLE NUMBER OF PAIRS BETWEEN 2 & 100"
130 PRINT "RE-";
140 GOTO 80
150 PRINT "TYPE DATA PAIRS ONE AT A TIME SEPARATED BY A COMMA"
160 IF I$ = "NO" THEN 190
170 PRINT "THEN PRESS RETURN, TYPE THE NEXT PAIR OF VALUES, RETURN ETC"
180 PRINT "YOU WILL HAVE THE CHANCE TO CORRECT TYPING ERRORS LATER"
190 FOR I = 1 TO N
200    INPUT X(I), Y(I)
210 NEXT I
220 REM *** CHECK THAT DATA ARE CORRECT
230 GOSUB 630
240 REM COUNT NUMBER OF AGREEMENTS & NUMBER OF DISAGREEMENTS
250 LET A = 0
260 FOR I = 1 TO N - 1
270    FOR J = I + 1 TO N
280      LET A = A + SGN(X(I) - X(J)) * SGN(Y(I) - Y(J))
290    NEXT J
300 NEXT I
310 REM *** CALCULATE NUMBER OF COMPARISONS
320 LET C = N * (N - 1) / 2
330 PRINT
340 LET K = A / C
350 PRINT "KENDALL'S RANK CORRELATION COEFFICIENT ="; K
360 PRINT
370 REM CONVERT K INTO  NO. OF STANDARD DEVIATIONS ON NORMAL CURVE
380 LET S = ABS(K) * SQR(4.5 * N * (N - 1) / (2 * N + 5))
390 REM CALC AREA UNDER NORMAL CURVE
400 GOSUB 1500
410 PRINT "THE ONE TAILED PROBABILITY OF A BETTER CORRELATION ARISING"
420 LET P = 1
430 IF (F - 0.05) * (F - 0.95) <= 0 THEN 450
440 LET P = 10
450 PRINT "BY CHANCE FROM RANDOM RANKS ="; INT((1-F)*P*100+0.5)/P; "%"
460 PRINT
470 PRINT "WOULD YOU LIKE ANOTHER RUN?"
480 GOSUB 1390
490 IF Q$ = "NO" THEN 600
500 LET I$ = "NO"
510 PRINT "TYPE NEW FOR A RUN WITH COMPLETELY NEW DATA"
520 PRINT "  OR OLD TO EDIT AND RERUN THE EXISTING DATA"
530 INPUT Q$
540 IF Q$ = "NEW" THEN 80
550 IF Q$ = "OLD" THEN 580
560 PRINT "REPLY '"; Q$; "' NOT UNDERSTOOD"
570 GOTO 510
580 GOSUB 680
590 GOTO 250
600 PRINT "END OF JOB"
610 STOP
```

```
620 REM SUBROUTINE TO CHECK THAT DATA ARE CORRECT & ALTER IF NECESSARY
630 PRINT "ARE THE DATA VALUES ENTERED CORRECT?";
640 REM A4 SHOULD BE SET TO THE NUMBER OF LINES ON THE VDU
650 LET A4 = 20
660 GOSUB 1390
670 IF Q$ = "YES" THEN 1370
680 PRINT "HERE IS A LIST OF THE CURRENT DATA"
690 PRINT "LINE NUMBER", "X", "Y"
700 FOR I = 1 TO N
710    PRINT I, X(I), Y(I)
720    IF INT(I / (A4 - 1)) * (A4 - 1) <> I THEN 760
730    PRINT "WOULD YOU LIKE TO CONTINUE LISTING";
740    GOSUB 1390
750    IF Q$ = "NO" THEN 770
760 NEXT I
770 PRINT "TYPE R TO REPLACE";
780 IF I$ = "NO" THEN 800
790 PRINT " AN EXISTING LINE OF DATA"
800 IF N = 100 THEN 850
810 PRINT TAB(5);" A TO ADD";
820 IF I$ = "NO" THEN 840
830 PRINT " AN EXTRA LINE"
840 IF N = 1 THEN 880
850 PRINT TAB(5); " D TO DELETE";
860 IF I$ = "NO" THEN 880
870 PRINT " AN EXISTING LINE"
880 PRINT TAB(5); " L TO LIST";
890 IF I$ = "NO" THEN 910
900 PRINT " THE DATA"
910 PRINT "  OR C TO CONTINUE";
920 IF I$ = "NO" THEN 940
930 PRINT " THE CALCULATION"
940 INPUT Q$
950 IF Q$ = "R" THEN 1050
960 IF N = 100 THEN 990
970 IF Q$ = "A" THEN 1170
980 IF N = 1 THEN 1000
990 IF Q$ = "D" THEN 1220
1000 IF Q$ = "L" THEN 680
1010 IF Q$ = "C" THEN 1370
1020 PRINT "REPLY '"; Q$; "' NOT UNDERSTOOD."
1030 GOTO 770
1040 REM REPLACE LINE
1050 PRINT "TYPE THE LINENUMBER OF THE LINE TO BE REPLACED";
1060 INPUT I
1070 IF I <> INT(I) THEN 1090
1080 IF (I - 1) * (I - N) <= 0 THEN 1120
1090 PRINT "LINENUMBER MUST BE AN INTEGER IN THE RANGE 1 -"; N
1100 PRINT "RE-";
1110 GOTO 1050
1120 PRINT "TYPE THE CORRECT LINE TO REPLACE THE ONE WHICH IS WRONG:"
1130 PRINT "X, Y"
1140 INPUT X(I), Y(I)
1150 GOTO 1200
1160 REM ADD A NEW LINE
1170 LET N = N + 1
1180 PRINT "TYPE THE ADDITIONAL LINE OF DATA AS SHOWN:   X,Y"
1190 INPUT X(N), Y(N)
1200 PRINT "OK"
1210 GOTO 770
1220 REM DELETE A LINE
```

146

```
1230 PRINT "TYPE THE LINENUMBER OF THE LINE TO BE DELETED"
1240 INPUT J
1250 IF (J - 1) * (J - N) > 0 THEN 1270
1260 IF J = INT(J) THEN 1290
1270 PRINT "LINENUMBER MUST BE AN INTEGER IN THE RANGE 1 -"; N
1280 GOTO 1230
1290 FOR I = J + 1 TO N
1300    LET X(I - 1) = X(I)
1310    LET Y(I - 1) = Y(I)
1320 NEXT I
1330 LET N = N - 1
1340 PRINT "OK"
1350 IF J > N THEN 770
1360 GOTO 680
1370 RETURN
1380 REM SUBROUTINE TO CHECK REPLIES
1390 IF I$ = "NO" THEN 1410
1400 PRINT " TYPE YES OR NO & PRESS RETURN."
1410 PRINT
1420 INPUT Q$
1430 IF Q$ = "YES" THEN 1470
1440 IF Q$ = "NO" THEN 1470
1450 PRINT "REPLY '"; Q$; "' NOT UNDERSTOOD.";
1460 GOTO 1400
1470 RETURN
1480 REM CALC CUMULATIVE AREA UNDER NORMAL CURVE
1490 REM CONSTANTS SET FOR 8 FIGURE ACCURACY
1500 LET X9 = -S * 7.0710678E-1
1510 LET F = 0
1520 IF X9 >= 9.5 THEN 1680
1530 LET F = 1
1540 IF X9 <= -4.5 THEN 1680
1550 LET T = 1 - 7.5 / (ABS(X9) + 3.75)
1560 LET Y = 0
1570 FOR I = 1 TO 12
1580    READ C
1590    LET Y = Y * T + C
1600 NEXT I
1610 RESTORE
1620 DATA 3.14753E-05, -0.000138746, -6.41279E-06, 0.00178663
1630 DATA -0.00823169, 0.0241519, -0.0547992, 0.102602
1640 DATA -0.163572, 0.226008, -0.273422, 0.14559
1650 LET F = 0.5 * EXP(-X9 * X9) * Y
1660 IF X9 >= 0 THEN 1680
1670 LET F = 1 - F
1680 RETURN
1690 END
```

ments' in the data are then calculated (lines 240–300) as the first step in calculating τ. The method of evaluation is different from that used in the manual calculation. Each combination of two data pairs is compared in turn. Consider the comparison of the Ith and Jth pairs. If the difference in the Ith and Jth ranks or score from judge X has the same sign as the equivalent difference for judge Y then the two judges agree about the relative positions of I

and J, and $+1$ is added to the agreement total A. If the differences are of opposite sign then the two judges disagree and -1 is added to A. If there is a tie then zero is added to A.

The number of comparisons required is calculated (lines 310–320) using $\frac{1}{2}n(n-1)$, and τ is calculated (line 340) using Equation 8, and printed (line 350).

The significance of τ is calculated using the slight approximation that τ is normally distributed with a mean of zero and a standard deviation of $\sqrt{[(2n+5)/(4.5n(n-1))]}$. The τ value is thus converted (line 380) into a number of standard deviations from the mean on a standardised normal curve. The area under the normal curve up to this number of standard deviations is then calculated using an order 11 empirical polynomial in a subroutine (lines 1480–1680). The area thus obtained is the probability of obtaining a worse correlation by chance. By subtraction from one the probability of obtaining a better correlation by chance is obtained. The probability is printed (lines 410–450) rounded to give a whole number if it is in the range 5%–95%, or with one decimal figure if it is outside this range. This method of calculating the area is based on that used by the Numerical Algorithms Group (NAG). (This subroutine is also used by the program which generates a table of areas under a normal curve given in Appendix 4.)

The program then asks if another run is required (line 470), and if so the option is offered of editing the existing data and re-running, or alternatively typing in a completely new set of data.

Exercises

8.1 Plot scatter diagrams for the following sets of data, and state whether the data appear to be approximately linearly related. Calculate Pearson's r to check this.
(a) x 0   0   1   1
    y 0   1   0   1
(b) x 0   1   2   3   4   5
    y 5.4 6.6 9.1 10.4 12.6 13.8

8.2 The examination marks of 10 students in mathematics and biology are:

Mathematics   69 74 88 47 66
              41 95 55 50 47
Biology   66 69 71 48 63
              60 85 45 51 32

Calculate Pearson's r and check whether the level correlation is significant at the 1% level.

8.3 The times in minutes taken by each of eight people to perform the same routine job on an assembly line were recorded early in the morning and again later in the afternoon.

Early morning 10.0 8.0 9.1   7.5
               8.6 9.2 11.3 15.1
Late afternoon 11.5 8.7 9.9 11.1
               9.9 9.1 13.6 13.2

Calculate Pearson's r, and state whether it is significant at the 5% level.

8.4 Two observers ranked the same five samples of eau-de-Cologne into order based on the strength of the perfume, on two different occasions. Their results are shown below:

	Observer 1				
First testing	1	5	4	3	2
Second testing	2	3	5	4	1
	Observer 2				
First testing	2	1	5	4	3
Second testing	3	2	5	4	1

Use Spearman's correlation coefficient on the replicate results from each observer to determine which of the two observers was the most self-consistent.

8.5 Use the data from Question 8.4 to see how well the two observers agreed on their second testing. Is the result significant at the 5% level?

8.6 Two TV critics were each asked to rank eight television programs. Use Kendall's rank correlation to establish whether their assessments (Table 8.8) agree significantly.

8.7 Using the data in Table 8.9, calculate Pearson's correlation coefficient to see if a relationship exists between the amount of summer rainfall in the months June–August and the annual crop of sugar beet.

Table 8.8

Program	Ranks from critic 1	Ranks from critic 2
A	1	2
B	2	3
C	3	1
D	4	5
E	5	4
F	6	6
G	7	8
H	8	7

Table 8.9

Year	Summer rainfall (cm)	Crop (in 100 000 tonne)
1970	19.8	62
1971	26.2	76
1972	17.4	60
1973	22.1	74
1974	23.9	44
1975	13.8	62
1976	13.5	68
1977	20.3	75

Table 8.10

Patient	Amount of drug absorbed	
	Day 1	Day 2
1	24.4	23.4
2	42.0	18.9
3	36.1	27.5
4	29.6	14.5
5	42.5	30.6
6	16.6	15.1
7	35.4	27.6
8	13.6	13.0
9	38.9	23.2
10	19.6	16.1

Table 8.11

Weight (kg) x	Food consumption (kcal) y
62	2850
65	2350
71	2700
75	2850
81	3300
85	3150
86	3200
90	2850
93	3300
95	3000

Table 8.12

	Kidney weight (g)	Heart weight (g)
1	333	271
2	357	439
3	361	328
4	305	326
5	269	276
6	340	305
7	369	404
8	312	262
9	268	255
10	354	350

8.8 Test the hypothesis that a patient who absorbs a drug well on one occasion will do so on another occasion, using the data given in Table 8.10 to calculate Pearson's r.

8.9 The weights and average food consumption for 10 obese adolescent females are given in Table 8.11. Plot the points on a scatter diagram and assess whether there appears to be correlation between these factors. Then calculate the value of Pearson's r.

8.10 Use Pearson's r to determine whether there is any correlation between the kidney weights and heart weights of adult men. The data given in Table 8.12 were obtained from fatal road accident victims.

9

Straight Line Fitting (Least Squares)

Many practical problems yield a series of pairs of data values. It is usual for each data pair that one value x is chosen by the experimenter while the other value y is obtained experimentally. A new value of x is then chosen and the appropriate value of y measured. Some examples include:

(i) Selecting the temperature x, and measuring the length y of a bar of metal.

(ii) Fixing the length x of a simple pendulum and measuring its periodic time y (time for one complete oscillation).

(iii) Driving a car at a constant speed x and measuring the fuel consumption as the number of miles per gallon y.

(iv) Treating equivalent plots of land with a given weight of fertilizer x and measuring the crop yield y.

(v) Measuring the results from a new analytical method x against results from a standard method y.

In such cases, the value which is chosen is always given the letter x, and is often referred to as the independent variable. The value which is measured is given the letter y, and is called the dependent variable since its value depends on x. The objective is to establish what, if any, relationship exists between the x and y values so obtained. Two methods of establishing if a relationship exists are plotting a scatter diagram, or calculating Pearson's correlation coefficient.

Scatter Diagram

A graph is plotted with each (x, y) pair represented by a point. This is called a scatter diagram. If the graph points lie within a narrow band, the variables x and y are said to be correlated, and a relationship exists between them. If no band exists there is an absence of correlation between x and y, and no relationship exists between them. The scatter diagrams (Fig. 9.1) are typical of results from (i), (ii), (iii) and (iv) above. Fig. 9.1 shows four examples of correlation between x and y. In Examples (i), (ii) and (iv) the value of y *increases* as x increases and the correlation is said to be 'direct'. In Example (iii) the value of y *decreases* as x increases, and the correlation is said to be 'inverse'.

Example (i) (Fig. 9.1) indicates a linear (straight line) relationship between x and y. Straight lines are easy to recognise and simple to express mathematically:

$$y = mx + c$$

where m is the slope of the line and c is the intercept on the y axis (see Fig. 9.2). The experimental points in Example (i) do not lie exactly on a straight line for several reasons. These may include: (a) human observational errors, for example parallax or carelessness; (b) random errors because of environmental changes; and (c) errors inherent in the apparatus for example play in mechanical parts, the stray inductance or capacitance of connecting leads in electrical apparatus. Fitting a straight line through the points has the advantage that it tends to remove errors of this sort. It should be noted that systematic errors are not removed in this way. Common systematic errors include balances or meters not correctly

151

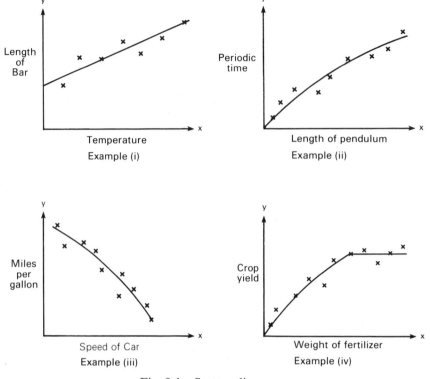

Fig. 9.1   Scatter diagrams.

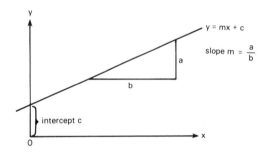

Fig. 9.2

zeroed, thermometers, pippettes and burettes incorrectly calibrated. The method by which the best values for the slope m and the intercept c are chosen is given in a later section.

The scatter diagram in Example (ii) (Fig. 9.1) suggests a curve rather than a straight line. The theoretical equation for a simple pendulum is:

$$T = 2\pi \sqrt{\dfrac{l}{g}}$$

where T is the periodic time, l is the length of the pendulum and g is the acceleration due to gravity (9.8 m s^{-2} or 32 ft s^{-2}). Using x and y for l and T respectively.

$$y = \dfrac{2\pi}{\sqrt{g}} \cdot \sqrt{x}$$

Comparing this with the straight line equation $y = mx + c$ it can be seen that by plotting a graph of y against $\sqrt{x}$ should give a line of slope $m = 2\pi/\sqrt{g}$ and intercept $c = 0$ (see Fig. 9.3).

There are no theoretical equations for Examples (iii) and (iv) (Fig. 9.1). In such cases one might obtain a straight line by changing the data before plotting, for example by plotting logarithms of one or both axes, by squaring or cubing one of the axes or by plotting reciprocals for one of the axes. If a straight line is not obtained it could be that the interrelation between x and y cannot be rearranged into a straight line form. Fitting a quadratic or a higher order polynomial to the

152

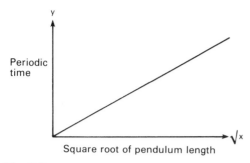

Fig. 9.3

curve may be sufficient for smoothing or interpolation. This is discussed and a program provided in Chapter 12.

Pearson's Correlation Coefficient

This is fully described in Chapter 8 on correlations. Pearson's coefficient r lies in the range $+1$ (perfect direct linear correlation), through zero (no linear correlation), to -1 (perfect inverse linear correlation). As a rule of thumb, a fairly good straight line is indicated by values above 0.95 or below -0.95.

Derivation of Least-Squares Equations

It is quite possible to use the equations at the end of this section without following their derivation. For those interested and for completeness, the derivation is given below.

Consider n points on a graph (x_1, y_1), $(x_2, y_2), \ldots, (x_n, y_n)$. The x_i values have been chosen by the experimenter (independent variable) while the y_i values have been measured (dependent variable) and are thus subject to errors. The 'best' straight line through these points has the equation

$$y = mx + c$$

The objective is to determine the values for slope m and intercept c. For a given values x_i the corresponding y value on the 'best' straight line is $mx_i + c$. The difference Δ between this calculated y value and the experi-

mentally observed y_i value is called the error in y_i (or the y residual), and has the value difference

$$\Delta_i = y_i - (mx_i + c)$$

The least-squares criterion used to determine the 'best' straight line is to make the sum of the differences squared as small as possible, i.e. minimise the expression E

$$E = \sum_{i=1}^{i=n} \Delta_i^2 = \sum_{i=1}^{i=n} (y_i - mx_i - c)^2 \qquad (1)$$

The standard way of obtaining the maxima and minima of an expression is to differentiate the expression and solve it for the case(s) where the derived function equals zero. Equation 1 contains two variables m and c, hence it is necessary to minimise E with respect to m, and also with respect to c. To do this requires the simultaneous solution of both partial derivatives $\partial E/\partial m$ and $\partial E/\partial c$ equal to zero. The partial derivative $\partial E/\partial m$ is obtained by differentiating the expression for E with respect to m and assuming that c is a constant. Similarly the partial derivative $\partial E/\partial c$ is obtained by differentiating E with respect to c and assuming that m is a constant.

Equation 1 may be expanded

$$E = \Sigma y_i^2 + m^2 \Sigma x_i^2 + nc^2 - \\ - 2m \Sigma x_i y_i - 2c \Sigma y_i + 2mc \Sigma x_i$$

partially differentiating with respect to m gives:

$$\frac{\partial E}{\partial m} = 0 + 2m \Sigma x_i^2 + 0 - \\ - 2 \Sigma x_i y_i - 0 + 2c \Sigma x_i \qquad (2)$$

similarly differentiating with respect to c gives:

$$\frac{\partial E}{\partial c} = 0 + 0 + 2nc - \\ - 0 - 2 \Sigma y_i + 2m \Sigma x_i \qquad (3)$$

Equating 2 to zero and rearranging

$$\Sigma x_i y_i = c \Sigma x_i + m \Sigma x_i^2 \qquad (4)$$

Equating 3 to zero and rearranging

$$\Sigma y_i = nc + m \Sigma x_i \qquad (5)$$

153

Equations 4 and 5 give two linearly independent equations in two unknowns. There are many ways of solving them to obtain m and c. One simple approach is:

(a) multiply each term in Equation 4 by n;
(b) multiply each term in Equation 5 by Σx_i;
(c) subtract these two equations from each other thus eliminating the c terms—the resulting equation involves only m;
(d) Substitute the equation for m back in Equation 4 to obtain an equation involving only c.

The correct solutions are

$$\text{Slope } m = \frac{n\Sigma x_i y_i - \Sigma x_i \cdot \Sigma y_i}{n\Sigma x_i^2 - (\Sigma x_i)^2} \qquad (6)$$

and

$$\text{Intercept } c = \frac{\Sigma x_i^2 \cdot \Sigma y_i - \Sigma x_i \cdot \Sigma x_i y_i}{n\Sigma x_i^2 - (\Sigma x_i)^2} \qquad (7)$$

The above equations for slope and intercept are widely used. It is worth noting that the denominators of both equations 6 and 7 involve the subtraction of two large positive quantities from each other. These equations are therefore potentially inaccurate if only a limited number of significant figures are carried as on calculators and computers. Any subtraction between two numbers which are almost equal results in a reduction in the number of significant figures of accuracy which may be claimed.

The way to reduce this problem of loss of accuracy is to take advantage of the fact that the least-squares line must pass through the point $(\bar{x}, \bar{y})$ where $\bar{x}$ and $\bar{y}$ are the mean values, i.e. $\Sigma x_i/n$ and $\Sigma y_i/n$. This can be seen from Equation 5 divided throughout by n

$$\frac{\Sigma y_i}{n} = c + \frac{m \Sigma x_i}{n}$$

thus

$$\bar{y} = c + m\bar{x}$$

i.e.

$$\bar{y} = m\bar{x} + c \qquad (8)$$

More accuracy will be retained by reducing the magnitude of terms which are summed. This may be achieved by subtracting $\bar{x}$ from all the x values and $\bar{y}$ from all the y values. This is equivalent to moving the origin of the graph from the point $(0, 0)$ to the point $(\bar{x}, \bar{y})$. This change has no effect on the slope but allows it to be calculated more accurately. The change makes the intercept zero, but the original intercept is readily calculated from Equation 8 as

$$c = \bar{y} - m\bar{x} \qquad (9)$$

The equation for the improved accuracy for the slope is

$$m = \frac{\Sigma X_i Y_i}{\Sigma X_i^2} \qquad (10)$$

where

$$X_i = (x_i - \bar{x}) \text{ and } Y_i = (y_i - \bar{y}).$$

Equation 10 can easily be obtained from Equation 6 by shifting the origin, i.e replacing x and y by X and Y, and remembering that ΣX_i must be zero.

A further marginal improvement to the accuracy of the intercept can be obtained by re-writing Equation 9

$$c = \frac{\Sigma y_i}{n} - \frac{\Sigma X_i Y_i}{\Sigma X_i^2} \cdot \frac{\Sigma x_i}{n}$$

$$c = \frac{\Sigma X_i^2 \cdot \Sigma y_i - \Sigma x_i \cdot \Sigma X_i Y_i}{n\Sigma X_i^2} \qquad (11)$$

Equation 11 is slightly superior to Equation 9 particularly when c is close to zero.

Equations 10 and 11 should always be used on a computer in preference to Equations 6 and 7. The more accurate equations require that all of the x and y values are stored in arrays, the values for $\bar{x}$ and $\bar{y}$ calculated, and then the sums of ΣX_i, ΣY_i, ΣX_i^2 and $\Sigma X_i Y_i$ collected. This involves slightly more arithmetic and appreciably more memory which is usually available on a computer but it is not available on a calculator. The improvement in accuracy is illustrated in the next section.

A Numerical Example of the Accuracy of the Different Methods

The object of this section is to illustrate the improved accuracy of Equations 10 and 11 over Equations 6 and 7 (see Table 9.1). The data have deliberately been chosen to give inaccurate values for the intercept with Equation 7.

Data	x	y
	10 000	10 001
	11 001	11 002
	12 002	12 003

Clearly the best straight line has $m = 1$ and $c = 1$. On a computer carrying six significant figures, Equation 7 yields a meaningless and incorrect value for the intercept while Equation 11 gives an intercept which is wrong by less than 1 in 999. With a computer carrying 12 significant figures, both methods yield the exact answer.

How Good a Fit is the Best Straight Line?

A problem with an automatic (computerised) method of fitting a straight line is that it will ALWAYS give a best line, even if the fit is totally unreasonable. If a graph or scatter diagram is plotted then wildly erroneous points will be readily detected and either re-measured or disregarded. The method of least squares uses all of the data, and the calculated values are affected by the erroneous value(s). Furthermore no reasonable human being would attempt to fit a straight line to a circle, to four points at the corners of a square, to a cosine wave and numerous other cases which are non-linear. The computer has no sense of what is unreasonable, and will produce a solution if so instructed. Some warning is needed that the computer is carrying out an unreasonable task. The program described next reports the following indicators:

(i) Pearson's correlation coefficient r;
(ii) standard deviation of the points from the least-squares line;
(iii) the individual errors Δ_i between each observed y value and corresponding point on the straight line;
(iv) the sum of the errors squared $\Sigma \Delta_i^2$.

Pearson's correlation coefficient should be close to $+1$ or -1 for a reasonable straight line. This is mentioned earlier in this chapter, and is more fully described in Chapter 8. It will detect the totally unreasonable cases mentioned above giving a value of r close to or equal to zero. It is less good at detecting a gentle curve and little use at detecting erroneous points.

A low standard deviation corresponds to a good fit. A high value is caused by a poor fit, and may be due to a bad scatter of points, an

Table 9.1

	Calculated value for m	Calculated value for c
Equations 6 and 7		
Computer carrying 6 significant figures	1.000 00	0.613 170
Computer carrying 12 significant figures	1.000 000 000 00	1.000 000 000 00
Equations 10 and 11		
Computer carrying 6 significant figures	1.000 00	0.999 004
Computer carrying 12 significant figures	1.000 000 000 00	1.000 000 000 00

unreasonable case as described above, or by one or more erroneous values. Unfortunately the magnitude of the standard deviation is directly proportional to the magnitude of the y_i values (i.e. doubling all of the y_i values gives double the standard deviation). Interpretation of the value of the standard deviation is therefore subjective.

The individual errors are invaluable in detecting wildly erroneous points, since these points have disproportionately large errors. If a poor fit is obtained then the individual errors should be examined to see if they follow some pattern. For example if the points actually followed a curve (see Fig. 9.4).

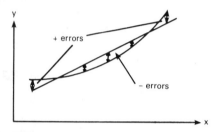

Fig. 9.4

If the list of individual errors suggests that an occasional point is erroneous, then this should be remeasured if possible or alternatively deleted. The program should then be re-run on the amended data. Discarding points should always be undertaken with care.

Confidence Limits for the Slope

It is often important to known how reliably the slope has been calculated, that is to quote a range of values within which the true slope is 95% likely to lie. Thus the slope should be quoted as:

$$\text{Slope} \pm t \cdot \text{Standard error of the slope} \quad (12)$$

where t is the 95% Student's t value (Appendix 6) with $(n-2)$ degrees of freedom. (The number of degrees of freedom is the number of points n less one for the slope and less

another since the straight line must pass through the point $(\bar{x}, \bar{y})$.)

First the standard deviation of the points from the least-squares line is calculated in the same same way as the standard deviation of y except that for each point $\bar{y}$ is replaced by the corresponding point on the straight line, which is $mx + c$.

Standard deviation of points from least-squares line

$$= \sqrt{\frac{[\Sigma[y_i - (mx_i + c)]^2]}{n - 2}} \quad (13)$$

The standard error of the slope can be derived from first principles and is:

Standard error of slope

$$= \frac{\text{Standard deviation of points from line}}{\sqrt{[\Sigma(x_i - \bar{x})^2]}} \quad (14)$$

Using the values derived from Equations 13 and 14 in Equation 12, limits of accuracy may be placed on the slope. These are shown graphically in Fig. 9.5.

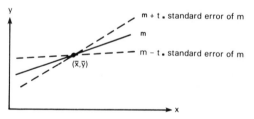

Fig. 9.5

Confidence Limits for a y Value Derived from the Line

In addition to the error in the slope discussed previously, there is also some additional error in the height of the line up the y axis. By definition the least-squares line must pass through $(\bar{x}, \bar{y})$ which is the mean of the observed data points. However $\bar{y}$ may differ from the true value because of errors in the individual y_i values.

156

From the central limit theorem, the standard error of $\bar{y}$ is

$$\frac{\text{Standard deviation of points from least-squares line}}{\sqrt{n}}$$

$$=\frac{\sqrt{\{\Sigma[y_i-(mx_i+c)]^2/(n-2)\}}}{\sqrt{n}} \qquad (15)$$

In an analogous manner to Equation 12 the true mean y value is 95% certain to lie in the range

$$\bar{y}\pm t\cdot\text{Standard error of }\bar{y} \qquad (16)$$

This is shown graphically in Fig. 9.6.

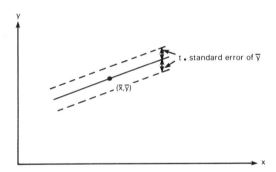

Fig. 9.6

From the error in the slope, and the error in $\bar{y}$, the standard error of a y value derived from the least-squares line at x may be expressed:

Standard error of calculated y value $=$

$$\frac{\text{Standard error of points}}{\text{from least-squares line}} \times \sqrt{\left[\frac{1}{n}+\frac{(x-\bar{x})^2}{\Sigma(x_i-\bar{x})^2}\right]} \qquad (17)$$

Hence the confidence limits for the calculated y value are

$$y\pm t\cdot\text{Standard error of calculated }y\text{ value}$$

In particular the confidence limits for the

intercept of the least-squares line on the y axis may be calculated by setting $x=0$. The value of t should have $(n-2)$ degrees of freedom, and should be chosen for the required probability level (usually 95%).

It is apparent from Equation 17 that the most reliable estimates of y are obtained for x values close to $\bar{x}$, and further x is from $\bar{x}$ the larger the possible error. This is shown in the graph (Fig. 9.7).

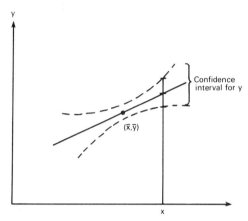

Fig. 9.7

Confidence Limits for an x Value Derived from the Line

In a number of methods of instrumental analysis such as a flame photometer or an absorption spectrophotometer, a number of solutions of known concentration of the material to be analysed are first used to obtain instrumental readings. A least-squares line is then fitted to these data to calibrate the instrument. A solution of unknown concentration is then introduced into the instrument giving a reading, which is converted into a concentration using the least-squares calibration graph. The problem of producing confidence limits is one of producing confidence limits for an x value from the least-squares line and a y value.

The standard error of x can be derived from first principles:

Standard error of calculated x value $=$

$$= \left[\frac{\text{Standard error of points from least-squares line}}{\text{Slope}} \right]$$

$$\cdot \sqrt{\left[1 + \frac{1}{n} + \frac{(y - \bar{y})^2}{\text{slope}^2 \cdot \Sigma(x_i - \bar{x})^2} \right]} \quad (18)$$

Provided that ($t \times$ standard error of points from line/slope) is small, i.e. less than about 0.3, the confidence limits for the calculated x values are estimated as:

$$x \pm t \cdot \text{Standard error of calculated } x \text{ value} \quad (19)$$

Closer confidence limits may be obtained by taking l instrumental readings on the unknown solution. If the average of these is used instead of the y value then Equation 18 becomes:

Standard error of calculated x value $=$

$$= \left[\frac{\text{Standard error of points from least-squares line}}{\text{Slope}} \right]$$

$$\cdot \sqrt{\frac{1}{l} + \frac{1}{n} + \frac{(y - \bar{y})^2}{\text{slope}^2 \cdot \Sigma(x_i - \bar{x})^2}} \quad (20)$$

Example

In the analysis of iron(III), ammonium thiocyanate is added and the intensity of colour produced is measured by a spectrophotometer. Four standard solutions containing 0.20, 0.40, 0.60 and 0.80 mg l^{-1} gave absorbances of 0.238, 0.506, 0.782 and 0.978 respectively. Calculate the average x and y values and the slope of the calibration curve, and use this to estimate the concentration of an iron(III) solution which gave an absorbance of 0.582. Calculate the standard deviation of points from the least-squares line and use this to estimate the standard error of the slope and $\bar{y}$. Finally calculate the confidence limits for the concentration of the solution whose absorbance was 0.582.

Calculation of Slope

Sum of x values	$= 2.00$
Mean x	$= 0.50$
Sum of y values	$= 2.504$
Mean y	$= 0.626$
Sum of $(x - x \text{ mean})^2$	$= 0.20$
Sum of $(y - y \text{ mean})^2$	$= 0.3132$
Sum of $(x - x \text{ mean}) \cdot (y - y \text{ mean})$	$= 0.2496$

Slope $= 0.2496/0.20 = 1.248$

Correlation coefficient r

$$= 0.2496 \Big/ \sqrt{(0.20 \times 0.3132)} = 0.997$$

Calculation of Concentration

For a straight line $y = mx + c$. Since we have not calculated the intercept c, but have calculated the means $\bar{x}$, $\bar{y}$ (the centroidal point), the alternative straight line equation may be used:

$$y - \bar{y} = m \, (x - \bar{x})$$

$$\text{hence } x = \bar{x} + (y - \bar{y})/m$$

$$= 0.50 + (0.582 - 0.626)/1.248$$

$$= 0.4647$$

Concentration of unknown $= 0.465$ mg l^{-1} iron(III)

Calculation of the Standard Deviation of the Points from the Line

x	y	y calculated from line	$(y - y_{calc})^2$
0.20	0.238	0.2516	0.000 185 0
0.40	0.506	0.5012	0.000 023 0
0.60	0.782	0.7508	0.000 973 4
0.80	0.978	1.0004	0.000 501 8
			$\Sigma 0.001 683 2$

158

Using Equation 13

Standard deviation of points from least-squares line

$$= \sqrt{\frac{0.001\ 683}{4-2}} = 0.029\ 01$$

Using Equation 14 standard error of the slope

$$= \frac{0.029\ 01}{\sqrt{0.2}} = 0.064\ 87$$

Using Equation 15

Standard error of $\bar{y}$

$$= \frac{0.029\ 01}{\sqrt{4}} = 0.014\ 5$$

Using Appendix 6 the 95% confidence t value for two degrees of freedom is 4.303

Hence using Equation 12

95% confidence limits for slope

$$= 1.248 \pm 4.303 \times 0.064\ 87$$

$$= 1.248 \pm 0.279$$

It is worth noting that the confidence limits on the slope are very wide even though the correlation coefficient is very close to 1. The reason for the wide confidence limits is the small number of points on the calibration curve.

Using Equation 16

95% confidence limits for $\bar{y}$

$$= 0.626 \pm 4.303 \times 0.0145$$

$$= 0.626 \pm 0.062$$

Confidence Limits for Concentration of Solution

Using Equation 18

Standard error of calculated x value

$$= \frac{0.029\ 01}{1.248} \cdot \sqrt{\left[1 + \frac{1}{4} + \frac{(0.582 - 0.626)^2}{1.248^2 \times 0.20}\right]}$$

$$= 0.029\ 20$$

Using Equation 19, confidence limit for calculated x value

$$= 0.0465 \pm 4.303 \times 0.029\ 20$$

$$= 0.465 \pm 0.126$$

$$= 0.47 \pm 0.13$$

The limits of accuracy are so bad that the analysis is useless since the result is subject to about 27% error. Any self-respecting analyst would repeat the experiment with more points on the calibration curve. This has a big effect on the value of t, as well as the n that appears in the equations. Several measurements on the unknown solution should be made, and their average value used as y. This means that Equation 20 should be used instead of Equation 18.

How to Apply Least Squares when both x and y are Equally Prone to Error

At the beginning of this chapter it was assumed that the x values were *chosen* by the experimenter and considered to be free from error, hence x is called the independent variable. The corresponding y values are *measured*, making y the dependent variable. The y measurements are subject to errors. The method of least squares assumes that all of the error is associated with the y values.

In some circumstances it is not obvious which variable is dependent and which is independent since both are measured, and both are subject to errors. Some simple examples are:

(i) The relationship between the intelligence quotient IQ (x) of an individual and the weight of his brain (y).
(ii) The relationship between the height (x) and weight (y) of human beings.
(iii) The relationship between liver weight (x) and kidney weight (y) of guinea pigs.

Plainly one cannot take an individual, adjust his IQ to a certain value, and then measure his brain weight, whereas one can adjust the length of a pendulum and measure its corresponding periodic time. In examples (i), (ii) and (iii) above both the x and y values are subject to errors

The conventional way of dealing with data of this sort is as follows:

1. Assume that all of the error is in the y terms (and none in the x), and fit the least-squares line. The graph (Fig. 9.8) shows the errors in the y values making this assumption, i.e. the least-squares line of regression of y on x.

2. Assume that all of the error occurs in the x terms (and none in the y terms), and fit the least-squares line. This can be done using the program for least squares by exchanging the x and y values that are typed in. The graph (Fig. 9.9) shows the errors in the x values for the least squares.

3. The two lines of regression are then compared. There are three possible cases:

(a) The slopes of both lines are identical. Both lines pass exactly through all of the points and consequently Pearson's correlation coefficient is $+1$ or -1.

(b) One line is horizontal and the other is vertical—that is the angle between the two lines is 90°. Pearson's correlation coefficient is zero, and there is no relationship whatsoever between x and y.

(c) The more usual case lies between these two extremes. If the angle between the two regression lines is small then x and y are related, and for example when estimating brain weight (y) from IQ (x) the first regression line (y on x) should be used. Conversely when estimating IQ (x) from brain weight (y) the second regression line (x on y) should be used.

Pearson's Correlation Coefficient

It should be noted that the value of Pearson's correlation coefficient is the same for regression of y on x and x on y. This property makes Pearson's correlation coefficient useful for determining whether a linear relationship exists between x and y. The program for Pearson's correlation given in Chapter 8 determines whether the calculated value is significant, i.e. whether a linear relationship between x and y exists at the 5% significance level.

The slope of x on y can be calculated from the slope of y on x and Pearson's correlation coefficient as shown below.

Slope of y on x

$$= \frac{\Sigma X_i \, Y_i}{\Sigma X_i^2} \text{ where } X_i = (x_i - \bar{x})$$
$$\text{and } Yi = (y_i - \bar{y})$$

Slope of x on y

$$= \frac{\Sigma X_i \, Y_i}{\Sigma Y_i^2}$$

Pearson's r

$$= \frac{\Sigma X_i \, Y_i}{\sqrt{\Sigma X_i^2 \cdot \Sigma Y_i^2}}$$

this is equivalent to Equation 5 Chapter 8.

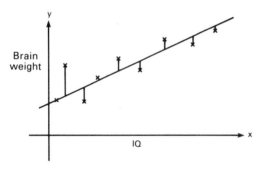

Fig. 9.8

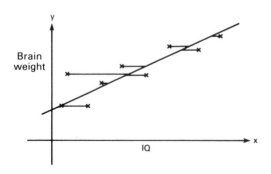

Fig. 9.9

160

$$r^2 = \frac{\Sigma X_i\,Y_i \cdot \Sigma X_i\,Y_i}{\Sigma X_i^2 \cdot \Sigma Y_i^2}$$

$$= (\text{slope of } y \text{ on } x) \cdot (\text{slope of } x \text{ on } y)$$

Hence slope of x on y

$$= \frac{r^2}{\text{slope of } y \text{ on } x}$$

Description of the Least-Squares Program (See Program 9.1)

First the program prints a title, and asks if full instructions are required. The reply is checked by a subroutine (lines 1230–1320) and must be either YES or NO. Any other reply is rejected. Full instructions or shortened ones are

Program 9.1   Trial run.

```
PROGRAM TO FIT A STRAIGHT LINE THROUGH A SET OF POINTS
======= == === = ======== ==== ======= = === == ======

WOULD YOU LIKE FULL INSTRUCTIONS?   TYPE YES OR NO & PRESS RETURN.

? YES
TYPE IN X & Y VALUES SEPARATED BY A COMMA
THEN PRESS RETURN AND TYPE THE NEXT PAIR OF VALUES
YOU WILL BE GIVEN THE CHANCE TO EDIT INCORRECT DATA LATER

INPUT DATA
TERMINATE DATA WITH BOTH X & Y EQUAL TO 999999
 X,Y
? 10000, 10001
? 11001, 11002
? 12002, 12003
? 999999, 999999
ARE THE DATA VALUES ENTERED CORRECT?   TYPE YES OR NO & PRESS RETURN.

? YES

RESULTS OF CALCULATION:
------- -- -----------
SLOPE  = 1

INTERCEPT ON Y AXIS  = 0.999004

PEARSON'S CORRELATION COEFFICIENT  = 1

STANDARD DEVIATION OF POINTS FROM THE LINE  = 0

WOULD YOU LIKE A LIST OF ERRORS?   TYPE YES OR NO & PRESS RETURN.

? YES

ERROR IN Y VALUE FOR EACH DATA POINT
X               Y OBS          Y CALC          ERROR
10000           10001          10001           0
11001           11002          11002           0
12002           12003          12003           0

SUM OF ERRORS SQUARED   = 0

WOULD YOU LIKE ANOTHER RUN?   TYPE YES OR NO & PRESS RETURN.

? NO
END OF JOB

10 DIM Q$(10), I$(3), X(100), Y(100)
20 PRINT "PROGRAM TO FIT A STRAIGHT LINE THROUGH A SET OF POINTS"
30 PRINT "======= == === = ======== ==== ======= = === == ======"
```

161

```
40 PRINT
50 PRINT "WOULD YOU LIKE FULL INSTRUCTIONS?";
60 GOSUB 1250
70 LET I$ = Q$
80 IF Q$ = "NO" THEN 130
90 PRINT "TYPE IN X & Y VALUES SEPARATED BY A COMMA"
100 PRINT "THEN PRESS RETURN AND TYPE THE NEXT PAIR OF VALUES"
110 PRINT "YOU WILL BE GIVEN THE CHANCE TO EDIT INCORRECT DATA LATER"
120 PRINT
130 PRINT "INPUT DATA"
140 PRINT "TERMINATE DATA WITH BOTH X & Y EQUAL TO 999999"
150 PRINT " X,Y"
160 REM SET INITIAL VALUES TO ZERO
170 READ C, S2, S3, X2, Y2, S4, E2, N
180 DATA 1, 0, 0, 0, 0, 0, 0, 0
190 REM READ IN X & Y FOR EACH DATA POINT & STORE IN ARRAYS
200 FOR I = 1 TO 100
210     INPUT X(I), Y(I)
220     IF ABS(X(I) - 999999) + ABS(Y(I) - 999999) = 0 THEN 260
230     LET N = N + 1
240 NEXT I
250 PRINT "PROGRAM CAN ONLY HANDLE A MAXIMUM OF 100 VALUES"
260 IF N > 0 THEN 300
270 PRINT "PLEASE TYPE IN SOME DATA VALUES"
280 GOTO 200
290 REM CALL SUBROUTINE TO CHECK AND ALTER DATA IF NECESSARY
300 GOSUB 1340
310 PRINT
320 IF N > 2 THEN 380
330 PRINT
340 PRINT "RUN ON THIS DATA ABANDONED BECAUSE THERE ARE"
350 PRINT "NOT ENOUGH DATA POINTS"
360 GOTO 620
370 REM CALL SUBROUTINE TO WORK OUT LEAST-SQUARES FIT ETC.
380 GOSUB 870
390 PRINT "RESULTS OF CALCULATION;"
400 PRINT "------- -- ------------"
410 PRINT "SLOPE  ="; S
420 PRINT
430 PRINT "INTERCEPT ON Y AXIS  ="; Y1
440 PRINT
450 PRINT "PEARSON'S CORRELATION COEFFICIENT  ="; C
460 PRINT
470 PRINT "STANDARD DEVIATION OF POINTS FROM THE LINE  ="; S1
480 PRINT
490 PRINT "WOULD YOU LIKE A LIST OF ERRORS?";
500 GOSUB 1240
510 IF Q$ = "NO" THEN 620
520 REM CALCULATE & PRINT ERROR FOR EACH DATA POINT
530 PRINT
540 PRINT "ERROR IN Y VALUE FOR EACH DATA POINT"
550 PRINT "X", "Y OBS", "Y CALC", "ERROR"
560 FOR I = 1 TO N
570     PRINT X(I), Y(I), A2 + S * (X(I)-A1), (Y(I)-A2) - S * (X(I)-A1)
580 NEXT I
590 PRINT
600 PRINT "SUM OF ERRORS SQUARED  ="; E2
610 REM DECIDE WHETHER TO FINISH OR HAVE ANOTHER RUN
620 PRINT
630 PRINT "WOULD YOU LIKE ANOTHER RUN?";
```

```
640 GOSUB 1240
650 IF Q$ = "NO" THEN 820
660 LET I$ = "NO"
670 RESTORE
680 PRINT "TYPE NEW FOR A RUN WITH COMPLETELY NEW DATA"
690 PRINT "  OR OLD TO EDIT AND RERUN THE EXISTING DATA"
700 INPUT Q$
710 IF Q$ = "NEW" THEN 780
720 IF Q$ = "OLD" THEN 750
730 PRINT "REPLY '"; Q$; "' NOT UNDERSTOOD"
740 GOTO 680
750 READ C, S2, S3, X2, Y2, S4, E2
760 GOSUB 1390
770 GOTO 320
780 PRINT "TYPE IN A NEW SET OF DATA"
790 PRINT "==== == = === === == ===="
800 GOTO 120
810 REM TERMINATE JOB
820 PRINT "END OF JOB"
830 STOP
840 REM ***** SUBROUTINE TO CALCULATE LEAST SQUARES FIT, ETC. *****
850 REM ARRAYS X & Y CONTAIN N DATA POINTS
860 REM CALCULATE SUMS OF X & Y VALUES
870 FOR I = 1 TO N
880    LET S2 = S2 + X(I)
890    LET S3 = S3 + Y(I)
900 NEXT I
910 REM CALCULATE THE AVERAGE X & Y VALUES
920 LET A1 = S2 / N
930 LET A2 = S3 / N
940 FOR I = 1 TO N
950    LET X9 = X(I) - A1
960    LET Y9 = Y(I) - A2
970    LET X2 = X2 + X9 * X9
980    LET Y2 = Y2 + Y9 * Y9
990    LET S4 = S4 + X9 * Y9
1000 NEXT I
1010 PRINT
1020 IF X2 <> 0 THEN 1070
1030 PRINT "RUN TERMINATED BY THE PROGRAM BECAUSE ALL"
1040 PRINT "THE X-COORDINATES ARE THE SAME."
1050 GOTO 620
1060 REM CALCULATE SLOPE
1070 LET S = S4 / X2
1080 REM CALCULATE INTERCEPT ON Y AXIS
1090 LET Y1 = (X2 * S3 - S2 * S4) / (N * X2)
1100 IF Y2 = 0 THEN 1140
1110 REM CALCULATE CORRELATION COEFFICIENT
1120 LET C = S4 / SQR(X2 * Y2)
1130 REM CALCULATE SUM OF ERRORS SQUARED
1140 FOR I = 1 TO N
1150    LET E1 = (Y(I) - A2) - S * (X(I) - A1)
1160    LET E2 = E2 + E1 * E1
1170 NEXT I
1180 REM CALCULATE STANDARD DEVIATION
1190 LET S1 = SQR(E2 / (N - 2))
1200 IF N < 30 THEN 1220
1210 LET S1 = SQR(E2 / N)
1220 RETURN
1230 REM ***** SUBROUTINE TO CHECK REPLIES *****
```

```
1240 IF I$ = "NO" THEN 1260
1250 PRINT " TYPE YES OR NO & PRESS RETURN."
1260 PRINT
1270 INPUT Q$
1280 IF Q$ = "YES" THEN 1320
1290 IF Q$ = "NO" THEN 1320
1300 PRINT "REPLY '"; Q$; "' NOT UNDERSTOOD.";
1310 GOTO 1250
1320 RETURN
1330 REM *****SUBROUTINE TO CHECK DATA ARE CORRECT & ALTER IF NECESSARY
1340 PRINT "ARE THE DATA VALUES ENTERED CORRECT?";
1350 REM A4 SHOULD BE SET TO THE NUMBER OF LINES ON THE VDU
1360 LET A4 = 20
1370 GOSUB 1240
1380 IF Q$ = "YES" THEN 2080
1390 PRINT "HERE IS A LIST OF THE CURRENT DATA"
1400 PRINT "LINE NUMBER", "X", "Y"
1410 FOR I = 1 TO N
1420    PRINT I, X(I), Y(I)
1430    IF INT(I / (A4 - 1)) * (A4 - 1) <> I THEN 1470
1440    PRINT "WOULD YOU LIKE TO CONTINUE LISTING?";
1450    GOSUB 1240
1460    IF Q$ = "NO" THEN 1480
1470 NEXT I
1480 PRINT "TYPE R TO REPLACE";
1490 IF I$ = "NO" THEN 1510
1500 PRINT " AN EXISTING LINE OF DATA"
1510 IF N = 100 THEN 1560
1520 PRINT TAB(5); " A TO ADD";
1530 IF I$ = "NO" THEN 1550
1540 PRINT " AN EXTRA LINE"
1550 IF N = 1 THEN 1590
1560 PRINT TAB(5); " D TO DELETE";
1570 IF I$ = "NO" THEN 1590
1580 PRINT " AN EXISTING LINE"
1590 PRINT TAB(5); " L TO LIST";
1600 IF I$ = "NO" THEN 1620
1610 PRINT " THE DATA"
1620 PRINT "  OR C TO CONTINUE";
1630 IF I$ = "NO" THEN 1650
1640 PRINT " THE CALCULATION"
1650 INPUT Q$
1660 IF Q$ = "R" THEN 1760
1670 IF N = 100 THEN 1700
1680 IF Q$ = "A" THEN 1880
1690 IF N = 1 THEN 1710
1700 IF Q$ = "D" THEN 1930
1710 IF Q$ = "L" THEN 1390
1720 IF Q$ = "C" THEN 2080
1730 PRINT "REPLY '"; Q$; "' NOT UNDERSTOOD."
1740 GOTO 1480
1750 REM REPLACE LINE
1760 PRINT "TYPE THE LINENUMBER OF THE LINE TO BE REPLACED";
1770 INPUT I
1780 IF I <> INT(I) THEN 1800
1790 IF (I - 1) * (I - N) <= 0 THEN 1830
1800 PRINT "LINENUMBER MUST BE AN INTEGER IN THE RANGE 1 -"; N
1810 PRINT "RE-";
1820 GOTO 1760
1830 PRINT "TYPE THE CORRECT LINE TO REPLACE THE ONE WHICH IS WRONG"
```

```
1840 PRINT "X, Y"
1850 INPUT X(I), Y(I)
1860 GOTO 1910
1870 REM ADD A NEW LINE
1880 LET N = N + 1
1890 PRINT "TYPE THE ADDITIONAL LINE OF DATA AS SHOWN:    X,Y"
1900 INPUT X(N), Y(N)
1910 PRINT "OK"
1920 GOTO 1480
1930 REM DELETE A LINE
1940 PRINT "TYPE THE LINENUMBER OF THE LINE TO BE DELETED"
1950 INPUT J
1960 IF (J - 1) * (J - N) > 0 THEN 1980
1970 IF J = INT(J) THEN 2000
1980 PRINT "LINENUMBER MUST BE AN INTEGER IN THE RANGE 1 -"; N
1990 GOTO 1940
2000 FOR I = J + 1 TO N
2010    LET X(I - 1) = X(I)
2020    LET Y(I - 1) = Y(I)
2030 NEXT I
2040 LET N = N - 1
2050 PRINT "OK"
2060 IF J > N THEN 1480
2070 GOTO 1390
2080 RETURN
2090 END
```

printed in the first run, but only shortened instructions are printed in subsequent runs.

The data input loop extends from lines 200–240. A pair of x and y values is typed on a line, and the RETURN key pressed. Further lines are typed in a similar way, and the end of data input is indicated by typing the dummy line 999999,999999 and RETURN. The X and Y arrays are dimensioned as 100 in the first line of the program, and limit the number of (x, y) points to a maximum of 100.

After the data input is complete a check is performed (lines 260–280) to ensure that at least one valid (x, y) value has been typed. Then a subroutine is entered (lines 1330–2080) which asks if the data entered are correct. If the answer is YES the subroutine is skipped, but if the answer is NO the current data are listed and instructions given explaining how to replace or delete existing lines, add new lines, re-list the data or continue the calculation. This subroutine is described in more detail in Chapter 8. When the data are correct a check is made to ensure that there are at least two data points (lines 320–360) since an infinite number of straight lines can be fitted through a single point.

The least-squares fitting is performed in a subroutine (lines 840–1220). This has deliberately been coded as a subroutine to allow it to be easily implemented in other programs. Equations 10 and 11 are used to prevent loss of accuracy as discussed earlier.

The results: slope, intercept, Pearson's correlation coefficient and the standard deviation of the points from the line are then printed in lines 390–470. The user is given the option of printing a table showing the 'error' in the y coordinate of each term (that is the difference between each input y coordinate and the straight line). The sum of the errors squared is also printed.

Finally the user is offered another run. If this is required then either a completely NEW set of data may be input, or the OLD (existing) data may be edited and re-run.

Exercises

9.1 Given 20 pairs of $(x_i \; y_i)$ values, the following quantities were calculated: $\Sigma x_i = 200$, $\Sigma y_i = 110$, $\Sigma x_i^2 = 2200$, $\Sigma y_i^2 = 655$, and $\Sigma x_i \; y_i = 1075$. Find the linear regression equations of x on y, and of y on x. Which would be more useful in the following cases?

(a) x is the telephone bill and y the number of employees for each of 20 companies.

(b) x is the age and y the reaction times of 20 children.

9.2 The amount of pocket money received by four school children from each of the classes aged 11, 12, 13, 14, 15 and 16 are given in Table 9.2.

(a) Plot these data on a scatter diagram.

(b) Calculate the mean pocket money for each age.

(c) Find the slope and intercept for the appropriate least-squares line (age/mean pocket money).

9.3 From a survey of 10 hospitals, the data in Table 9.3 were collected. Fit a least-squares line to the data, and estimate the cost per patient per day in a 500 bed hospital.

9.4 The monthly electricity bills (y) of 20 families of size (x) are given in Table 9.4. By fitting a least-squares line, estimate the mean monthly electricity bill for a family of size 3. Calculate the average bill for households of 3 people, and explain which gives the better estimate.

9.5 The age at which people get married was studied by considering a sample of 24 couples. The age of the wife (x) and husband (y) are given in Table 9.5. Calculate the regression lines of x on y and of y on x. Use these to predict the age of the spouse of a man and woman aged 25.

9.6 The number of Civil Servants required to collect Value Added Tax in 10 towns of different sizes is given below

Population of town (thousands):
6, 8, 9, 12, 17, 19, 21, 23, 25, 26
Number of Civil Servants:
9, 12, 14, 19, 25, 27, 32, 33, 35, 35

Calculate the regression line for predicting the number of Civil Servants needed for different towns, and use this to predict the number required for a town of 15 000 inhabitants.

Table 9.2

Age (years)	Pocket money (pence)
11	20, 35, 40, 70
12	30, 50, 60, 70
13	35, 40, 70, 80
14	35, 50, 90, 100
15	50, 85, 110, 150
16	50, 75, 100, 200

Table 9.3

Hospital	Number of beds	Cost per patient per day (£)
1	810	39
2	190	54
3	310	46
4	1100	40
5	230	51
6	400	45
7	760	37
8	240	51
9	610	42
10	400	46

Table 9.4

x	y	x	y	x	y	x	y	x	y
1	6	2	6	3	10	4	9	5	10
1	6	2	8	3	10	4	11	5	15
1	8	2	9	3	12	4	13	5	18
1	8	2	9	3	16	4	19	5	21

Table 9.5

x	y	x	y	x	y	x	y
17	20	19	25	18	20	21	26
22	23	16	26	25	27	17	26
19	21	28	23	25	28	21	32
21	29	20	43	26	28	26	29
54	58	29	33	21	25	29	28
27	29	25	25	23	25	23	20

9.7 The linear growth of *Aspergillus niger* was measured when grown on a nutrient medium containing glucose and mineral elements. The results are shown in Table 9.6. Calculate the slope (linear growth rate) and the correlation coefficient.

Table 9.6

Time (days after inoculation)	Diameter of colony (mm)
3	9.3
4	16.8
5	22.8
6	28.5
7	33.6
8	36.6
9	42.8

9.8 Calculate the least squares regression line of crop yield against the amount of fertilizer added, using the data in Table 9.7.

Table 9.7

Amount of fertilizer (ounce/ square yard)	Crop yield (cwt/acre)
0	160
1	165
2	172
3	178
4	181
5	186
6	188
7	189
8	192

10

Some Biological Applications of Linear Regression

Bacterial Growth Curves

Background

Higher plants and animals usually have a well defined life cycle involving growth in size (with cell division by mitosis) and reproduction (generally involving meiosis in certain tissues). In contrast, microorganisms do not show such a life cycle. Cells usually increase their size, weight and quantity of cellular contents to a predetermined size, then undergo cell division, resulting in an increased number of cells. In practice the direct study of the growth of a bacterial cell is difficult because of its small size, and growth studies are usually carried out on a colony of cells grown in a culture. There are four common methods of measuring the growth of a bacterial culture:

1. direct counting of the number of cells in a Petroff–Hauser counting chamber, using a microscope,
2. plating cells on an agar culture medium, allowing the individual cells to grow into colonies large enough to be seen by eye, and then counting the colonies,
3. by centrifuging, drying and weighing to obtain the 'dry weight' of the cells present, and
4. an estimate of the number of cells present can be obtained by measuring the turbidity of the culture in its nutrient medium using a spectrophotometer.

The first method is the most accurate, and gives the total number of cells, both viable and dead. The second method is slow, and measures just the number of viable cells. The third method provides a fast estimate of the total mass (both living and dead), whilst the last method is quick, does not interfere with the course of the experiment, and gives the total cell content.

The requirements for growth are the presence of nutrient, water, trace elements and carbon dioxide. The presence of oxygen is required by strict aerobes, must be absent for strict anaerobes, and may be present for facultative anaerobes. Other factors which affect the rate of growth are temperature, pH, osmotic pressure and surface tension. Each particular species of bacteria has its own preferred conditions for optimum growth.

Phases of Growth

A typical plot showing the phases of growth of a bacterial culture is shown in Fig. 10.1.

The lag phase is variable in duration, and represents the time taken by the cells to adjust to the environment in which they are growing. Factors which contribute to this are the number of viable cells present, the 'physiological age' of the cells, and also their previous conditioning to a similar environment.

After the lag phase, binary fission of the cells occurs at regular intervals, and the growth of the cell mass is exponential (some-

168

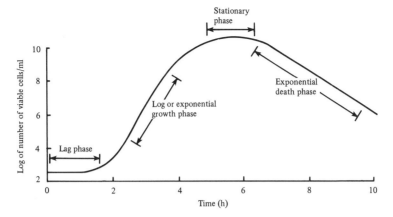

Fig. 10.1   Typical growth phases for bacteria.

times called logarithmic). This is characterised by each cell dividing into two after a fixed time interval called the generation time, g. Thus the rate of growth at any time is proportional to the number of viable cells present. This is expressed mathematically:

$$\frac{dM}{dt} = kM \qquad (1)$$

Where M is the mass of the viable cells present, and dM/dt is the rate of change of M with time. Equation 1 is integrated from time 0 to t and rearranged giving:

$$\ln M_t = \ln M_0 + kT \qquad (2)$$

where M_t is the mass of viable cells at time t, and M_0 is the mass at time zero. It can be seen from Fig. 10.2 that Equation 2 represents a straight line of slope k and intercept $\ln M_0$. It is because of this relationship that growth in this phase is sometimes called 'logarithmic'.

By definition the generation time, g, is the time taken for one cell to grow and divide into two cells. Thus at time $t = g$ it follows that $M_t = 2M_0$, and hence from Equation 2

$$\ln 2M_0 = \ln M_0 + kg$$

$$\ln 2 + \ln M_0 = \ln M_0 + kg$$

$$\ln 2 = kg$$

$$g = \frac{\ln 2}{k} = \frac{0.693}{k}$$

Eventually the growth rate slows down and stops. This region of the growth curve is called the stationary phase, and the number of live cells is at its maximum. The change from exponential growth to stationary growth may be due to a limited supply of nutrients,

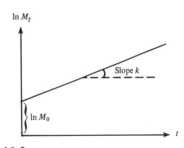

Fig. 10.2

increasing amounts of toxic waste products, the diffusion of oxygen or carbon dioxide becoming limiting, or the pH of the culture falling. For a given culture medium and species of bacteria, the maximum cell concentration which can be attained is fixed, and does not depend on the number of cells inoculated at the beginning. During the stationary phase cell division occurs at the same rate as cell death.

In the final stage, toxic materials (excreted by the cells or released from dead cells by autolysis) accumulate to such a level that exponential (logarithmic) death occurs, and the viable cell population falls drastically. Some microorganisms, such as mycobacteria, still contain viable organisms months after the stationary phase has been reached, while others, including many pathogens such as pneumococci, die rapidly and become sterile.

Description of the Program to Evaluate Bacterial Growth

The program (Program 10.1) prints a heading (lines 20–30) and then asks if full instructions are required (line 50). The answer must be YES or NO, and is checked in a subroutine (lines 1160–1240). Full or abbreviated instructions are printed as requested in the first run, but shortened messages are given in subsequent runs. A number of constants are set to 1 or 0 in lines 80–90.

A loop is used to input the data values (lines 240–330). A pair of values for the time and the number of cells are entered on one line separated by a comma. Up to 100 pairs of values may be entered in this way, the limit being set by the DIMensions of the X, Y and Z arrays in line 10. Generally there are less than 100 data pairs and the end of the data is indicated by typing a dummy line of $-1, -1$. A check is performed on the values typed for the number of cells, to ensure that it is greater than zero, since this is impossible and would cause failure later when taking logarithms of the values. The number of cells is not checked to make sure that it is an integer, so that 'dry weight' or 'turbidity' data may be used, rather than restricting the program to cell counts. The natural logarithm of the number of cells is evaluated (line 310), and the number of valid data pairs is counted (line 320).

On leaving the input loop, a check is carried out (lines 360–380) to make sure that some valid data exists, and then (lines 390–400) a subroutine (lines 1640–2480) is called to permit listing and correction of the input data. Lines of data may be added, replaced or deleted, and the subroutine is similar to that described more fully in Chapter 8. It differs in that new data are checked to ensure that the

number of cells exceeds zero, and also in that it calculates the logarithm of the number of cells. When the data are correct, another check is carried out (lines 410–460) to make sure that at least three valid points remain, since the program would fail when calculating the standard deviation (through dividing by zero in line 1600) if only two points existed.

Next a subroutine (lines 1250–1630) is called to fit a least-squares line to the time and logarithm (cells) data. The subroutine is described in Chapter 9. The slope, intercept, correlation coefficient and standard deviation are printed (lines 500–530), and the user is asked if a table of residuals is required. The reply must be YES or NO, and is checked by the subroutine (lines 1160–1240). The residuals are the differences between the input data values for logarithm (cells) and the corresponding point on the least-squares line. The residuals are evaluated using:

$$y - \bar{y} = m(x - \bar{x})$$

since this gives better numerical accuracy than:

$$y = mx + c$$

A message (line 670) asks if a graph is required. The answer must be YES or NO, and is checked by a subroutine (lines 1160–1240). If a graph is required, the maximum and minimum logarithm (cells) values are calculated (lines 700–780) and used to scale the graph. A check is carried out to make sure that the log terms are not all equal, and a graph is plotted assuming that the time values (x-axis) are equally spaced.

The user is then asked if another run is required (lines 950–960), and a subroutine is called (lines 1160–1240) to obtain an answer of YES or NO. If the answer is NO, a message is printed (line 1140), indicating the end of the job. If the answer is YES, the user is invited (lines 1010–1070) to choose between typing a completely NEW set of data, or editing and re-running the OLD existing data. The latter option may be useful if the first point(s) appear to be in the lag-region, or if the last point(s) are in the stationary phase. These

Program 10.1   Trial run.

```
EVALUATION OF BACTERIAL GROWTH
========== == =================

WOULD YOU LIKE FULL INSTRUCTIONS
TYPE YES OR NO AND PRESS RETURN
? YES

THE PROGRAM FITS A LEAST-SQUARES LINE TO BACTERIAL GROWTH-
DATA (TIME & NUMBER OF CELLS).  THE PROGRAM TAKES THE
LOGARITHM OF THE NUMBER OF CELLS FOR A LOG GROWTH PLOT

TYPE IN TIME AND NO. OF CELLS SEPARATED BY A COMMA
PRESS RETURN AND TYPE IN NEXT PAIR OF VALUES
YOU WILL HAVE A CHANCE TO CORRECT THE INPUT DATA LATER
TERMINATE DATA WITH -1,-1
TIME,CELLS

? 2,125
? 12,135
? 22,142
? 32,159
? 42,182
? 52,202
? 62,239
? 72,264
? 82,292
? 92,309
? 102,337
? 112,346
? -1,-1
ARE THE DATA VALUES ENTERED CORRECT? TYPE YES OR NO AND PRESS RETURN
? YES


SLOPE = 0.00443383
INTERCEPT ON LOG(Y) AXIS = 2.07797
CORRELATION COEFFICIENT = 0.992774
STANDARD DEVIATION = 0.0202657

WOULD YOU LIKE A LIST OF RESIDUALS?
TYPE YES OR NO AND PRESS RETURN
? YES

RESIDUAL FOR EACH DATA POINT
X          Y          LOG(Y)         RESIDUAL
2          125        2.09691        0.0100714
12         135        2.13033        -0.000842956
22         142        2.15229        -0.0232267
32         159        2.2014         -0.0184559
42         182        2.26007        -0.00412032
52         202        2.30535        -0.00317859
62         239        2.3784         0.0255299
72         264        2.4216         0.0243973
82         292        2.46538        0.0238385
92         309        2.48996        0.00407546
102        337        2.52763        -0.00259123
112        346        2.53908        -0.0354835

WOULD YOU LIKE A GRAPH?
TYPE YES OR NO AND PRESS RETURN
? YES
```

```
TIME                         LOG(CELLS)
(MINS)    2.09691                                              2.53908
          +!----------------------------------------------------!
2         !*
          !
12        !    *
          !
22        !      *
          !
32        !         *
          !
42        !            *
          !
52        !               *
          !
62        !                  *
          !
72        !                     *
          !
82        !                        *
          !
92        !                          *
          !
102       !                             *
          !
112       !                             *
          !


WOULD YOU LIKE ANOTHER RUN?
TYPE YES OR NO AND PRESS RETURN
? NO

END OF PROGRAM


10  DIM Q$(10), I$(3), X(100), Y(100), Z(100)
20  PRINT TAB(8); "EVALUATION OF BACTERIAL GROWTH"
30  PRINT TAB(8); "========== == ================="
40  PRINT
50  PRINT "WOULD YOU LIKE FULL INSTRUCTIONS"
60  GOSUB 1180
70  LET I$ = Q$
80  READ C, S2, S3, X2, Y2, S4, E2, N
90  DATA 1, 0, 0, 0, 0, 0, 0, 0
100 IF I$ = "YES" THEN 130
110 PRINT "TYPE IN THE DATA. TERMINATE WITH -1,-1"
120 GOTO 230
130 PRINT
140 PRINT "THE PROGRAM FITS A LEAST-SQUARES LINE TO BACTERIAL GROWTH-"
150 PRINT "DATA (TIME & NUMBER OF CELLS).  THE PROGRAM TAKES THE"
160 PRINT "LOGARITHM OF THE NUMBER OF CELLS FOR A LOG GROWTH PLOT"
170 PRINT
180 PRINT "TYPE IN TIME AND NO. OF CELLS SEPARATED BY A COMMA"
190 PRINT "PRESS RETURN AND TYPE IN NEXT PAIR OF VALUES"
200 PRINT "YOU WILL HAVE A CHANCE TO CORRECT THE INPUT DATA LATER"
210 PRINT "TERMINATE DATA WITH -1,-1"
220 PRINT "TIME,CELLS"
230 PRINT
240 REM START LOOP TO INPUT DATA
250 FOR I = 1 TO 100
260    INPUT X(I), Y(I)
```

```
270    IF ABS(Y(I) + 1) + ABS(X(I) + 1) = 0 THEN 360
280    IF Y(I) > 0 THEN 310
290    PRINT "NUMBER OF CELLS MUST BE POSITIVE - RETYPE LINE CORRECTLY"
300    GOTO 260
310    LET Z(I) =  LOG(Y(I))/LOG(10)
320    LET N = N + 1
330 NEXT I
340 PRINT "PROGRAM CAN ONLY HANDLE A MAX OF 100 VALUES"
350 REM CHECK THAT THERE IS SOME DATA
360 IF N > 0 THEN 400
370 PRINT "PLEASE TYPE IN SOME DATA VALUES"
380 GOTO 250
390 REM CALL SUBROUTINE TO CHECK EDIT & PRINT DATA
400 GOSUB 1650
410 REM CHECK THAT THERE ARE ENOUGH DATA POINTS
420 IF N>2 THEN 480
430 PRINT
440 PRINT "RUN ON THIS DATA ABANDONED BECAUSE THERE ARE"
450 PRINT "NOT ENOUGH DATA POINTS"
460 GOTO 940
470 REM CALL SUBROUTINE TO WORK OUT LEAST-SQUARES FIT ETC.
480 GOSUB 1280
490 PRINT
500 PRINT "SLOPE ="; S
510 PRINT "INTERCEPT ON LOG(Y) AXIS ="; Y1
520 PRINT "CORRELATION COEFFICIENT ="; C
530 PRINT "STANDARD DEVIATION ="; S1
540 PRINT
550 PRINT "WOULD YOU LIKE A LIST OF RESIDUALS?"
560 GOSUB 1170
570 PRINT
580 IF Q$ ="NO" THEN 660
590 REM CALCULATE & PRINT RESIDUAL FOR EACH DATA POINT
600 PRINT "RESIDUAL FOR EACH DATA POINT"
610 PRINT "X","Y","LOG(Y)","RESIDUAL"
620 FOR I = 1 TO N
630    PRINT X(I),Y(I),Z(I),(Z(I)-A2)-S*(X(I)-A1)
640 NEXT I
650 REM SEE IF GRAPH IS REQUIRED
660 PRINT
670 PRINT "WOULD YOU LIKE A GRAPH?"
680 GOSUB 1170
690 IF Q$ ="NO" THEN 940
700 REM CALCULATE MIN M1 & MAX M2 VALUES OF LOG(Y)
710 LET M1 = Z(1)
720 LET M2 = Z(1)
730 FOR I = 2 TO N
740    IF Z(I) >= M1 THEN 760
750    LET M1 = Z(I)
760    IF Z(I) <= M2 THEN 780
770    LET M2 = Z(I)
780 NEXT I
790 REM CHECK IF ALL Y VALUES THE SAME & IF SO CANNOT PLOT GRAPH
800 IF M1 <> M2 THEN 840
810 PRINT "Y VALUES ALL THE SAME GIVING A HORIZONTAL LINE"
820 GOTO 940
830 REM PRINT GRAPH
840 PRINT
850 PRINT "TIME                                    LOG(CELLS)"
860 PRINT "(MINS)";TAB(12);M1;TAB(62);M2
870 PRINT TAB(12);
```

```
880 PRINT "+!------------------------------------------------!"
890 FOR I = 1 TO N
900    LET A = INT( ((Z(I) - M1) * 50) / (M2 - M1) + 0.5 )
910    PRINT X(I);TAB(12);"!";TAB(A + 13);"*"
920    PRINT TAB(12);"!"
930 NEXT I
940 PRINT
950 PRINT "WOULD YOU LIKE ANOTHER RUN?"
960 GOSUB 1170
970 PRINT
980 IF Q$ ="NO" THEN 1140
990 LET I$ = "NO"
1000 RESTORE
1010 PRINT "TYPE NEW FOR A RUN WITH COMPLETELY NEW DATA"
1020 PRINT "  OR OLD TO EDIT AND RERUN THE EXISTING DATA"
1030 INPUT Q$
1040 IF Q$ = "NEW" THEN 1110
1050 IF Q$ = "OLD" THEN 1080
1060 PRINT "REPLY '";Q$;"' NOT UNDERSTOOD"
1070 GOTO 1010
1080 READ C, S2, S3, X2, Y2, S4, E2
1090 GOSUB 1700
1100 GOTO 420
1110 PRINT "TYPE IN A NEW SET OF DATA"
1120 PRINT "==== == = === === == ===="
1130 GOTO 80
1140 PRINT "END OF PROGRAM"
1150 STOP
1160 REM ***** SUBROUTINE TO CHECK REPLIES *****
1170 IF I$ = "NO" THEN 1190
1180 PRINT "TYPE YES OR NO AND PRESS RETURN"
1190 INPUT Q$
1200 IF Q$ ="YES" THEN 1240
1210 IF Q$ ="NO"  THEN 1240
1220 PRINT "REPLY '"; Q$; "' NOT UNDERSTOOD. RE-";
1230 GOTO 1180
1240 RETURN
1250 REM ***** SUBROUTINE TO CALCULATE LEAST SQUARES FIT ETC *****
1260 REM ARRAYS X & Z CONTAIN N DATA POINTS
1270 REM CALCULATE SUMS OF X & Z VALUES
1280 FOR I = 1 TO N
1290    LET S2 = S2 + X(I)
1300    LET S3 = S3 + Z(I)
1310 NEXT I
1320 REM CALCULATE AVERAGE X & Z VALUES
1330 LET A1 = S2 / N
1340 LET A2 = S3 / N
1350 FOR I = 1 TO N
1360    LET X9 = X(I) - A1
1370    LET Y9 = Z(I) - A2
1380    LET X2 = X2 + X9 * X9
1390    LET Y2 = Y2 + Y9 * Y9
1400    LET S4 = S4 + X9 * Y9
1410 NEXT I
1420 PRINT
1430 IF X2 <> 0 THEN 1480
1440 PRINT "RUN TERMINATED BY THE PROGRAM BECAUSE ALL THE X"
1450 PRINT "COORDINATES ARE THE SAME, GIVING AN INFINITE SLOPE"
1460 GOTO 940
1470 REM CALCULATE THE SLOPE
1480 LET S = S4 / X2
```

```
1490 REM CALCULATE INTERCEPT ON Y AXIS
1500 LET Y1 = (X2 * S3 - S2 * S4) / (N * X2)
1510 IF Y2 = 0 THEN 1550
1520 REM CALCULATE CORRELATION COEFFICIENT
1530 LET C = S4 / SQR(X2 * Y2)
1540 REM CALCULATE SUM OF ERRORS SQUARED
1550 FOR I = 1 TO N
1560    LET E1 = (Z(I)-A2)-S*(X(I)-A1)
1570    LET E2 = E2 + E1 * E1
1580 NEXT I
1590 REM CALCULATE STANDARD DEVIATION
1600 LET S1 = SQR(E2 / (N - 2))
1610 IF N < 30 THEN 1630
1620 LET S1 = SQR(E2 / N)
1630 RETURN
1640 REM *****SUBROUTINE TO CHECK DATA ARE CORRECT & ALTER IF NECESSARY*
1650 PRINT "ARE THE DATA VALUES ENTERED CORRECT?";
1660 REM A4 SHOULD BE SET TO THE NUMBER OF LINES ON THE VDU
1670 LET A4 = 20
1680 GOSUB 1170
1690 IF Q$ = "YES" THEN 2480
1700 PRINT "HERE IS A LIST OF THE CURRENT DATA"
1710 PRINT "LINE NUMBER", "X", "Y"
1720 FOR I = 1 TO N
1730    PRINT I, X(I), Y(I)
1740    IF INT(I / (A4 - 1)) * (A4 - 1) <> I THEN 1780
1750    PRINT "WOULD YOU LIKE TO CONTINUE LISTING?";
1760    GOSUB 1170
1770    IF Q$ = "NO" THEN 1790
1780 NEXT I
1790 PRINT "TYPE R TO REPLACE";
1800 IF I$ = "NO" THEN 1820
1810 PRINT " AN EXISTING LINE OF DATA"
1820 IF N = 100 THEN 1870
1830 PRINT TAB(5); " A TO ADD";
1840 IF I$ = "NO" THEN 1860
1850 PRINT " AN EXTRA LINE"
1860 IF N = 1 THEN 1900
1870 PRINT TAB(5); " D TO DELETE";
1880 IF I$ = "NO" THEN 1900
1890 PRINT " AN EXISTING LINE"
1900 PRINT TAB(5); " L TO LIST";
1910 IF I$ = "NO" THEN 1930
1920 PRINT " THE DATA"
1930 PRINT "   OR C TO CONTINUE";
1940 IF I$ = "NO" THEN 1960
1950 PRINT " THE CALCULATION"
1960 INPUT Q$
1970 IF Q$ = "R" THEN 2070
1980 IF N = 100 THEN 2010
1990 IF Q$ = "A" THEN 2230
2000 IF N = 1 THEN 2020
2010 IF Q$ = "D" THEN 2320
2020 IF Q$ = "L" THEN 1700
2030 IF Q$ = "C" THEN 2480
2040 PRINT "REPLY '";Q$;"' NOT UNDERSTOOD."
2050 GOTO 1790
2060 REM REPLACE LINE
2070 PRINT "TYPE THE LINENUMBER OF THE LINE TO BE REPLACED";
2080 INPUT I
2090 IF I <> INT(I) THEN 2110
```

```
2100 IF (I - 1) * (I - N) <= 0 THEN 2140
2110 PRINT "LINENUMBER MUST BE AN INTEGER IN THE RANGE 1 -"; N
2120 PRINT "RE-";
2130 GOTO 2070
2140 PRINT "TYPE THE CORRECT LINE TO REPLACE THE ONE WHICH IS WRONG"
2150 PRINT "X, Y"
2160 INPUT X(I), Y(I)
2170 IF Y(I) > 0 THEN 2200
2180 PRINT "NUMBER OF CELLS MUST BE POSITIVE - RETYPE CORRECTLY"
2190 GOTO 2140
2200 LET Z(I) = LOG(Y(I)) / LOG(10)
2210 GOTO 2300
2220 REM ADD A NEW LINE
2230 LET N = N + 1
2240 PRINT "TYPE THE ADDITIONAL LINE OF DATA AS SHOWN;   X,Y"
2250 INPUT X(N), Y(N)
2260 IF Y(N) > 0 THEN 2290
2270 PRINT "NUMBER OF CELLS MUST BE POSITIVE - RETYPE CORRECTLY"
2280 GOTO 2240
2290 LET Z(N) = LOG(Y(N)) / LOG(10)
2300 PRINT "OK"
2310 GOTO 1790
2320 REM DELETE A LINE
2330 PRINT "TYPE THE LINENUMBER OF THE LINE TO BE DELETED"
2340 INPUT J
2350 IF (J - 1) * (J - N) > 0 THEN 2370
2360 IF J = INT(J) THEN 2390
2370 PRINT "LINENUMBER MUST BE AN INTEGER IN THE RANGE 1 -"; N
2380 GOTO 2330
2390 FOR I = J + 1 TO N
2400    LET X(I - 1) = X(I)
2410    LET Y(I - 1) = Y(I)
2420    LET Z(I - 1) = Z(I)
2430 NEXT I
2440 LET N = N - 1
2450 PRINT "OK"
2460 IF J > N THEN 1790
2470 GOTO 1700
2480 RETURN
2490 END
```

points could easily be deleted and a new line fitted to the remaining data. This is necessary because the slope of the line in the region of exponential growth is required to calculate the generation time.

Mean Single Survivor Time

There are many methods for testing the effectiveness of disinfectants or antibiotics against microorganisms. Biological systems are inherently subject to variation, and some of the classical methods of determining the effectiveness produce a very low level of reproducibility between different laboratories. The mean single survivor time is a newer and more reliable method of comparison which evaluates the time taken to reduce the population of microorganisms to one.

Background

Most classical methods for the evaluation of disinfectants and antibiotics involve the calculation of the phenol coefficient. This is simply the killing power of the disinfectant towards the bacterial organism under test, compared

176

with that of phenol under similar conditions. The Rideal–Walker test (British Standard Specification 541:1934) and the Chick–Martin test (British Standard Specification 808:1938) are both used for quantitative comparisons of different disinfectants.

In principle the bacterial culture is treated with several known concentrations of the disinfectant. After a specified exposure time of perhaps 30 min, a very small sample is removed from each of the test mixtures, using a sterile platinum loop, and is used to inoculate a tube or plate of nutrient medium. After incubation for 24 h, the cultures are examined to find the last sample which showed no growth, and the first which does show growth. The mean of the concentrations of disinfectant in these two cultures is calculated. The same procedure is followed using several different concentrations of phenol instead of the disinfectant. The phenol coefficient is obtained by dividing the mean of the phenol concentrations which do and do not show growth by the mean of the disinfectant concentrations which do and do not show growth.

Though widely used, these methods give variations of up to 360%. The main reasons for this lack of reproducibility are (i) the sampling method—the platinum loop does not deliver the same amount of culture each time, (ii) the organisms may clump together, hence the sample withdrawn may not be representative, and (iii) since one tube either grows or does not grow, the effect of one particularly resistant organism will have a marked effect on the result. The fact that the result is based on one single tube is plainly undesirable. The test is also criticised for its lack of realism in the test conditions, since comparison of water-soluble phenol with a non-water-soluble disinfectant which may require a detergent to dissolve it is comparing unlike materials. The choice of organisms used in the test is also criticised. Improvements in these phenol coefficient tests are incorporated in the Kelsey-Sykes test, but a more reliable extinction test is described in the following section.

The Mean Single Survivor Test

An extinction test based on the mean single survivor time was devised by H. Berry and H. S. Bean (*Journal of Pharmacy and Pharmacology*, 1954, **6**, 649) to determine bactericidal activity. The principle is to inoculate a culture of the organisms with an appropriate concentration of the bactericidal agent. After varying time intervals, several small samples of solution are withdrawn, grown on plates, and the number of surviving organisms in each sample estimated. The number of organisms surviving in each of the plates produced at a given time will vary, but the mean number of survivors λ decreases logarithmically with increased time of contact with the bactericidal agent.

$$\ln \lambda = -kT + c \qquad (1)$$

where T is the contact time, k is a positive constant, and c is a constant.

In practise it is difficult to estimate the number of surviving organisms in each sample, and the following procedure is generally adopted. The inoculated culture is split into a large number (for example 100) tubes. After a certain time, T, a batch of say 20 tubes is examined, and the number of tubes which show no growth, G, is counted. At subsequent times, further batches of tubes are withdrawn and examined in the same way. This method has the advantage that several replicate samples are used.

The value of λ is calculated from G using the Poisson distribution. If λ is the mean number of viable (surviving) organisms in each tube then the proportion of tubes with no living organisms is given by $e^{-\lambda}$ (see Chapter 4 and Chapter 5, Example 6). By observation, the proportion of tubes with no living organisms in each batch is, G, divided by the number of tubes in the batch

$$e^{-\lambda} = \frac{G}{\text{Batch size}}$$

rearranging

$$\lambda = -\ln(G/\text{Batch size}) \qquad (2)$$

substituting Equation 2 into Equation 1:

$$\ln[-\ln(G/\text{Batch size})] = -kT + c \quad (3)$$

$$y = mx + c$$

This equation has the form of a straight line, hence a graph of $\ln\{-\ln(G/\text{Batch size})\}$ against time, T, yields an approximate straight line, with slope $-k$ and intercept c. (But for random and experimental errors, an exact straight line would be obtained.) (See Fig. 10.3.)

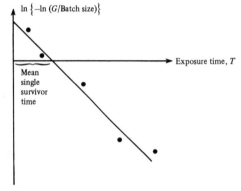

Fig. 10.3

The mean single survivor time is the time at which the mean number of organisms surviving in each tube is one, i.e. $\lambda = 1$. If $\lambda = 1$ then $\ln\{-\ln(G/\text{Batch size})\} = \ln(\lambda) = 0$. This corresponds to the point where the graph cuts the time axis. The mean single survivor time calculated in this way is generally more reliable than other extinction methods because it is based on all of the readings rather than on a single reading. The necessity for plotting a graph may be obviated if a computer is used.

The computer can fit the 'best' least-squares line through the points (see Chapter 9) and calculates the slope and the intercept on the ln–ln axis. The mean single survivor time may be calculated as $-\text{intercept}/\text{slope}$.

Example 1

A culture was inoculated with a known concentration of disinfectant, and batches of 20 tubes were taken at each of seven time intervals. These gave the results shown in Table 10.1. Plotting a graph as described above gives a mean single survivor time of 33 min.

Description of the Program (Program 10.2)

After printing a heading (lines 20–30) the user is asked (line 50) if full instructions are required. The answer which must be YES or NO is checked in a subroutine (lines 2170–2260), and full or abbreviated instructions given subsequently. The user is then requested (line 80) to type in the number of tubes tested in each batch. This value is checked (lines 110–140) to ensure that it is an integer, and in the range 2–50 inclusive. If the value is unacceptable, the user is requested to retype the value correctly (lines 150–160).

A message (lines 200–260) invites the user to type in up to 50 values for time and the number of tubes which showed no growth. Generally there are less than 50 data pairs, and the end of the data is indicated by typing the dummy values -1, -1. A subroutine

Table 10.1

Contact time, T (min)	20	30	40	50	60	70	80
Number of tubes with no growth, G	0	6	10	14	17	18	20
Fraction of tubes with no growth, F	excluded	0.30	0.50	0.70	0.85	0.90	excluded
$\ln\{-\ln(F)\}$	—	0.186	-0.367	-1.031	-1.817	-2.250	—

178

Program 10.2 Trial run.

```
MEAN SINGLE SURVIVOR CALCULATION
==== ====== ======== ===========

WOULD YOU LIKE FULL INSTRUCTIONS?
 TYPE YES OR NO & PRESS RETURN.

? YES
TYPE IN THE NUMBER OF TUBES TESTED IN EACH BATCH & PRESS RETURN

? 20

TYPE IN TIME (MIN.) AND NO. OF TUBES WHICH SHOWED
NO GROWTH, WITH A COMMA BETWEEN TERMS. THEN PRESS RETURN.
YOU WILL HAVE A CHANCE TO CORRECT WRONGLY TYPED DATA LATER
TERMINATE DATA WITH -1, -1

TIME, NUMBER OF TUBES WHICH SHOWED NO GROWTH
? 20, 0
READINGS WHERE ALL THE TUBES SHOW GROWTH, OR NO TUBES SHOW
GROWTH ARE NOT USED BY THE PROGRAM TO CALCULATE THE SLOPE
HENCE THIS PAIR OF VALUES IS IGNORED

? 30, 6
? 40, 10
? 50, 14
? 60, 17
? 70, 18
? 80, 20
READINGS WHERE ALL THE TUBES SHOW GROWTH, OR NO TUBES SHOW
GROWTH ARE NOT USED BY THE PROGRAM TO CALCULATE THE SLOPE
HENCE THIS PAIR OF VALUES IS IGNORED

? -1, -1
ARE THE DATA VALUES ENTERED CORRECT?  TYPE YES OR NO & PRESS RETURN.

? YES

STANDARD DEVIATION OF POINTS FROM LINE = 0.0929386
BASED ON 5 VALID DATA POINTS

MEAN SINGLE SURVIVOR TIME = 33.3 MINUTES

ALL BUT ONE BUGS DEAD - END OF JOB


10 DIM T(50), G(50), Y(50), Q$(10), I$(3)
20 PRINT TAB(5); "MEAN SINGLE SURVIVOR CALCULATION"
30 PRINT TAB(5); "==== ====== ======== ==========="
40 PRINT
50 PRINT "WOULD YOU LIKE FULL INSTRUCTIONS?"
60 GOSUB 2190
70 LET I$ = Q$
80 PRINT "TYPE THE NUMBER OF TUBES TESTED IN EACH BATCH & PRESS RETURN"
90 PRINT
100 INPUT N1
110 IF N1 = INT(N1) THEN 140
120 PRINT "THERE MUST BE A WHOLE NUMBER OF TUBES!"
130 GOTO 160
```

```
140 IF (N1 - 2) * (N1 - 50) <= 0 THEN 180
150 PRINT "NUMBER OF TUBES MUST BE IN THE RANGE 2 - 50"
160 PRINT "RETYPE CORRECT VALUE AND PRESS RETURN."
170 GOTO 90
180 PRINT
190 IF I$ = "YES" THEN 220
200 PRINT "TYPE IN THE DATA"
210 GOTO 250
220 PRINT "TYPE IN TIME (MIN.) AND NO. OF TUBES WHICH SHOWED"
230 PRINT "NO GROWTH, WITH A COMMA BETWEEN TERMS. THEN PRESS RETURN."
240 PRINT "YOU WILL HAVE A CHANCE TO CORRECT WRONGLY TYPED DATA LATER"
250 PRINT "TERMINATE DATA WITH -1, -1"
260 PRINT
270 PRINT "TIME, NUMBER OF TUBES WHICH SHOWED NO GROWTH"
280 LET N = 0
290 LET E1 = 0
300 FOR I = 1 TO 50
310   REM CALL SUBROUTINE TO INPUT AND CHECK A PAIR OF VALUES
320   GOSUB 1000
330   IF E1 = 1 THEN 390
340   LET N = N + 1
350 NEXT I
360 PRINT
370 PRINT "ARRAYS FULL - PROGRAM CAN HANDLE A MAXIMUM OF 50 POINTS"
380 REM CHECK THAT AT LEAST ONE VALID PAIR EXISTS
390 IF N > 0 THEN 430
400 PRINT "PLEASE ENTER SOME VALID DATA"
410 GOTO 290
420 REM CALL SUBROUTINE TO EDIT DATA IF NECESSARY
430 GOSUB 1320
440 REM CHECK THAT THERE ARE AT LEAST 3 VALID DATA POINTS
450 PRINT
460 IF N > 2 THEN 490
470 PRINT "JOB TERMINATED - NOT ENOUGH VALID DATA POINTS."
480 GOTO 2270
490 READ S2, S3, S4, T2, Y2
500 DATA 0, 0, 0, 0, 0
510 REM CALCULATE AVERAGE VALUES & WORK OUT LOG-LOG VALUES
520 FOR I = 1 TO N
530   LET P = G(I) / N1
540   LET Y(I) = LOG(-LOG(P))
550   LET S2 = S2 + T(I)
560   LET S3 = S3 + Y(I)
570 NEXT I
580 LET A1 = S2 / N
590 LET A2 = S3 / N
600 FOR I = 1 TO N
610   LET T9 = T(I) - A1
620   LET Y9 = Y(I) - A2
630   LET T2 = T2 + T9 * T9
640   LET Y2 = Y2 + Y9 * Y9
650   LET S4 = S4 + T9 * Y9
660 NEXT I
670 REM CHECK T-COORDINATES ARE NOT ALL THE SAME.
680 IF T2 <> 0 THEN 780
690 PRINT "RUN TERMINATED BY PROGRAM BECAUSE"
700 PRINT "THE TIME VALUES ARE ALL THE SAME"
710 GOTO 2270
720 REM CHECK THAT THE LN-LN VALUES ARE NOT ALL THE SAME
730 IF Y2 <> 0 THEN 780
740 PRINT "RUN TERMINATED BY THE PROGRAM BECAUSE"
```

```
750 PRINT "THE TUBE VALUES ARE ALL THE SAME"
760 GOTO 2270
770 REM WORK OUT COEFFICIENT FOR SLOPE OF STRAIGHT LINE THROUGH POINTS
780 LET S = S4 / T2
790 REM WORK OUT COEFFICIENT FOR INTERCEPT OF STRAIGHT LINE
800 LET Y1 = (T2 * S3 - S2 * S4) / (N * T2)
810 REM WORK OUT TIME AT LN.LN(P) = 0
820 LET T1 = -(T2 * S3 - S2 * S4) / (N * S4)
830 REM WORK OUT STANDARD DEVIATION
840 LET E2 = 0
850 FOR I = 1 TO N
860    LET E = (Y(I) - A2) - S * (T(I) - A1)
870    LET E2 = E2 + E * E
880 NEXT I
890 LET S1 = SQR(E2 / (N - 2))
900 IF N < 30 THEN 920
910 LET S1 = SQR(E2 / N)
920 PRINT "STANDARD DEVIATION OF POINTS FROM LINE ="; S1
930 PRINT "BASED ON"; N; "VALID DATA POINTS"
940 PRINT
950 PRINT "MEAN SINGLE SURVIVOR TIME ="; INT(T1*10+0.5)/10; "MINUTES"
960 PRINT
970 PRINT "ALL BUT ONE BUGS DEAD - END OF JOB"
980 GOTO 2270
990 REM ***** SUBROUTINE TO INPUT AND CHECK A PAIR OF DATA VALUES   *****
1000 INPUT T(I), G(I)
1010 IF E1 = 1 THEN 1050
1020 REM CHECK FOR TERMINATOR
1030 IF ABS(T(I) + 1) + ABS(G(I) + 1) = 0 THEN 1290
1040 REM CHECK IF DATA VALUES ARE IMPOSSIBLE
1050 IF T(I) >= 0 THEN 1080
1060 PRINT "TIME VALUE CANNOT BE NEGATIVE."
1070 GOTO 1160
1080 IF G(I) = INT(G(I)) THEN 1110
1090 PRINT "NUMBER OF TUBES MUST BE A WHOLE NUMBER!"
1100 GOTO 1160
1110 IF G(I) >= 0 THEN 1140
1120 PRINT "THE NUMBER OF TUBES CANNOT BE NEGATIVE!"
1130 GOTO 1160
1140 IF G(I) <= N1 THEN 1210
1150 PRINT "NUMBER OF TUBES IS TOO LARGE. THE MAXIMUM IS"; N1
1160 PRINT "LAST PAIR OF VALUES IGNORED - RETYPE CORRECTLY"
1170 PRINT
1180 GOTO 1000
1190 REM CHECK IF DATA POINT IS TO BE EXCLUDED FROM THE LEASTSQUARES
1200 REM FIT BECAUSE NO TUBES, OR ALL THE TUBES SHOWED NO GROWTH
1210 IF (G(I) - N1) * G(I) <> 0 THEN 1300
1220 PRINT "READINGS WHERE ALL THE TUBES SHOW GROWTH, OR NO TUBES SHOW"
1230 PRINT "GROWTH ARE NOT USED BY THE PROGRAM TO CALCULATE THE SLOPE"
1240 PRINT "HENCE THIS PAIR OF VALUES IS IGNORED"
1250 PRINT
1260 IF E1 = 0 THEN 1000
1270 LET E1 = 999
1280 GOTO 1300
1290 LET E1 = 1
1300 RETURN
1310 REM *****SUBROUTINE TO CHECK DATA ARE CORRECT & ALTER IF NECESSARY*
1320 PRINT "ARE THE DATA VALUES ENTERED CORRECT?";
1330 REM A4 SHOULD BE SET TO THE NUMBER OF LINES ON THE VDU
1340 LET A4 = 20
1350 GOSUB 2180
```

```
1360 IF Q$ = "YES" THEN 2160
1370 PRINT "HERE IS A LIST OF THE CURRENT DATA"
1380 PRINT "LINE NUMBER", "TIME", "NUMBER OF TUBES WITH NO GROWTH"
1390 FOR I = 1 TO N
1400    PRINT I, T(I), G(I)
1410    IF INT(I / (A4 - 1)) * (A4 - 1) <> I THEN 1450
1420    PRINT "WOULD YOU LIKE TO CONTINUE LISTING";
1430    GOSUB 2180
1440    IF Q$ = "NO" THEN 1460
1450 NEXT I
1460 PRINT "TYPE L TO LIST";
1470 IF I$ = "NO" THEN 1490
1480 PRINT " THE DATA"
1490 IF N = 50 THEN 1540
1500 PRINT TAB(5); " A TO ADD";
1510 IF I$ = "NO" THEN 1530
1520 PRINT " AN EXTRA LINE"
1530 IF N = 1 THEN 1600
1540 PRINT TAB(5); " D TO DELETE";
1550 IF I$ = "NO" THEN 1570
1560 PRINT " AN EXISTING LINE"
1570 PRINT TAB(5); " R TO REPLACE";
1580 IF I$ = "NO" THEN 1600
1590 PRINT " AN EXISTING LINE OF DATA"
1600 PRINT "  OR C TO CONTINUE";
1610 IF I$ = "NO" THEN 1630
1620 PRINT "THE CALCULATION"
1630 INPUT Q$
1640 IF Q$ = "L" THEN 1370
1650 IF N = 50 THEN 1680
1660 IF Q$ = "A" THEN 1900
1670 IF N = 1 THEN 1700
1680 IF Q$ = "D" THEN 2020
1690 IF Q$ = "R" THEN 1740
1700 IF Q$ = "C" THEN 2160
1710 PRINT "REPLY '";Q$;"' NOT UNDERSTOOD."
1720 GOTO 1460
1730 REM REPLACE LINE
1740 PRINT "TYPE THE LINENUMBER OF THE LINE TO BE REPLACED";
1750 INPUT I
1760 IF I <> INT(I) THEN 1780
1770 IF (I - 1) * (I - N) <= 0 THEN 1810
1780 PRINT "LINENUMBER MUST BE AN INTEGER IN THE RANGE 1 -"; N
1790 PRINT "RE-";
1800 GOTO 1740
1810 PRINT "TYPE THE CORRECT LINE TO REPLACE THE ONE WHICH IS WRONG;"
1820 PRINT "TIME, NUMBER OF TUBES WITH NO GROWTH"
1830 REM CALL SUBROUTINE TO INPUT & CHECK A PAIR OF VALUES
1840 GOSUB 1000
1850 IF E1 = 1 THEN 1990
1860 LET E1 = 1
1870 LET J = I
1880 GOTO 2080
1890 REM ADD A NEW LINE
1900 LET N = N + 1
1910 PRINT "TYPE THE ADDITIONAL LINE OF DATA AS SHOWN;"
1920 PRINT "TIME, NUMBER OF TUBES WITH NO GROWTH"
1930 REM CALL SUBROUTINE TO INPUT AND CHECK A PAIR OF VALUES
1940 LET I = N
1950 GOSUB 1000
1960 LET N = N - INT(E1 / 990)
```

```
1970 LET El = 1
1980 GOTO 1460
1990 PRINT "OK"
2000 GOTO 1460
2010 REM DELETE A LINE
2020 PRINT "TYPE THE LINENUMBER OF THE LINE TO BE DELETED"
2030 INPUT J
2040 IF (J - 1) * (J - N) > 0 THEN 2060
2050 IF J = INT(J) THEN 2080
2060 PRINT "LINENUMBER MUST BE AN INTEGER IN THE RANGE 1 -"; N
2070 GOTO 2020
2080 FOR I = J + 1 TO N
2090    LET T(I - 1) = T(I)
2100    LET G(I - 1) = G(I)
2110 NEXT I
2120 LET N = N - 1
2130 PRINT "OK"
2140 IF J > N THEN 1460
2150 GOTO 1370
2160 RETURN
2170 REM ***** SUBROUTINE TO CHECK REPLIES *****
2180 IF I$ = "NO" THEN 2200
2190 PRINT " TYPE YES OR NO & PRESS RETURN."
2200 PRINT
2210 INPUT Q$
2220 IF Q$ = "YES" THEN 2260
2230 IF Q$ = "NO" THEN 2260
2240 PRINT "REPLY '";Q$;"' NOT UNDERSTOOD.";
2250 GOTO 2190
2260 RETURN
2270 END
```

(lines 990–1300) is used to input and check each pair of data values to ensure that:

   (i) the time value is not negative,
  (ii) the number of tubes is a whole number,
 (iii) the number of tubes is positive or zero,
 (iv) the number of tubes does not exceed the batch size.

If the data are unacceptable on any of these four checks, a message explains why the pair of values has been rejected, and requests the user to retype the correct values.

In addition, readings where all of the tubes showed growth, or none of the tubes showed growth cannot be used, since the ln–ln term cannot be evaluated. Although such terms may be valid readings, they are ignored and an explanatory message is printed.

If 50 data pairs are entered, a message (line 370) explains that the arrays are full, and the calculation continues with the data already input. Generally there are less than 50 data entries. A check is performed (lines 380–410) to ensure that some valid data has been input, and if not, the user is requested to type in more data.

Next (lines 420–430) a subroutine is called (lines 1310–2160) which prints a message asking if the data values entered are correct. The reply to this must be either YES or NO, and is checked in another subroutine (lines 2170–2260). If the reply is NO the data are listed, and instructions are given explaining how to add, replace or delete lines of data, to re-list the data or to continue the calculation. This subroutine is broadly similar to the checking subroutine which is fully described in Chapter 8, but it contains modifications specific to this program to allow the detailed checking of any changed data.

Once the data are correct the program continues at lines 440–480 and checks that at least three valid points exist, since with only two points the program would fail through dividing by zero at line 890 when calculating the standard deviation.

The input G values are converted to ln–ln values, and the average time and ln–ln values are calculated (lines 510–590). The sums of a number of terms $\Sigma(T_i - \bar{T})^2$, $\Sigma(Y_i - \bar{Y})^2$ and $\Sigma(T_i - \bar{T})$ $(Y_i - \bar{Y})$ are collected (lines 600–660) for use in the least squares solution. (T_i are the time values, Y_i are the ln–ln values, $\bar{T}$ and $\bar{Y}$ are the mean of the T_i and Y_i values respectively.)

A check is then performed (lines 670–710) to ensure that all of the time values are not the same. This would correspond to a vertical straight line. Then a check is performed (lines 720–760) to ensure that all of the Y_i values are not the same. This would mean that all of the G values are the same, and would correspond to a horizontal straight line.

The slope of the best least squares line is calculated (line 780), and the intercept on the ln–ln axis is calculated (line 800). The mean single survivor time is calculated in line 820 and the standard deviation of the points from the line is calculated (lines 830–910). If there are less than 30 points the divisor used is $(n-2)$, otherwise the divisor is n, where n is the number of data pairs included in the least squares fit. The results are printed out (lines 920–950), followed by a finishing message (line 970).

11

Comparison of Regression Lines—Analysis of Variance

A commonly occurring situation in biological and other scientific work requires the comparison of two or more regression lines calculated from different sets of data. A decision is required as to whether they differ sufficiently to be distinct, or whether the differences are sufficiently small to allow the data to be pooled and treated as one common line.

Some examples requiring the comparison of entire lines are:

(1) comparison of the regression lines for crop yield against the amount of fertilizer applied for two different fertilizers,
(2) comparison of the regression lines for body weight against lung capacity for male and female caucasians aged between 20 and 25.

In both of the above examples the lines are to be considered different if either the slopes are significantly different, or if the slopes are similar but the heights of the lines are significantly different.

In other cases only the comparison of the slopes of the lines is of interest. This is usually the case with studies of enzyme kinetics, and is always the case with growth curves of microorganisms, where the height of the curve depends on the number of organisms present in the culture. Furthermore, in these cases, the height or the curve at a given time may also be affected by an initial lag phase. Both the comparison of entire lines and the comparison of slopes are treated by analysis of variance, in which the sums of residuals squared are

calculated in two ways and compared using an F-test.

Comparison of Two Regression Lines

In cases where there are only two lines, a t-test may be used to compare them. If there are more than two lines then they should not be compared by the t-test for the reasons explained at the beginning of Chapter 7. In these cases an analysis of variance using an F-test is performed as described later.

The method of comparing two lines using a t-test is as follows.

(1) Calculate the slopes of the two individual lines.
(2) Pool all of the data and calculate the standard error of the slope of the common line using Equation 14 from Chapter 9.
(3) Calculate t as the difference in slopes from (1) divided by the standard error from (2).
(4) The calculated t-value is compared with a significance table for t-values (Appendix 6), using a two-tailed probability. The number of degrees of freedom v is equal to the total number of points minus four. If the calculated value exceeds, for example, the 5% table value, then there is less than a 5% probability that lines so different could occur by chance if the lines should really be the same. This would provide

some evidence that the lines are different.

A *t*-test is valid only if the data for the two lines are sufficiently similar to be pooled, and is never valid if there are more than two lines. A more general comparison of lines is provided by analysis of variance using an *F*-test. This test compares the lines to determine whether they differ so significantly that the data cannot legitimately be pooled.

Comparison of Two or More Regression Lines

For simplicity in the following discussion only two lines will be compared, but the method is completely general and any number of lines may be compared.

  (i) A least squares line is calculated for the first set of data, giving coefficients for the slope and intercept. The slope so obtained is substituted into the equation $y - \bar{y} = m(x - \bar{x})$ to give calculated y values for each x value. (This gives greater numerical accuracy than $y = mx + c$.) The differences between the calculated y values and the observed y values are called the residuals, and the sum of all the residuals squared is collected.

 (ii) The steps outlined in (i) are repeated for the second (and subsequent) set(s) of data.

(iii) The sum of the residuals squared from all of the individual lines is calculated.

(iv) The number of degrees of freedom arising from the individual lines is calculated as (number of points in line 1–2)+(number of points in line 2–2)+ . . . .

 (v) The sum of residuals squared from (iii) is divided by the number of degrees of freedom from (iv) to give a mean square. This corresponds to the amount of random experimental error in each measurement, and is

thus a measure of the precision of the experiment.

 (vi) Then a single least squares line is calculated for the pooled data comprising all of the data for all of the individual lines. The slope, intercept and sum of residuals squared are calculated as in (i).

(vii) The difference between the sum of residuals squared for a single line (vi) and the individual lines (iii) is calculated. This provides a measure of how much worse a single line fits all of the data than the several individual lines fit the separate data.

(viii) The number of degrees of freedom for the difference (vii) is calculated as (number of individual lines−1). In the case of comparing two lines, this number of degrees of freedom is one, and step (ix) can be omitted.

 (ix) The difference in residuals squared from (vii) is divided by the number of degrees of freedom from (viii), to give another mean square.

  (x) The ratio of the mean square from (ix) divided by the mean square from (v) is evaluated to give an *F* value. This *F* value is compared with tables of *F* values (Appendix 7). If the calculated value exceeds the table value for say the 5% level of significance, then there is less than a 5% chance that a difference between the lines such as this could have arisen by chance if the data really belong to one straight line. This suggests that the differences are systematic not random, and that the data should not be pooled.

Example 1

An agricultural research station compared the effect of two different fertilizers on the crop yield and obtained the results given in Table 11.1. The problem is to determine whether the two fertilizers produce significantly different crop yields.

186

Table 11.1

Amount of fertilizer applied (ounces per square yard)	Crop yield (cwt per acre)	
	Fertilizer 1	Fertilizer 2
0	160	162
1	165	168
2	172	174
3	178	180
4	181	181
5	186	185
6	188	187
7	189	186
8	192	188

(i) A least squares line is fitted to the data for fertilizer 1 and the slope is found to be 4.0, and the intercept 163.0. Using these the sum of residuals squared is calculated to be 50.0.

(ii) A least squares line for fertilizer 2 data has a slope of 3.15, an intercept of 166.4 and a sum of residuals squared of 74.65.

(iii) The sum of the residuals squared from both the separate lines is calculated:

$$50.0 + 74.65 = 124.65$$

(iv) The number of degrees of freedom arising from the individual lines is calculated:

$$(9-2) + (9-2) = 14$$

(v) The sum of residuals squared is divided by the number of degrees of freedom to give a mean square

$$124.65/14 = 8.904$$

(vi) A least squares line is fitted to the pooled data, and gives a slope of 3.575, an intercept of 164.7 and a sum of residuals squared of 146.325.

(vii) The difference in residuals squared between the pooled line and the separate lines is calculated:

$$146.325 - 124.65 = 21.675$$

(viii) The number of degrees of freedom for this difference is

$$2 - 1 = 1$$

(ix) The difference in residuals squared is divided by the number of degrees of freedom to give a mean square

$$21.675/1 = 21.675$$

(x) F is calculated as

$$21.675/8.904 = 2.434.$$

This value should be compared with the significance values for F given in Appendix 7 with the number of degrees of freedom $v_1 = 1$ and $v_2 = 14$:

Significance level	Table F value
10%	3.10
5%	4.60
1%	8.86

This calculated value of $F = 2.434$ is seen to be less than the 10% table value, and hence there is more than a 10% chance that lines so different could occur by chance if the two fertilizers were equally good. Thus one cannot conclude that the fertilizers are significantly different.

Description of Program to Compare Several Straight Lines

The program (Program 11.1) prints a heading (lines 20–30) and then asks (line 50) if full instructions are required. The answer must be YES or NO, and is checked in a subroutine (lines 880–970). Full or abbreviated instructions are printed as appropriate for the first run, but the instructions are always shortened for a second or subsequent run, or after using the subroutine to edit the data.

Program 11.1   Trial run.

```
PROGRAM TO COMPARE SEVERAL STRAIGHT LINES
======= == ======= ======= ======== =====

WOULD YOU LIKE FULL INSTRUCTIONS?   TYPE YES OR NO & PRESS RETURN.

? YES

THIS PROGRAM COMPARES FROM 2 TO 8 LINES TO DETERMINE WHETHER
THEY DIFFER SIGNIFICANTLY, OR WHETHER THE DATA MAY BE POOLED

HOW MANY LINES WOULD YOU LIKE TO COMPARE
? 2
TYPE IN X & Y VALUES SEPARATED BY A COMMA
THEN PRESS RETURN AND TYPE THE NEXT PAIR OF VALUES
YOU WILL BE GIVEN THE CHANCE TO EDIT INCORRECT DATA LATER

INPUT DATA FOR LINE 1
TERMINATE DATA WITH BOTH X & Y EQUAL TO 999999
 X,Y
? 0,160
? 1,165
? 2,172
? 3,178
? 4,181
? 5,186
? 6,188
? 7,189
? 8,192
? 999999,999999
ARE THE DATA VALUES ENTERED CORRECT?   TYPE YES OR NO & PRESS RETURN.

? YES


RESULTS OF CALCULATION FOR LINE 1
------- -- ----------- --- ---- -

NUMBER OF VALID DATA POINTS = 9
SLOPE  = 4
INTERCEPT ON Y AXIS  = 163
SUM OF DIFFERENCES SQUARED  = 50

WOULD YOU LIKE A LIST OF DIFFERENCES?   TYPE YES OR NO & PRESS RETURN.

? NO

INPUT DATA FOR LINE 2
TERMINATE DATA WITH BOTH X & Y EQUAL TO 999999
 X,Y
? 0,162
? 1,168
? 2,174
? 3,180
? 4,181
? 5,185
? 6,187
? 7,186
? 8,188
? 999999,999999
ARE THE DATA VALUES ENTERED CORRECT?   TYPE YES OR NO & PRESS RETURN.

? YES
```

188

```
RESULTS OF CALCULATION FOR LINE 2
------- -- ----------- --- ---- -

NUMBER OF VALID DATA POINTS = 9
SLOPE  = 3.15
INTERCEPT ON Y AXIS  = 166.4
SUM OF DIFFERENCES SQUARED  = 74.65


WOULD YOU LIKE A LIST OF DIFFERENCES?  TYPE YES OR NO & PRESS RETURN.

? NO


RESULTS OF CALCULATION FOR THE COMBINED LINE
------- -- ----------- --- --- -------- ----

NUMBER OF VALID DATA POINTS = 18
SLOPE  = 3.575
INTERCEPT ON Y AXIS  = 164.7
SUM OF DIFFERENCES SQUARED  = 146.325


WOULD YOU LIKE A LIST OF DIFFERENCES?  TYPE YES OR NO & PRESS RETURN.

? YES


DIFFERENCE IN Y VALUE FOR EACH DATA POINT
X            Y OBS        Y CALC        DIFFERENCE
0            160          164.7         -4.7
1            165          168.275       -3.275
2            172          171.85        0.15
3            178          175.425       2.575
4            181          179           2
5            186          182.575       3.425
6            188          186.15        1.85
7            189          189.725       -0.725
8            192          193.3         -1.3
0            162          164.7         -2.7
1            168          168.275       -0.275001
2            174          171.85        2.15
3            180          175.425       4.575
4            181          179           2
5            185          182.575       2.425
6            187          186.15        0.85
7            186          189.725       -3.725
8            188          193.3         -5.3


F VALUE = 2.43441
NUMBER OF DEGREES OF FREEDOM: NU(1)= 1    NU(2)= 14


PROBABILITY THAT SUCH A DIFFERENCE IN VARIANCES COULD OCCUR
BY CHANCE IF THE PARENT POPULATIONS HAVE THE SAME VARIANCE
IS 14.1 %


WOULD YOU LIKE ANOTHER RUN?  TYPE YES OR NO & PRESS RETURN.

? NO
END OF JOB
```

```
10 DIM Q$(10), I$(3), X(400), Y(400)
20 PRINT "PROGRAM TO COMPARE SEVERAL STRAIGHT LINES"
30 PRINT "======= == ======= ======= ======== ====="
40 PRINT
50 PRINT "WOULD YOU LIKE FULL INSTRUCTIONS?";
60 REM CALL SUBROUTINE TO OBTAIN AN ANSWER OF YES OR NO
70 GOSUB   900
80 LET I$ = Q$
90 IF Q$ = "NO" THEN 130
100 PRINT
110 PRINT "THIS PROGRAM COMPARES FROM 2 TO 8 LINES TO DETERMINE WHETHER"
120 PRINT "THEY DIFFER SIGNIFICANTLY, OR WHETHER THE DATA MAY BE POOLED"
130 PRINT
140 PRINT "HOW MANY LINES WOULD YOU LIKE TO COMPARE"
150 INPUT L
160 IF L <> INT(L) THEN 180
170 IF (L - 1) * (L - 9) < 0 THEN 200
180 PRINT "NUMBER OF LINES MUST BE AN INTEGER BETWEEN 2 & 8"
190 GOTO   140
200 LET E3 = 0
210 LET V2 = 0
220 LET N1 = 0
230 LET V1 = L - 1
240 REM START LOOP TO PERFORM LEASTSQUARES FOR SEPARATE & COMBINED LINES
250 FOR J1 = 1 TO L + 1
260    REM SET INITIAL VALUES TO ZERO
270    READ S2, S3, X2, S4, E2
280    DATA 0, 0, 0, 0, 0
290    REM CALL SUBROUTINE TO INPUT A SET OF N (X,Y) DATA VALUES
300    IF J1 > L THEN 410
310    GOSUB   990
320    REM CALL SUBROUTINE TO CHECK AND ALTER DATA IF NECESSARY
330    GOSUB  1550
340    PRINT
350    IF N >= 2 THEN 410
360    PRINT
370    PRINT "RUN ON THIS DATA ABANDONED BECAUSE THERE ARE"
380    PRINT "NOT ENOUGH DATA POINTS"
390    GOTO   760
400    REM CALL SUBROUTINE TO WORK OUT LEAST-SQUARES FIT ETC.
410    GOSUB  1240
420    REM CHECK FOR ERROR FLAGGED BY SUBROUTINE
430    IF N < 0 THEN 760
440    REM CALC. SUM OF DIFFERENCES SQUARED E3 & DEGREES OF FREEDOM V2
450    REM FOR THE INDIVIDUAL LINES
460    IF J1 = L + 1 THEN 500
470    LET E3 = E3 + E2
480    LET V2 = V2 + (N - 2)
490    REM CALL SUBROUTINE TO PRINT OUT RESULTS OF REGRESSION CALCS.
500    GOSUB  2340
510    LET N1 = N + N1
520    IF J1 <> L THEN 570
530    REM RESET NO. OF POINTS & STARTING PLACE IN ARRAYS PRIOR TO
540    REM FITTING THE COMBINED LINE
550    LET N = N1
560    LET N1 = 0
570    RESTORE
580 NEXT J1
590 REM CALCULATE F IF DATA PERMITS
600 IF V2 > 0 THEN 640
610 PRINT "RUN TERMINATED - ALL THE INDIVIDUAL LINES HAVE ONLY TWO"
```

```
620 PRINT "POINTS, AND HENCE FIT EXACTLY"
630 GOTO 760
640 IF E3 > 0 THEN 680
650 PRINT "RUN TERMINATED - ALL OF THE INDIVIDUAL LINES PASS EXACTLY"
660 PRINT "THROUGH THEIR DATA"
670 GOTO 760
680 LET F = ((E2 - E3) / V1) / (E3 / V2)
690 PRINT
700 PRINT "F VALUE ="; F
710 PRINT "NUMBER OF DEGREES OF FREEDOM; NU(1)="; V1; "   NU(2)="; V2
720 PRINT
730 REM CALL SUBROUTINE TO WORK OUT PROBABILITY FOR THIS F
740 GOSUB 2590
750 REM DECIDE WHETHER TO FINISH OR HAVE ANOTHER RUN
760 PRINT
770 PRINT "WOULD YOU LIKE ANOTHER RUN?";
780 REM CALL SUBROUTINE TO OBTAIN AN ANSWER OF YES OR NO
790 GOSUB 890
800 IF Q$ = "NO" THEN 860
810 PRINT "TYPE IN A NEW SET OF DATA"
820 PRINT "==== == = === === == ===="
830 LET I$ = "NO"
840 GOTO 140
850 REM TERMINATE JOB
860 PRINT "END OF JOB"
870 STOP
880 REM ***** SUBROUTINE TO CHECK REPLIES *****
890 IF I$ = "NO" THEN 910
900 PRINT " TYPE YES OR NO & PRESS RETURN."
910 PRINT
920 INPUT Q$
930 IF Q$ = "YES" THEN 970
940 IF Q$ = "NO" THEN 970
950 PRINT "REPLY '"; Q$; "' NOT UNDERSTOOD.";
960 GOTO 900
970 RETURN
980 REM ***** SUBROUTINE TO INPUT A SET OF N (X,Y) DATA PAIRS *****
990 IF I$ = "NO" THEN 1040
1000 IF J1 > 1 THEN 1040
1010 PRINT "TYPE IN X & Y VALUES SEPARATED BY A COMMA"
1020 PRINT "THEN PRESS RETURN AND TYPE THE NEXT PAIR OF VALUES"
1030 PRINT "YOU WILL BE GIVEN THE CHANCE TO EDIT INCORRECT DATA LATER"
1040 PRINT
1050 PRINT "INPUT DATA FOR LINE"; J1
1060 PRINT "TERMINATE DATA WITH BOTH X & Y EQUAL TO 999999"
1070 PRINT " X,Y"
1080 REM SET NUMBER OF POINTS TO ZERO
1090 LET N = 0
1100 REM READ IN X & Y FOR EACH DATA POINT & STORE IN ARRAYS
1110 FOR I = N1 + 1 TO N1 + 50
1120    INPUT X(I), Y(I)
1130    IF ABS(X(I) - 999999) + ABS(Y(I) - 999999) = 0 THEN 1170
1140    LET N = N + 1
1150 NEXT I
1160 PRINT "PROGRAM CAN ONLY HANDLE A MAXIMUM OF 50 VALUES IN EACH LINE
1170 IF N > 0 THEN 1200
1180 PRINT "PLEASE TYPE IN SOME DATA VALUES FOR LINE"; J
1190 GOTO 1110
1200 RETURN
1210 REM ***** SUBROUTINE TO CALCULATE LEAST SQUARES FIT, ETC.*****
1220 REM ARRAYS X & Y CONTAIN N DATA POINTS
```

```
1230 REM CALCULATE SUMS OF X & Y VALUES
1240 FOR I = N1 + 1 TO N1 + N
1250   LET S2 = S2 + X(I)
1260   LET S3 = S3 + Y(I)
1270 NEXT I
1280 REM CALCULATE THE AVERAGE X & Y VALUES
1290 LET A1 = S2 / N
1300 LET A2 = S3 / N
1310 FOR I = N1 + 1 TO N1 + N
1320   LET X9 = X(I) - A1
1330   LET Y9 = Y(I) - A2
1340   LET X2 = X2 + X9 * X9
1350   LET S4 = S4 + X9 * Y9
1360 NEXT I
1370 PRINT
1380 IF X2 <> 0 THEN 1450
1390 PRINT "RUN TERMINATED BY THE PROGRAM BECAUSE ALL"
1400 PRINT "THE X-COORDINATES ARE THE SAME FOR LINE"; J
1410 REM SET ERROR FLAG
1420 LET N = -1
1430 GOTO  1530
1440 REM CALCULATE SLOPE
1450 LET S = S4 / X2
1460 REM CALCULATE INTERCEPT ON Y AXIS
1470 LET Y1 = (X2 * S3 - S2 * S4) / (N * X2)
1480 REM CALCULATE SUM OF DIFFERENCES SQUARED
1490 FOR I = N1 + 1 TO N1 + N
1500   LET E1 = (Y(I) - A2) - S * (X(I) - A1)
1510   LET E2 = E2 + E1 * E1
1520 NEXT I
1530 RETURN
1540 REM *****SUBROUTINE TO CHECK DATA ARE CORRECT & ALTER IF NECESSARY*
1550 PRINT "ARE THE DATA VALUES ENTERED CORRECT?";
1560 REM A4 SHOULD BE SET TO THE NUMBER OF LINES ON THE VDU
1570 LET A4 = 20
1580 REM CALL SUBROUTINE TO OBTAIN ANSWER OF YES OR NO
1590 GOSUB  890
1600 IF Q$ = "YES" THEN 2320
1610 PRINT "HERE IS A LIST OF THE CURRENT DATA"
1620 PRINT "LINE NUMBER", "X", "Y"
1630 FOR I = 1 TO N
1640   PRINT I, X(N1 + I), Y(N1 + I)
1650   IF INT(I / (A4 - 1)) * (A4 - 1) <> I THEN 1700
1660   PRINT "WOULD YOU LIKE TO CONTINUE LISTING?";
1670   REM CALL SUBROUTINE TO OBTAIN ANSWER OF YES OR NO
1680   GOSUB  890
1690   IF Q$ = "NO" THEN 1710
1700 NEXT I
1710 PRINT "TYPE R TO REPLACE";
1720 IF I$ = "NO" THEN 1740
1730 PRINT " AN EXISTING LINE OF DATA"
1740 IF N = 50 THEN 1790
1750 PRINT TAB(5); " A TO ADD";
1760 IF I$ = "NO" THEN 1780
1770 PRINT " AN EXTRA LINE"
1780 IF N = 1 THEN 1820
1790 PRINT TAB(5); " D TO DELETE";
1800 IF I$ = "NO" THEN 1820
1810 PRINT " AN EXISTING LINE"
1820 PRINT TAB(5); " L TO LIST";
1830 IF I$ = "NO" THEN 1850
```

```
1840 PRINT " THE DATA"
1850 PRINT "  OR C TO CONTINUE";
1860 IF I$ = "NO" THEN 1890
1870 PRINT " THE CALCULATION"
1880 LET I$ = "NO"
1890 INPUT Q$
1900 IF Q$ = "R" THEN 2000
1910 IF N = 50 THEN 1940
1920 IF Q$ = "A" THEN 2120
1930 IF N = 1 THEN 1950
1940 IF Q$ = "D" THEN 2170
1950 IF Q$ = "L" THEN 1610
1960 IF Q$ = "C" THEN 2320
1970 PRINT "REPLY '"; Q$; "' NOT UNDERSTOOD."
1980 GOTO  1710
1990 REM REPLACE LINE
2000 PRINT "TYPE THE LINENUMBER OF THE LINE TO BE REPLACED";
2010 INPUT I
2020 IF I <> INT(I) THEN 2040
2030 IF (I - 1) * (I - N) <= 0 THEN 2070
2040 PRINT "LINENUMBER MUST BE AN INTEGER IN THE RANGE 1 -"; N
2050 PRINT "RE-";
2060 GOTO  2000
2070 PRINT "TYPE THE CORRECT LINE TO REPLACE THE ONE WHICH IS WRONG"
2080 PRINT "X, Y"
2090 INPUT X(N1 + I), Y(N1 + I)
2100 GOTO  2150
2110 REM ADD A NEW LINE
2120 LET N = N + 1
2130 PRINT "TYPE THE ADDITIONAL LINE OF DATA AS SHOWN;   X,Y"
2140 INPUT X(N1 + N), Y(N1 + N)
2150 PRINT "OK"
2160 GOTO  1710
2170 REM DELETE A LINE
2180 PRINT "TYPE THE LINENUMBER OF THE LINE TO BE DELETED"
2190 INPUT J
2200 IF (J - 1) * (J - N) > 0 THEN 2220
2210 IF J = INT(J) THEN 2240
2220 PRINT "LINENUMBER MUST BE AN INTEGER IN THE RANGE 1 -"; N
2230 GOTO  2180
2240 FOR I = N1 + J + 1 TO N1 + N
2250    LET X(I - 1) = X(I)
2260    LET Y(I - 1) = Y(I)
2270 NEXT I
2280 LET N = N - 1
2290 PRINT "OK"
2300 IF J > N THEN 1710
2310 GOTO  1610
2320 RETURN
2330 REM ***** SUBROUTINE TO PRINT RESULTS OF REGRESSION *****
2340 IF J1 = L + 1 THEN 2380
2350 PRINT "RESULTS OF CALCULATION FOR LINE"; J1
2360 PRINT "------- -- ----------- --- ---- -"
2370 GOTO  2400
2380 PRINT "RESULTS OF CALCULATION FOR THE COMBINED LINE"
2390 PRINT "------- -- ----------- --- --- -------- ----"
2400 PRINT
2410 PRINT "NUMBER OF VALID DATA POINTS ="; N
2420 PRINT "SLOPE  ="; S
2430 PRINT "INTERCEPT ON Y AXIS  ="; Y1
2440 PRINT "SUM OF DIFFERENCES SQUARED  ="; E2
```

```
2450 PRINT
2460 PRINT "WOULD YOU LIKE A LIST OF DIFFERENCES?";
2470 REM CALL SUBROUTINE TO OBTAIN AN ANSWER OF YES OR NO
2480 GOSUB   890
2490 IF Q$ = "NO" THEN 2570
2500 REM CALCULATE & PRINT DIFFERENCE FOR EACH DATA POINT
2510 PRINT
2520 PRINT "DIFFERENCE IN Y VALUE FOR EACH DATA POINT"
2530 PRINT "X", "Y OBS", "Y CALC", "DIFFERENCE"
2540 FOR I = 1 TO N
2550 PRINT X(N1+I),Y(N1+I),A2+S*(X(N1+I)-A1),(Y(N1+I)-A2)-S*(X(N1+I)-A1)
2560 NEXT I
2570 RETURN
2580 REM ***** SUBROUTINE TO CALCULATE F PROBABILITY *****
2590 LET E = 0
2600 IF V1 = 2 * INT(V1 / 2 + 0.1) THEN 2640
2610 IF V2 = 2 * INT(V2 / 2 + 0.1) THEN 2690
2620 GOTO   2870
2630 REM CALCULATE PROBABILITY IF V1 EVEN
2640 LET U = 1 / (1 + V2 / (F * V1))
2650 LET P1 = V1 + 1
2660 LET Q = V2 - 2
2670 GOTO   2730
2680 REM CALCULATE PROBABILITY IF V2 EVEN
2690 LET E = 1
2700 LET U = 1 / (1 + F * V1 / V2)
2710 LET P1 = V2 + 1
2720 LET Q = V1 - 2
2730 LET S = 0
2740 LET W = 1
2750 LET K = 2
2760 LET S = S + W
2770 LET W = W * U * (K + Q) / K
2780 LET K = K + 2
2790 IF K < P1 THEN 2760
2800 LET Z = SQR(1 - U)
2810 IF E = 0 THEN 2840
2820 LET P = 100 * S * (Z ^ V1)
2830 GOTO   3100
2840 LET P = 100 * S * (Z ^ V2)
2850 GOTO   3100
2860 REM CALCULATE PROBABILITY IF V1 & V2 BOTH ODD
2870 LET U = 1 / (1 + F * V1 / V2)
2880 LET X = 1 - U
2890 LET S = 0
2900 LET W = 1
2910 LET K = 2
2920 LET P1 = V2
2930 GOTO   2970
2940 LET S = S + W
2950 LET W = W * U * K / (K + 1)
2960 LET K = K + 2
2970 IF K < P1 THEN 2940
2980 LET W = W * V2
2990 LET K = 3
3000 LET P1 = V1 + 1
3010 LET Q = V2 - 2
3020 GOTO   3060
3030 LET S = S - W
3040 LET W = W * X * (K + Q) / K
3050 LET K = K + 2
```

```
3060 IF K < P1 THEN 3030
3070 LET T1 = ATN(SQR(F * V1 / V2))
3080 LET R1 = S * SQR(X * U)
3090 LET P = 100 * (1 - 2 * (T1 + R1) / 3.14159)
3100 IF P <= 50 THEN 3130
3110 LET P = 100 - P
3120 REM ROUND OFF ANSWER
3130 IF P < 5 THEN 3150
3140 LET P = INT(P * 10 + 0.5) * 0.1
3150 IF (P - 5) * (P - 0.5) > 0 THEN 3170
3160 LET P = INT(P * 100 + 0.5) * 0.01
3170 IF P > 0.05 THEN 3190
3180 LET P = INT(P * 1000 + 0.5) * 0.001
3190 PRINT "PROBABILITY THAT SUCH A DIFFERENCE IN VARIANCES COULD OCCUR"
3200 PRINT "BY CHANCE IF THE PARENT POPULATIONS HAVE THE SAME VARIANCE"
3210 IF P < 0.01 THEN 3240
3220 PRINT "IS"; P; "%"
3230 GOTO 3250
3240 PRINT "IS LESS THAN 0.01 %"
3250 PRINT
3260 RETURN
3270 END
```

The user is asked to type the number of lines to be compared (lines 140–150), and this number is checked to ensure that it is an integer between 2 and 8 (lines 160–190). The upper limit of 8 lines is an empirical limit set by the program, and some information on how to change this is given later.

A loop to perform the main calculations extends from lines 250 to 580, and is executed once for each of the separate lines and once more for the combined lines. The steps carried out in the loop are as follows:

1. A number of totals are set to zero (lines 260–280).
2. A subroutine (lines 980–1200) is called to input up to 50 (x,y) data pairs for each line. The input of data for each line is terminated by typing a dummy pair of values of 999999, 999999. In the subroutine the number of valid data pairs is counted. If no values are input, an error message is printed and a request made to input some data.
3. Next a subroutine (lines 1540–2320) is called to check and edit the data for the line just entered. This allows the data to be listed, existing values to be deleted or replaced, or additional lines of data to be added. A more detailed description of this subroutine is given in Chapter 8.
4. A check is then carried out (lines 350–390) to verify that at least two data points exist before attempting to fit a straight line.
5. A subroutine (lines 1210–1530) is called to fit a least squares line to the data. The slope and the sum of differences squared are usually calculated, but an error flag is set if all the x values are the same since this corresponds to a vertical line (with an infinite slope). On returning from the subroutine the run is aborted (lines 420–430) if this error flag has been set.
6. The total of the sums of differences squared for the individual lines, and the sum of the number of degrees of freedom v_2 for the individual lines are collected (lines 470–480).
7. A subroutine (lines 2330–2570) is called to print the results of the least squares calculation.

After fitting the separate least squares lines and the combined line, two checks are performed, the value of F is calculated and printed, and the number of degrees of freedom, v_1 and v_2, are printed (lines 590–710).

The first check (lines 600–630) is to make sure that the degrees of freedom v_2 exceeds zero; that is, at least one of the individual lines has more than two points. The second check (lines 640–670) is to make sure that the sum of the individual sums of differences squared is not zero, since this corresponds to all of the individual lines passing exactly through their data—giving no error.

Following this, a subroutine (lines 2580–3260) is called to calculate the probability of such an F value. This is similar to the program for F and t-tests in comparison of two samples (Chapter 6). Underflow and overflow problems may be encountered on some computers with this calculation if the number of degrees of freedom is large, and the range of numbers on the computer is small.

Finally the user is asked if another run is required (lines 770–800).

As provided, the program can handle up to 50 points in each of eight separate lines, and this accounts for the DIMension of the X and Y arrays in line 10. If the user wishes to change the sizes 50 or 8 then the following lines may need attention: 10, 110, 170, 180, 1110, 1160, 1740, and 1910.

Comparison of Slopes (Regression Coefficients) for Two or More Lines

This is a general method by which the slopes of any number of lines may be compared using analysis of variance. Since only the slopes of the lines are of interest, and not their heights or lateral positions, each line is translated so that their mid points coincide. This is achieved by subtracting the mean values $\bar{x}_1$ and $\bar{y}_1$ from the data values for line one, subtracting $\bar{x}_2$ and $\bar{y}_2$ from the data for line two, and so on. The effect of this is shown in Fig. 11.1.

The steps in the calculation are identical to those outlined (steps (i) to (x)) in the previous section for comparing two or more regression lines except for step (vi):

> (vi) A least squares line is calculated for the pooled *modified* data, and the sum of residuals squared calculated.

These steps are expanded in Example 2.

Example 2—Growth Curves of *E. coli*

An agar slope of *E. coli* was used to inoculate a 200 ml sample of sterile nutrient broth, which was grown overnight in an orbital incubator.

100 ml of the overnight sample was taken, and its absorbance measured (at 650 nm in a spectrophotometer, against a blank of sterile nutrient broth) every 10 min. This gives a measure of the number of bacteria present.

The remaining 100 ml of overnight sample was divided into two equal portions. One 50 ml portion was autoclaved to kill the cells, cooled to room temperature, and re-mixed with the other 50 ml portion. Growth curve data for this modified culture was obtained from absorbance measurements at 10 min intervals as before.

To obtain growth curves for the two sets of data, it is necessary to plot straight lines through log (absorbance) against time.

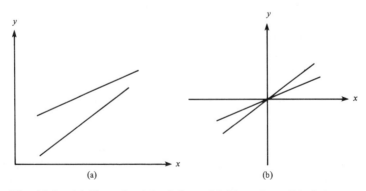

Fig. 11.1   (a) Plot of original data. (b) Plot of modified data.

Intuitively one would expect the sample with all live bacteria to grow more rapidly than the sample with half the bacteria dead. If, however, the rate of growth depends only on the number of live bacteria, the slopes of the two growth curves should be the same. If other factors such as a lag phase, the presence or absence of some ingredient in the nutrient broth, or of some other organism in the *E. coli* culture affect the rate of growth, then the slopes of the two growth curves will differ. The problem is to find if the slopes differ significantly (see Tables 11.2 and 11.3).

(i) A least squares line is fitted to the data for log (absorbance) against

Table 11.2   All Live Bacteria

Time (min)	Absorbance	Log (absorbance)
0	134	2.127
10	145	2.161
20	152	2.182
30	170	2.230
40	192	2.283
50	217	2.336
60	248	2.394
70	285	2.455
80	315	2.498
90	341	2.533

Table 11.3   50% Live Bacteria

Time (min)	Absorbance	Log (absorbance)
0	111	2.045
10	123	2.090
20	148	2.170
30	158	2.199
40	171	2.233
50	182	2.260
60	204	2.310
70	236	2.373
80	266	2.425
90	290	2.462

time for 'all live bacteria', and the slope is found to be 0.004802, and the intercept 2.104. Using these, the sum of residuals squared is calculated to be 0.001829.

(ii) A similar least squares line for the '50% live bacteria' gave a slope of 0.004529, an intercept of 2.053 and a sum of residuals squared of 0.001635.

(iii) The sum of the residuals squared from both the separate lines is calculated

$$0.001829 + 0.001635 = 0.003464$$

(iv) The number of degrees of freedom arising from the individual lines is calculated

$$(10-2) + (10-2) = 16$$

(v) The sum of residuals squared is divided by the number of degrees of freedom to give a mean square

$$0.003464/16 = 0.0002165$$

(vi) A least squares line is calculated for the pooled modified data ($\bar{x}_1 = 45$, $\bar{y}_1 = 2.3199$, $\bar{x}_2 = 45$, $\bar{y}_2 = 2.2567$) and gives a slope of 0.00466 (intercept $= 0$) and a sum of residuals squared of 0.003770.

(vii) The difference in residuals squared between the pooled modified data and the separate lines is calculated

$$0.003770 - 0.003464 = 0.000306$$

(viii) The number of degree of freedom for this difference is

$$2-1 = 1$$

(ix) The difference in residuals squared is divided by the number of degrees of freedom to give a mean square

$$0 \cdot 000306/1 = 0 \cdot 000306$$

(x) F is calculated as

$$0 \cdot 000306/0 \cdot 0002165 = 1.41$$

This value should be compared with the significance values for F given in

Appendix 7, with the number of degrees of freedom $v_1 = 1$ and $v_2 = 16$. The calculated F value exceeds the table value for a 20% probability, hence there is more than a 20% chance that slopes as different could have occurred by chance if they are really the same. Thus one cannot conclude that the slopes of the lines are significantly different.

Description of Program to Compare Several Slopes

The program (Program 11.2) prints a heading (lines 20–30) and then asks (line 50) whether full or abbreviated instructions are required. The answer must be YES or NO, and is checked in a subroutine (lines 870–960).

The user is then asked to type the number of slopes to be compared (lines 100–140). The answer is checked (lines 150–190) to ensure that it is an integer, and is in the range 2–8. The upper limit is empirical and is set by the program for the same reasons as those outlined in the program to compare several straight lines.

The calculations for all of the separate lines are performed in a loop from lines 270 to 560. In this loop a number of totals are re-set (lines 280–300), and a subroutine (lines 970–1190) is called to input the (x,y) data values. Then a subroutine (lines 1530–2310) is called to check that the data for the line just entered are correct, and if necessary to add, delete or replace lines of data. This subroutine is more fully described in Chapter 8. A check is then performed (lines 360–400) to ensure that there

Program 11.2   Trial run.

```
PROGRAM TO COMPARE SEVERAL SLOPES
======= == ======= ======= ======

WOULD YOU LIKE FULL INSTRUCTIONS?   TYPE YES OR NO & PRESS RETURN.

? YES

THIS PROGRAM COMPARES FROM 2 TO 8 LINES TO DETERMINE WHETHER
THEIR SLOPES DIFFER SIGNIFICANTLY.

HOW MANY SLOPES WOULD YOU LIKE TO COMPARE
? 2
TYPE IN X & Y VALUES SEPARATED BY A COMMA
THEN PRESS RETURN AND TYPE THE NEXT PAIR OF VALUES
YOU WILL BE GIVEN THE CHANCE TO EDIT INCORRECT DATA LATER

INPUT DATA FOR LINE 1
TERMINATE DATA WITH BOTH X & Y EQUAL TO 999999
 X,Y
? 0, 2.127
? 10,2.161
? 20,2.182
? 30,2.230
? 40,2.283
? 50,2.336
? 60,2.394
? 70,2.455
? 80,2.498
? 90,2.533
? 999999,999999
ARE THE DATA VALUES ENTERED CORRECT?   TYPE YES OR NO & PRESS RETURN.

? YES
```

RESULTS OF CALCULATION FOR LINE 1
------- -- ----------- --- ---- -

NUMBER OF VALID DATA POINTS = 10
SLOPE = 0.00480182
INTERCEPT ON Y AXIS = 2.10382
SUM OF DIFFERENCES SQUARED = 0.00182888

WOULD YOU LIKE A LIST OF DIFFERENCES? TYPE YES OR NO & PRESS RETURN.

? NO

INPUT DATA FOR LINE 2
TERMINATE DATA WITH BOTH X & Y EQUAL TO 999999
 X,Y
? 0, 2.045
? 10,2.090
? 20,2.170
? 30,2.199
? 40,2.233
? 50,2.260
? 60,2.310
? 70,2.373
? 80,2.425
? 90,2.462
? 999999,999999
ARE THE DATA VALUES ENTERED CORRECT? TYPE YES OR NO & PRESS RETURN.

? YES

RESULTS OF CALCULATION FOR LINE 2
------- -- ----------- --- ---- -

NUMBER OF VALID DATA POINTS = 10
SLOPE = 0.00452909
INTERCEPT ON Y AXIS = 2.05289
SUM OF DIFFERENCES SQUARED = 0.00163461

WOULD YOU LIKE A LIST OF DIFFERENCES? TYPE YES OR NO & PRESS RETURN.

? NO
COMMON SLOPE = 0.00466545

F VALUE = 1.41747
NUMBER OF DEGREES OF FREEDOM: NU(1)= 1   NU(2)= 16

PROBABILITY THAT SUCH A DIFFERENCE IN VARIANCES COULD OCCUR
BY CHANCE IF THE PARENT POPULATIONS HAVE THE SAME VARIANCE
IS 25.1 %

WOULD YOU LIKE ANOTHER RUN? TYPE YES OR NO & PRESS RETURN.

? NO
END OF JOB

```
10 DIM Q$(10), I$(3), X(50), Y(50)
20 PRINT "PROGRAM TO COMPARE SEVERAL SLOPES"
30 PRINT "======= == ======= ======= ======"
40 PRINT
50 PRINT "WOULD YOU LIKE FULL INSTRUCTIONS?";
60 REM CALL SUBROUTINE TO OBTAIN AN ANSWER OF YES OR NO
70 GOSUB   890
80 LET I$ = Q$
90 IF Q$ = "NO" THEN 130
100 PRINT
110 PRINT "THIS PROGRAM COMPARES FROM 2 TO 8 LINES TO DETERMINE WHETHER"
120 PRINT "THEIR SLOPES DIFFER SIGNIFICANTLY."
130 PRINT
140 PRINT "HOW MANY SLOPES WOULD YOU LIKE TO COMPARE"
150 INPUT L
160 IF L <> INT(L) THEN 130
170 IF (L - 1) * (L - 9) < 0 THEN 200
180 PRINT "NUMBER OF SLOPES MUST BE AN INTEGER BETWEEN 2 & 8"
190 GOTO   140
200 LET E3 = 0
210 LET V2 = 0
220 LET V1 = L - 1
230 LET T1 = 0
240 LET T2 = 0
250 LET T3 = 0
260 REM START LOOP TO PERFORM LEASTSQUARES FOR SEPARATE & COMBINED LINES
270 FOR J1 = 1 TO L
280    REM SET INITIAL VALUES TO ZERO
290    READ S2, S3, X2, S4, E2
300    DATA 0, 0, 0, 0, 0
310    REM CALL SUBROUTINE TO INPUT A SET OF N (X,Y) DATA VALUES
320    GOSUB   980
330    REM CALL SUBROUTINE TO CHECK AND ALTER DATA IF NECESSARY
340    GOSUB   1540
350    PRINT
360    IF N >= 2 THEN 420
370    PRINT
380    PRINT "RUN ON THIS DATA ABANDONED BECAUSE THERE ARE"
390    PRINT "NOT ENOUGH DATA POINTS"
400    GOTO   750
410    REM CALL SUBROUTINE TO WORK OUT LEAST-SQUARES FIT ETC.
420    GOSUB   1230
430    REM CHECK FOR ERROR FLAGGED BY SUBROUTINE
440    IF N < 0 THEN 750
450    REM CALC. SUM OF DIFFERENCES SQUARED E3 & DEGREES OF FREEDOM V2
460    REM FOR THE INDIVIDUAL LINES
470    LET E3 = E3 + E2
480    LET V2 = V2 + (N - 2)
490    REM COLLECT TOTALS
500    LET T1 = T1 + X2
510    LET T2 = T2 + S4
520    LET T3 = T3 + S4 * S4 / X2
530    REM CALL SUBROUTINE TO PRINT OUT RESULTS OF REGRESSION CALCS.
540    GOSUB   2330
550    RESTORE
560 NEXT J1
570 PRINT "COMMON SLOPE = "; T2 / T1
580 REM CALCULATE F IF DATA PERMITS
590 IF V2 > 0 THEN 630
600 PRINT "RUN TERMINATED - ALL THE INDIVIDUAL LINES HAVE ONLY TWO"
610 PRINT "POINTS, AND HENCE FIT EXACTLY"
```

```
620 GOTO   750
630 IF E3 > 0 THEN 670
640 PRINT "RUN TERMINATED - ALL OF THE INDIVIDUAL LINES PASS EXACTLY"
650 PRINT "THROUGH THEIR DATA"
660 GOTO   750
670 LET F = ((T3 - T2 * T2 / T1) / V1) / (E3 / V2)
680 PRINT
690 PRINT "F VALUE ="; F
700 PRINT "NUMBER OF DEGREES OF FREEDOM; NU(1)="; V1; "  NU(2)="; V2
710 PRINT
720 REM CALL SUBROUTINE TO WORK OUT PROBABILITY FOR THIS F
730 GOSUB   2540
740 REM DECIDE WHETHER TO FINISH OR HAVE ANOTHER RUN
750 PRINT
760 PRINT "WOULD YOU LIKE ANOTHER RUN?";
770 REM CALL SUBROUTINE TO OBTAIN AN ANSWER OF YES OR NO
780 GOSUB   880
790 IF Q$ = "NO" THEN 850
800 PRINT "TYPE IN A NEW SET OF DATA"
810 PRINT "==== == = === === == ===="
820 LET I$ = "NO"
830 GOTO   140
840 REM TERMINATE JOB
850 PRINT "END OF JOB"
860 STOP
870 REM ***** SUBROUTINE TO CHECK REPLIES *****
880 IF I$ = "NO" THEN 900
890 PRINT " TYPE YES OR NO & PRESS RETURN."
900 PRINT
910 INPUT Q$
920 IF Q$ = "YES" THEN 960
930 IF Q$ = "NO" THEN 960
940 PRINT "REPLY '"; Q$; "' NOT UNDERSTOOD.";
950 GOTO   890
960 RETURN
970 REM ***** SUBROUTINE TO INPUT A SET OF N (X,Y) DATA PAIRS *****
980 IF I$ = "NO" THEN 1030
990 IF J1 > 1 THEN 1030
1000 PRINT "TYPE IN X & Y VALUES SEPARATED BY A COMMA"
1010 PRINT "THEN PRESS RETURN AND TYPE THE NEXT PAIR OF VALUES"
1020 PRINT "YOU WILL BE GIVEN THE CHANCE TO EDIT INCORRECT DATA LATER"
1030 PRINT
1040 PRINT "INPUT DATA FOR LINE"; J1
1050 PRINT "TERMINATE DATA WITH BOTH X & Y EQUAL TO 999999"
1060 PRINT " X,Y"
1070 REM SET NUMBER OF POINTS TO ZERO
1080 LET N = 0
1090 REM READ IN X & Y FOR EACH DATA POINT & STORE IN ARRAYS
1100 FOR I = 1 TO 50
1110    INPUT X(I), Y(I)
1120    IF ABS(X(I) - 999999) + ABS(Y(I) - 999999) = 0 THEN 1160
1130    LET N = N + 1
1140 NEXT I
1150 PRINT "PROGRAM CAN ONLY HANDLE A MAXIMUM OF 50 VALUES IN EACH LINE"
1160 IF N > 0 THEN 1190
1170 PRINT "PLEASE TYPE IN SOME DATA VALUES FOR LINE"; J
1180 GOTO   1100
1190 RETURN
1200 REM ***** SUBROUTINE TO CALCULATE LEAST SQUARES FIT, ETC.*****
1210 REM ARRAYS X & Y CONTAIN N DATA POINTS
1220 REM CALCULATE SUMS OF X & Y VALUES
```

```
1230 FOR I = 1 TO N
1240    LET S2 = S2 + X(I)
1250    LET S3 = S3 + Y(I)
1260 NEXT I
1270 REM CALCULATE THE AVERAGE X & Y VALUES
1280 LET A1 = S2 / N
1290 LET A2 = S3 / N
1300 FOR I = 1 TO N
1310    LET X9 = X(I) - A1
1320    LET Y9 = Y(I) - A2
1330    LET X2 = X2 + X9 * X9
1340    LET S4 = S4 + X9 * Y9
1350 NEXT I
1360 PRINT
1370 IF X2 <> 0 THEN 1440
1380 PRINT "RUN TERMINATED BY THE PROGRAM BECAUSE ALL"
1390 PRINT "THE X-COORDINATES ARE THE SAME FOR LINE"; J
1400 REM SET ERROR FLAG
1410 LET N = -1
1420 GOTO 1520
1430 REM CALCULATE SLOPE
1440 LET S = S4 / X2
1450 REM CALCULATE INTERCEPT ON Y AXIS
1460 LET Y1 = (X2 * S3 - S2 * S4) / (N * X2)
1470 REM CALCULATE SUM OF DIFFERENCES SQUARED
1480 FOR I = 1 TO N
1490    LET E1 = (Y(I) - A2) - S * (X(I) - A1)
1500    LET E2 = E2 + E1 * E1
1510 NEXT I
1520 RETURN
1530 REM *****SUBROUTINE TO CHECK DATA ARE CORRECT & ALTER IF NECESSARY*
1540 PRINT "ARE THE DATA VALUES ENTERED CORRECT?";
1550 REM A4 SHOULD BE SET TO THE NUMBER OF LINES ON THE VDU
1560 LET A4 = 20
1570 REM CALL SUBROUTINE TO OBTAIN ANSWER OF YES OR NO
1580 GOSUB 880
1590 IF Q$ = "YES" THEN 2310
1600 PRINT "HERE IS A LIST OF THE CURRENT DATA"
1610 PRINT "LINE NUMBER", "X", "Y"
1620 FOR I = 1 TO N
1630    PRINT I, X(I), Y(I)
1640    IF INT(I / (A4 - 1)) * (A4 - 1) <> I THEN 1690
1650    PRINT "WOULD YOU LIKE TO CONTINUE LISTING?";
1660    REM CALL SUBROUTINE TO OBTAIN ANSWER OF YES OR NO
1670    GOSUB 880
1680    IF Q$ = "NO" THEN 1700
1690 NEXT I
1700 PRINT "TYPE R TO REPLACE";
1710 IF I$ = "NO" THEN 1730
1720 PRINT " AN EXISTING LINE OF DATA"
1730 IF N = 50 THEN 1780
1740 PRINT TAB(5); " A TO ADD";
1750 IF I$ = "NO" THEN 1770
1760 PRINT " AN EXTRA LINE"
1770 IF N = 1 THEN 1810
1780 PRINT TAB(5); " D TO DELETE";
1790 IF I$ = "NO" THEN 1810
1800 PRINT " AN EXISTING LINE"
1810 PRINT TAB(5); " L TO LIST";
1820 IF I$ = "NO" THEN 1840
1830 PRINT " THE DATA"
```

```
1840 PRINT "   OR C TO CONTINUE";
1850 IF I$ = "NO" THEN 1880
1860 PRINT " THE CALCULATION"
1870 LET I$ = "NO"
1880 INPUT Q$
1890 IF Q$ = "R" THEN 1990
1900 IF N = 50 THEN 1930
1910 IF Q$ = "A" THEN 2110
1920 IF N = 1 THEN 1940
1930 IF Q$ = "D" THEN 2160
1940 IF Q$ = "L" THEN 1600
1950 IF Q$ = "C" THEN 2310
1960 PRINT "REPLY '"; Q$; "' NOT UNDERSTOOD."
1970 GOTO  1700
1980 REM REPLACE LINE
1990 PRINT "TYPE THE LINENUMBER OF THE LINE TO BE REPLACED";
2000 INPUT I
2010 IF I <> INT(I) THEN 2030
2020 IF (I - 1) * (I - N) <= 0 THEN 2060
2030 PRINT "LINENUMBER MUST BE AN INTEGER IN THE RANGE 1 -"; N
2040 PRINT "RE-";
2050 GOTO  1990
2060 PRINT "TYPE THE CORRECT LINE TO REPLACE THE ONE WHICH IS WRONG"
2070 PRINT "X, Y"
2080 INPUT X(I), Y(I)
2090 GOTO  2140
2100 REM ADD A NEW LINE
2110 LET N = N + 1
2120 PRINT "TYPE THE ADDITIONAL LINE OF DATA AS SHOWN;   X,Y"
2130 INPUT X(N), Y(N)
2140 PRINT "OK"
2150 GOTO  1700
2160 REM DELETE A LINE
2170 PRINT "TYPE THE LINENUMBER OF THE LINE TO BE DELETED"
2180 INPUT J
2190 IF (J - 1) * (J - N) > 0 THEN 2210
2200 IF J = INT(J) THEN 2230
2210 PRINT "LINENUMBER MUST BE AN INTEGER IN THE RANGE 1 -"; N
2220 GOTO  2170
2230 FOR I = J + 1 TO N
2240    LET X(I - 1) = X(I)
2250    LET Y(I - 1) = Y(I)
2260 NEXT I
2270 LET N = N - 1
2280 PRINT "OK"
2290 IF J > N THEN 1700
2300 GOTO  1600
2310 RETURN
2320 REM ***** SUBROUTINE TO PRINT RESULTS OF REGRESSION *****
2330 PRINT "RESULTS OF CALCULATION FOR LINE"; J1
2340 PRINT "------- -- ----------- --- ---- -"
2350 PRINT
2360 PRINT "NUMBER OF VALID DATA POINTS ="; N
2370 PRINT "SLOPE   ="; S
2380 PRINT "INTERCEPT ON Y AXIS  ="; Y1
2390 PRINT "SUM OF DIFFERENCES SQUARED  ="; E2
2400 PRINT
2410 PRINT "WOULD YOU LIKE A LIST OF DIFFERENCES?";
2420 REM CALL SUBROUTINE TO OBTAIN AN ANSWER OF YES OR NO
2430 GOSUB  880
2440 IF Q$ = "NO" THEN 2520
```

```
2450 REM CALCULATE & PRINT DIFFERENCE FOR EACH DATA POINT
2460 PRINT
2470 PRINT "DIFFERENCE IN Y VALUE FOR EACH DATA POINT"
2480 PRINT "X", "Y OBS", "Y CALC", "DIFFERENCE"
2490 FOR I = 1 TO N
2500    PRINT X(I), Y(I), A2 + S * (X(I)-A1), (Y(I)-A2)-S * (X(I)-A1)
2510 NEXT I
2520 RETURN
2530 REM ***** SUBROUTINE TO CALCULATE F PROBABILITY *****
2540 LET E = 0
2550 IF V1 = 2 * INT(V1 / 2 + 0.1) THEN 2590
2560 IF V2 = 2 * INT(V2 / 2 + 0.1) THEN 2640
2570 GOTO  2820
2580 REM CALCULATE PROBABILITY IF V1 EVEN
2590 LET U = 1 / (1 + V2 / (F * V1))
2600 LET P1 = V1 + 1
2610 LET Q = V2 - 2
2620 GOTO  2680
2630 REM CALCULATE PROBABILITY IF V2 EVEN
2640 LET E = 1
2650 LET U = 1 / (1 + F * V1 / V2)
2660 LET P1 = V2 + 1
2670 LET Q = V1 - 2
2680 LET S = 0
2690 LET W = 1
2700 LET K = 2
2710 LET S = S + W
2720 LET W = W * U * (K + Q) / K
2730 LET K = K + 2
2740 IF K < P1 THEN 2710
2750 LET Z = SQR(1 - U)
2760 IF E = 0 THEN 2790
2770 LET P = 100 * S * (Z ^ V1)
2780 GOTO  3050
2790 LET P = 100 * S * (Z ^ V2)
2800 GOTO  3050
2810 REM CALCULATE PROBABILITY IF V1 & V2 BOTH ODD
2820 LET U = 1 / (1 + F * V1 / V2)
2830 LET X = 1 - U
2840 LET S = 0
2850 LET W = 1
2860 LET K = 2
2870 LET P1 = V2
2880 GOTO  2920
2890 LET S = S + W
2900 LET W = W * U * K / (K + 1)
2910 LET K = K + 2
2920 IF K < P1 THEN 2890
2930 LET W = W * V2
2940 LET K = 3
2950 LET P1 = V1 + 1
2960 LET Q = V2 - 2
2970 GOTO  3010
2980 LET S = S - W
2990 LET W = W * X * (K + Q) / K
3000 LET K = K + 2
3010 IF K < P1 THEN 2980
3020 LET T1 = ATN(SQR(F * V1 / V2))
3030 LET R1 = S * SQR(X * U)
3040 LET P = 100 * (1 - 2 * (T1 + R1) / 3.14159)
3050 IF P <= 50 THEN 3080
```

```
3060 LET P = 100 - P
3070 REM ROUND OFF ANSWER
3080 IF P < 5 THEN 3100
3090 LET P = INT(P * 10 + 0.5) * 0.1
3100 IF (P - 5) * (P - 0.5) > 0 THEN 3120
3110 LET P = INT(P * 100 + 0.5) * 0.01
3120 IF P > 0.05 THEN 3140
3130 LET P = INT(P * 1000 + 0.5) * 0.001
3140 PRINT "PROBABILITY THAT SUCH A DIFFERENCE IN VARIANCES COULD OCCUR"
3150 PRINT "BY CHANCE IF THE PARENT POPULATIONS HAVE THE SAME VARIANCE"
3160 IF P < 0.01 THEN 3190
3170 PRINT "IS"; P; "%"
3180 GOTO  3200
3190 PRINT "IS LESS THAN 0.01 %"
3200 PRINT
3210 RETURN
3220 END
```

are at least two data points, and a subroutine (lines 1200–1520) is called to fit a least squares line. The run is aborted (line 440) if a vertical straight line is found, since this has an infinite slope. A number of totals are collected (lines 470–520) and the results of the regression are printed in a subroutine (lines 2320–2520). This whole procedure is repeated by the loop for the second and subsequent lines.

The slope of the common line is then calculated and printed (line 570). Two checks are performed (lines 580–660) to ensure that it is possible to calculate an F value. Should the number of degrees of freedom in the residual be zero, or should all of the lines pass exactly through their data, it is not possible to calculate F, and the user is asked if another run is required. Otherwise the value of F is calculated and printed in lines 670–700. A subroutine (lines 2530–3210) is then called to evaluate and print the probability of such an F value occurring by chance if the lines are drawn from parent populations whose slopes are really the same.

Finally the user is asked if another run is required (line 760), and the answer is checked by a subroutine (lines 870–960) to make sure that it is either YES or NO.

12

Curve Fitting (Polynomials)

Polynomials are often fitted to experimental data either to smooth the data or to allow interpolation. A polynomial equation has the form

$$y = a + bx + cx^2 + dx^3 + \ldots \qquad (1)$$

where $a, b, c, d \ldots$ are constants and are called the coefficients of the polynomial. The 'order' of the polynomial is the largest power of x in the equation. Thus for a quadratic equation $y = a + bx + cx^2$ the order is 2, and for a straight line $y = a + bx$ the order is 1.

When fitting a polynomial to n data points $(x_1 y_1), (x_2 y_2), (x_3 y_3), \ldots, (x_n, y_n)$ it is assumed that all of the errors occur in the measured y values and that there are no errors in the measured x values. For each value of x_r (with $r = 1, 2, \ldots, n$) there will be a calculated y value from the polynomial, and the difference between this and the observed y_r value is called the residual Δ_r in y_r. Thus

$$\Delta_r = y_r - (a + bx_r + cx_r^2 + dx_r^3 + \ldots)$$

The principle underlying the fitting of the polynomial is to make the sum of the residuals squared as small as possible by choosing appropriate values for $a, b, c, d \ldots$

$$\text{Sum of residuals squared} = E = \sum_{r=1}^{r=n} \Delta_r^2$$

The 'best' polynomial of a given order is that which minimises E. Increasing the order of the polynomial can only decrease the minimum value of E, but may result in less smoothing of the data. In the extreme case the order is equal to $n-1$, and the polynomial passes exactly through each point in a general set of data.

This is called the interpolating polynomial, for which $E = 0$ and there is no smoothing of the data.

The method for minimising E is to partially differentiate E with respect to each of the coefficients in turn, and equate each of the partial derivatives to zero. Since the order of the polynomial is m, this will result in $m+1$ simultaneous equations, which are called the normal equations. Solving the normal equations yields the required coefficients $a, b, c, d, \ldots$ This approach is identical to that used in Chapter 9 for fitting a straight line—that is a polynomial of order 1—except that some symbols are different:

$$y = mx + c \text{ becomes } y = a + bx$$

since the slope m becomes b and the intercept c becomes a. Re-writing Equations 2 and 3 from Chapter 9 for the partial derivatives gives:

$$\frac{\partial E}{\partial a} = 2na - 2\Sigma y_r + 2b\Sigma x_r \qquad (2)$$

and

$$\frac{\partial E}{\partial b} = 2b\Sigma x_r^2 - 2\Sigma x_r y_r + 2a\Sigma x_r \qquad (3)$$

Equating the partial derivatives to zero and rearranging

$$an + b\Sigma x_r = \Sigma y_r$$
$$a\Sigma x_r + b\Sigma x_r^2 = \Sigma x_r y_r$$

The normal equations above can be written in matrix form:

$$\begin{pmatrix} n & \Sigma x_r \\ \Sigma x_r & \Sigma x_r^2 \end{pmatrix} \begin{pmatrix} a \\ b \end{pmatrix} = \begin{pmatrix} \Sigma y_r \\ \Sigma x_r y_r \end{pmatrix} \qquad (4)$$

One method of calculating the coefficients a and b is to evaluate the inverse of the 2×2 matrix

$$\text{inverse of } \begin{pmatrix} n & \Sigma x_r \\ \Sigma x_r & \Sigma x_r^2 \end{pmatrix} = \begin{pmatrix} n & \Sigma x_r \\ \Sigma x_r & \Sigma x_r^2 \end{pmatrix}^{-1}$$

and then premultiplying both sides of Equation 4 by the inverse matrix to give

$$\begin{pmatrix} a \\ b \end{pmatrix} = \begin{pmatrix} n & \Sigma x_r \\ \Sigma x_r & \Sigma x_r^2 \end{pmatrix}^{-1} \begin{pmatrix} \Sigma y_r \\ \Sigma x_r y_r \end{pmatrix}$$

Fitting a polynomial of order 2, that is a quadratic equation $y = a + bx + cx^2$ yields the following normal equations:

$$an \quad + b\Sigma x_r + c\Sigma x_r^2 = \Sigma y_r$$

$$a\Sigma x_r + b\Sigma x_r^2 + c\Sigma x_r^3 = \Sigma x_r y_r$$

$$a\Sigma x_r^2 + b\Sigma x_r^3 + c\Sigma x_r^4 = \Sigma x_r^2 y_r$$

hence in matrix form

$$\begin{pmatrix} n & \Sigma x_r & \Sigma x_r^2 \\ \Sigma x_r & \Sigma x_r^2 & \Sigma x_r^3 \\ \Sigma x_r^2 & \Sigma x_r^3 & \Sigma x_r^4 \end{pmatrix} \begin{pmatrix} a \\ b \\ c \end{pmatrix} = \begin{pmatrix} \Sigma y_r \\ \Sigma x_r y_r \\ \Sigma x_r^2 y_r \end{pmatrix}$$

Premultiplying both sides of this equation by the inverse matrix gives

$$\begin{pmatrix} a \\ b \\ c \end{pmatrix} = \begin{pmatrix} n & \Sigma x_r & \Sigma x_r^2 \\ \Sigma x_r & \Sigma x_r^2 & \Sigma x_r^3 \\ \Sigma x_r^2 & \Sigma x_r^3 & \Sigma x_r^4 \end{pmatrix}^{-1} \begin{pmatrix} \Sigma y_r \\ \Sigma x_r y_r \\ \Sigma x_r^2 y_r \end{pmatrix}$$

In an analogous way a polynomial of order m yields an $m+1$ by $m+1$ matrix, and the coefficients may be obtained by

$$(5)$$

$$\begin{bmatrix} a \\ b \\ \cdot \\ \cdot \\ \cdot \\ \cdot \end{bmatrix} = \begin{bmatrix} n & \Sigma x_r & \dots & \Sigma x_r^m \\ \Sigma x_r & \Sigma x_r^2 & \dots & \Sigma x_r^{m+1} \\ \cdot & \cdot & & \cdot \\ \cdot & \cdot & & \cdot \\ \cdot & \cdot & & \cdot \\ \Sigma x_r^m & \Sigma x_r^{m+1} & \dots & \Sigma x_r^{2m} \end{bmatrix}^{-1} \begin{bmatrix} \Sigma x_r \\ \Sigma x_r y_r \\ \cdot \\ \cdot \\ \cdot \\ \Sigma x_r^m y_r \end{bmatrix}$$

Polynomials of order 1 or 2 (straight line or quadratic fits) can be obtained by manually building the matrix and solving the matrix equation. Though in theory it is possible to fit higher order polynomials in this way, the time taken to perform the calculations renders this impracticable. Furthermore, unless a large number of significant figures are carried throughout the calculation then the result may be wildly inaccurate.

The speed of a computer overcomes the excessive time taken for manual calculation of higher order polynomials. However, since a computer only carries a limited number of significant figures, the problems of accuracy remain. Even using double precision where the computer carries double the usual number of significant figures only allows reliable answers up to about an order of six on most computers.

It is important to identify the sources of inaccuracy as a first step towards overcoming the problem of loss of accuracy. It has been found that the major cause of inaccuracy is ill-conditioning of the solution of the normal equations with respect to the coefficients of the matrix. More simply a small change in the value of a particular term or terms in the matrix drastically changes the coefficients of the polynomial obtained as the solution. Ill-conditioning does not always occur, and is a property of the way in which the problem has been formulated. It follows that by formulating the problem in a different way it may be possible to avoid the difficulties of ill-conditioning.

A second smaller source of inaccuracy is termed instability, and is caused by the accumulation of rounding errors in performing the calculations. Instability is a direct result of the method of calculation.

Method of Reducing Loss in Accuracy

1. The accuracy of the coefficients is likely to be improved by performing the calculations carrying a larger number of significant figures.

207

Table 9.1 Equation 7

$$
\begin{bmatrix} c_0 \\ c_1 \\ c_2 \\ \cdot \\ \cdot \\ \cdot \\ c_m \end{bmatrix}
=
\begin{bmatrix}
\Sigma p_0(x_r)p_0(x_r) & \Sigma p_0(x_r)p_1(x_r) & \Sigma p_0(x_r)p_2(x_r)\ldots \\
\Sigma p_1(x_r)p_0(x_r) & \Sigma p_1(x_r)p_1(x_r) & \Sigma p_1(x_r)p_2(x_r)\ldots \\
\Sigma p_2(x_r)p_0(x_r) & \Sigma p_2(x_r)p_1(x_r) & \Sigma p_2(x_r)p_2(x_r)\ldots \\
\cdot \\
\cdot \\
\cdot \\
\Sigma p_m(x_r)p_0(x_r) & \Sigma p_m(x_r)p_1(x_r) & \Sigma p_m(x_r)p_2(x_r)\ldots
\end{bmatrix}^{-1}
\begin{bmatrix} \Sigma p_0(x_r)y_r \\ \Sigma p_1(x_r)y_r \\ \Sigma p_2(x_r)y_r \\ \cdot \\ \cdot \\ \cdot \\ \Sigma p_m(x_r)y_r \end{bmatrix}
\qquad (7)
$$

2. The magnitude of the x, y data may affect the accuracy of the coefficients, particularly if the mean of the x values is a long way from zero. This can be overcome by scaling the data into a small range close to zero—for example into the range $+2$ to -2, before building and solving the matrix. The resulting coefficients refer to the scaled data, but may be appropriately converted back to refer to the original unscaled data.

3. Results which are unreliable will be produced if the problem is 'ill-conditioned'. This is generally associated with values off the leading diagonal which are large relative to the values on the leading diagonal. Better results will be obtained if a method is found which keeps the off-diagonal terms small, and the ideal case is to have all off-diagonal terms zero. A method of achieving this was devised by G. E. Forsythe (*Journal of the Society for Industrial and Applied Mathematics*, 1957, **5**, 74) which has revolutionised the whole subject of polynomial curve fitting.

Orthogonal Polynomials

It is possible to re-write Equation 1 in terms of a set of polynomials p_i, where p_i is an arbitrary polynomial of order i. Thus

$$y = c_0 p_0(x) + c_1 p_1(x) + c_2 p_2(x) + $$
$$+ c_3 p_3(x) + \ldots + c_m p_m(x) \quad (6)$$

The polynomials $p_0, p_1, p_2, \ldots, p_m$ are derived directly from the x_r data and are thus known. The problem of fitting the best curve to the data becomes one of choosing the best coeffi-

cients $c_0, c_1, c_2, \ldots, c_m$ and from the values of these deriving the values of the coefficients a, b, c, $d \ldots$ used in Equation 1.

It might at first sight seem that there is considerably more work in evaluating the coeffients $c_0, c_1, c_2, c_3 \ldots$ than in evaluating the coefficients a, b, c, $d \ldots$ directly for no apparent benefit. Furthermore the relatively simple normal equations given in Equation 5 are replaced by the more complicated ones in Table 9.1.

The objective is to choose the polynomials p_i such that all terms which are off the leading diagonal in the matrix above will evaluate to zero. The matrix can then be inverted accurately in one line thus eliminating problems from ill-conditioning. This is clearly a major benefit. For example

$$
\begin{bmatrix}
2 & 0 & 0 & 0 \\
0 & 5 & 0 & 0 \\
0 & 0 & 4 & 0 \\
0 & 0 & 0 & 8
\end{bmatrix}^{-1}
=
\begin{bmatrix}
\frac{1}{2} & 0 & 0 & 0 \\
0 & \frac{1}{5} & 0 & 0 \\
0 & 0 & \frac{1}{4} & 0 \\
0 & 0 & 0 & \frac{1}{8}
\end{bmatrix}
$$

exactly

Polynomials with these properties are orthogonal over the data.

Equation 1 may be considered as a special case of Equation 6 where $p_0(x) = 1$, $p_1(x) = x$, $p_2(x) = x^2$, $p_3(x) = x^3, \ldots,$ $p_i(x) = x^i, \ldots,$ $p_m(x) = x^m$. This special case is not orthogonal, and ill-conditioning of the problem can and frequently does occur.

Forsythe's Orthogonal Polynomials

Forsythe showed that it was always possible to derive an orthogonal set of polynomials p_i

from the x_r data provided, and he gave a set of rules for doing this. The use of these orthogonal polynomials has revolutionised the whole subject of polynomial curve fitting, and has made it possible to fit much higher orders than hitherto.

The polynomials p_i are derived from each other, and are defined

$$p_0(x) = 1$$

$$p_1(x) = 2\ (x - \alpha_1)p_0(x)$$

$$p_2(x) = 2(x - \alpha_2)p_1(x) - \beta_1 p_0(x)$$

$$p_3(x) = 2(x - \alpha_3)p_2(x) - \beta_2 p_1(x)$$

.

.

.

$$p_{i+1}(x) = 2(x - \alpha_{i+1})p_i(x) - \beta_i p_{i-1}(x)$$

where

$$\alpha_{i+1} = \frac{\sum_{r=1}^{r=n} x_r p_i^2(x_r)}{\sum_{r=1}^{r=n} p_i^2(x_r)}$$

and

$$\beta_i = \frac{\sum_{r=1}^{r=n} p_i^2(x_r)}{\sum_{r=1}^{r=n} p_{i-1}^2(x_r)}$$

Use of the above procedure overcomes the problem of loss of accuracy due to ill-conditioning when solving the matrix equation. The orthogonal polynomials and the coefficients c_0, c_1, $c_2 \ldots c_m$ can be used to evaluate the polynomial fitted to the original data. It is often required to express the fitted polynomial in terms of a power series as in Equation 1. It is possible to convert the coefficients c_0, c_1, $c_2, \ldots, c_m$ back into the coefficients a, b, c, $d, \ldots$ used in Equation 1. Though a power series may be a more convenient expression, the conversion process may introduce arithmetic rounding errors.

The use of orthogonal polynomials has cured the problem of ill-conditioning, but instability due to accumulation of rounding

errors remains. Instability is reduced by carrying more significant figures in the calculation.

A polynomial in the form of a power series may be evaluated in an elegant way. Consider the polynomial

$$y = 2x^4 - 4x^3 + 6x^2 - 3x + 15 \qquad (8)$$

The correct way to evaluate this is

$$y = \{[(2x - 4)x + 6]x - 3\}x + 15$$

This is known as Horner's rule, although the method was first given by Isaac Newton in 1711. The principle is to start with the coefficient of the highest order term, multiply by x, add in the next coefficient, multiply by x, add the next coefficient, and continue multiplying by x and adding coefficients until all the coefficients have been used. Whether the calculation is performed by hand, or by computer, the calculation is faster and is likely to produce a more accurate answer than evaluating Equation 8 as it is written.

If further accuracy is required in evaluating the polynomial then the power series given in Equation 1 must be abandoned. The value of the polynomial at a point can be evaluated directly from Forsythe's α and β values, avoiding the necessity of converting α and β into a power series form. A further slight improvement in accuracy can be obtained by representing the fitting polynomial in the form of a series of Chebyshev polynomials. It is necessary to use the Chebyshev polynomial directly to evaluate the function, since conversion of Chebyshev polynomials into power series coefficients generally destroys the extra accuracy just gained. This technique is discussed in G. J. Hayes's book *Numerical Approximation to Functions and Data*, Athlone Press 1970.

Finally it should be noted that high order polynomials may produce unwanted spikes between data points and hence interpolation should be performed with care. Furthermore, polynomials other than order zero tend to plus or minus infinity for large plus or minus values of x, hence extrapolation beyond the range of data values supplied should not be undertaken.

Description of the Program to Fit Polynomials
(see Program 12.1)

Accuracy

It is strongly recommended that this program is only run on computers which support a version of BASIC which carries 12 or more significant decimal figures. Loss of accuracy will probably be significant—particularly with high order polynomials, if fewer figures are carried. On a number of mainframes this is accomplished by using a double precision version of BASIC. On microcomputers some versions of BASIC including CBASIC, Cromenco disc BASIC and Xitan disc BASIC (as used on the 380-Z), automatically carry sufficient significant figures. Some other versions allow double precision to be specified in the program. Microsoft BASIC-80 and TRS level II BASIC achieve this by adding the line:

5 DEFDBL A-Z

The results obtained from BASIC's carrying only six or eight figures may be erroneous.

Memory Requirements

The arrays DIMensioned at 100 in line 10 use a considerable amount of memory. The value of 100 permits up to 100 (x, y) data pairs to be stored. If memory is restricted, the value of 100 may be reduced to a smaller value in all of these arrays, but a few additional changes are also required: the value of 100 must be changed to the new value in lines 210, 290, 2940 and 3100.

The Program

First the program prints a heading (lines 70–80), and a message (line 100) asks if full instructions are required. The answer which must be YES or NO is input and checked in a subroutine (lines 3580–3670). Full or abbreviated instructions are printed as appropriate

throughout the program on the first run, but only abbreviated instructions are given on a second or subsequent run.

Instructions for inputing the data are given (lines 140–190), and the data are input in a loop from lines 210 to 280. The data are entered as an (x, y) pair and the appropriate weight on one line. For many purposes the weights are chosen as 1, but different weights may be chosen for individual points based on the reliability of the particular point. Weights are checked to ensure that they are greater than zero (lines 240–260).

If the maximum number of points which the arrays can hold have been entered then a message (line 290) is printed before the calculation proceeds.

A check is performed (lines 300–310) to ensure that at least two acceptable data pairs have been provided. If they have not, then the run is abandoned with a message (line 1070), and the user is asked if another run is required. Provided that sufficient data have been input, a subroutine (lines 2760–3570) is entered to check that the data are correct. If the data are correct then the subroutine is exited, but otherwise the data are listed and instructions are printed explaining how to replace or delete an existing line, or alternatively to add a new line. The operation of this subroutine is fully described in Chapter 8. When the data are correct, the calculation continues.

The maximum order of polynomial which the program tests for is set to 9 in line 370, and this value is re-set (lines 380–390) to the number of points minus two if there are 10 or fewer points. A message (lines 410–470) requests the order of the polynomial to be fitted, or zero if the program is to choose the 'best' polynomial based on the goodness of fit. The value typed is checked (lines 490–530) to ensure

(i) that it is an integer,
(ii) that it is not negative, and
(iii) that it does not exceed the maximum order.

If the value typed is rejected, a message requests the user to re-type the correct value,

Program 12.1 Trial run.

```
PROGRAM TO FIT A POLYNOMIAL TO A SET OF POINTS
======= == === = ========= == = === == ======

WOULD YOU LIKE FULL INSTRUCTIONS.  TYPE YES OR NO & PRESS RETURN.

? YES
TYPE IN A PAIR OF X & Y VALUES & WEIGHT SEPARATED BY COMMAS
THEN PRESS RETURN, TYPE THE NEXT PAIR OF VALUES ETC
TERMINATE DATA WITH  999999, 999999, 999999

STARTING DATA
 X, Y, WEIGHT
? 1, 10, 1
? 2, 49, 1
? 3, 142, 1
? 4, 313, 1
? 5, 586, 1
? 6, 985, 1
? 7, 1534, 1
? 8, 2257, 1
? 9, 3178, 1
? 999999, 999999, 999999

ARE THE DATA VALUES ENTERED CORRECT?  TYPE YES OR NO & PRESS RETURN.

? YES

TYPE IN THE ORDER REQUIRED IN THE RANGE 1 - 7  OF THE
ONE SPECIFIC POLYNOMIAL REQUIRED.
OR TYPE 0 IF ALL THE POLYNOMIALS FROM ORDER 0 - 7 ARE TO
BE EXAMINED, AND THE ONE WHICH FITS BEST REPORTED,
THEN PRESS RETURN.
? 0

MAXIMUM ORDER OF POLYNOMIAL TESTED FOR = 7
ORDER OF BEST POLYNOMIAL FOUND = 3

POLYNOMIAL ORDER    GOODNESS OF FIT
0          0.4918592889118
1          0.070900097402596
2          0.001515151515145
3          1.465494392505E-14
4          1.831867990632E-14
5          2.442490654179E-14
6          3.663735981269E-14
7          7.327471962657E-14

COEFFICIENTS OF THE BEST OR SPECIFIED ORDER POLYNOMIAL
(Y = A + B*X + C*X^2 + D*X^3 +...)
A=  0.999999993597
B=  2.000000000512
C=  2.999999999878
D=  4.000000000008

WOULD YOU LIKE A TABLE OF RESIDUALS
 TYPE YES OR NO & PRESS RETURN.

? NO
WOULD YOU LIKE ANOTHER RUN  TYPE YES OR NO & PRESS RETURN.
```

? NO

REMEMBER THAT YOU MUST NOT EXTRAPOLATE BEYOND THE
DATA POINTS, AND ALSO THAT INTERPOLATION BETWEEN
POINTS IS DANGEROUS WITH HIGH ORDER POLYNOMIALS.
END OF JOB


```
10 DIM P(100),R(100),T(100),U(100),V(100),W(100),X(100),Y(100),Z(100)
20 DIM A(10),B(10),C(10),D(11),F(10),G(10),L(10),Q(10),S(10)
30 DIM A$(2),I$(3),Q$(10)
40 REM ARRAY SIZES LIMIT PROGRAM TO A MAXIMUM OF 100 DATA POINTS.
50 REM THE NUMBER OF DATA POINTS SHOULD BE AT LEAST 2 GREATER THAN THE
60 REM MAXIMUM ORDER OF THE POLYNOMIAL.
70 PRINT "PROGRAM TO FIT A POLYNOMIAL TO A SET OF POINTS"
80 PRINT "======= == === = ========== == = === == ======"
90 PRINT
100 PRINT "WOULD YOU LIKE FULL INSTRUCTIONS.";
110 GOSUB 3600
120 LET I$ = Q$
130 IF I$ = "NO" THEN 160
140 PRINT "TYPE IN A PAIR OF X & Y VALUES & WEIGHT SEPARATED BY COMMAS"
150 PRINT "THEN PRESS RETURN, TYPE THE NEXT PAIR OF VALUES ETC"
160 PRINT "TERMINATE DATA WITH  999999, 999999, 999999"
170 PRINT
180 PRINT "STARTING DATA"
190 PRINT " X, Y, WEIGHT"
200 LET N = 0
210 FOR I = 1 TO 100
220    INPUT X(I), Y(I), W(I)
230    IF ABS(X(I) - 999999) + ABS(Y(I) - 999999) = 0 THEN 310
240    IF W(I) >= 0 THEN 270
250    PRINT "NEGATIVE WEIGHTS ARE IMPOSSIBLE - RETYPE LAST LINE"
260    GOTO 220
270    LET N = N + 1
280 NEXT I
290 PRINT "PROGRAM CAN ONLY HANDLE MAXIMUM OF 100 VALUES"
300 REM CHECK THAT THERE ARE AT LEAST 2 POINTS
310 IF N < 2 THEN 1070
320 PRINT
330 REM CALL SUBROUTINE TO CHECK THAT DATA ARE CORRECT
340 GOSUB 2770
350 PRINT
360 REM CALCULATE MAXIMUM ORDER BASED ON NUMBER OF DATA POINTS
370 LET N9 = 9
380 IF N - 2 >= 9 THEN 400
390 LET N9 = N - 2
400 IF I$ = "YES" THEN 430
410 PRINT "TYPE ORDER REQUIRED"
420 GOTO 480
430 PRINT "TYPE IN THE ORDER REQUIRED IN THE RANGE 1 -"; N9; " OF THE"
440 PRINT "ONE SPECIFIC POLYNOMIAL REQUIRED."
450 PRINT "OR TYPE 0 IF ALL THE POLYNOMIALS FROM ORDER 0 -";N9;"ARE TO"
460 PRINT "BE EXAMINED, AND THE ONE WHICH FITS BEST REPORTED,"
470 PRINT "THEN PRESS RETURN."
480 INPUT L
490 IF L <> INT(L) THEN 520
500 IF L < 0 THEN 520
510 IF L <= N9 THEN 550
520 PRINT "INCORRECT VALUE TYPED"
530 GOTO 430
```

212

```
540 REM SET THE MAXIMUM ORDER TO 9, IE M1 (MAXORDER+1) TO 10
550 LET M1 = 10
560 IF L <= 0 THEN 580
570 LET M1 = L + 1
580 LET I = N - 1
590 IF M1 <= I THEN 620
600 LET M1 = I
610 REM CALL SUBROUTINE TO FIT THE POLYNOMIAL
620 GOSUB 1180
630 LET M2 = M1 - 1
640 PRINT
650 IF L = 0 THEN 680
660 PRINT "ORDER OF POLYNOMIAL SPECIFIED ="; N2
670 GOTO 710
680 PRINT "MAXIMUM ORDER OF POLYNOMIAL TESTED FOR ="; M2
690 PRINT "ORDER OF BEST POLYNOMIAL FOUND ="; N2
700 PRINT
710 PRINT "POLYNOMIAL ORDER    GOODNESS OF FIT"
720 FOR I = 1 TO M1
730    PRINT I - 1; TAB(10); G(I)
740 NEXT I
750 PRINT
760 PRINT "COEFFICIENTS OF THE BEST OR SPECIFIED ORDER POLYNOMIAL"
770 PRINT "(Y = A + B*X + C*X^2 + D*X^3 +...)"
780 LET N3 = N2 + 1
790 FOR I = 1 TO N3
800    READ A$
810    PRINT A$; TAB(5); F(I)
820 NEXT I
830 DATA "A=", "B=", "C=", "D=", "E=", "F=", "G=", "H=", "I=", "J="
840 RESTORE
850 PRINT
860 PRINT "WOULD YOU LIKE A TABLE OF RESIDUALS"
870 GOSUB 3590
880 IF Q$ = "NO" THEN 980
890 PRINT "X              Y           Y(CALC)        DIFF"
900 LET R2 = 0
910 FOR I = 1 TO N
920    PRINT X(I), Y(I), Z(I), R(I)
930    LET R2 = R2 + R(I) ^ 2
940 NEXT I
950 PRINT
960 PRINT "SUM OF ERRORS SQUARED ="; R2
970 PRINT
980 PRINT "WOULD YOU LIKE ANOTHER RUN";
990 GOSUB 3590
1000 IF Q$ = "NO" THEN 1100
1010 LET I$ = "NO"
1020 PRINT "WOULD YOU LIKE TO TRY ANOTHER ORDER WITH THE SAME DATA"
1030 GOSUB 3590
1040 IF Q$ = "YES" THEN 310
1050 GOTO 160
1060 REM ENTER IF THERE ARE NOT ENOUGH POINTS
1070 PRINT "RUN TERMINATED - NOT ENOUGH DATA POINTS"
1080 GOTO 980
1090 REM TERMINATE JOB
1100 IF I$ = "NO" THEN 1140
1110 PRINT "REMEMBER THAT YOU MUST NOT EXTRAPOLATE BEYOND THE"
1120 PRINT "DATA POINTS, AND ALSO THAT INTERPOLATION BETWEEN"
1130 PRINT "POINTS IS DANGEROUS WITH HIGH ORDER POLYNOMIALS."
1140 PRINT "END OF JOB"
```

```
1150 STOP
1160 REM SUBROUTINE TO CALCULATE A WEIGHTED LEAST SQUARES POLYNOMIAL
1170 REM BY FORSYTHE"S METHOD USING ORTHOGONAL POLYNOMIALS.
1180 LET M3 = M1 - 1
1190 LET N2 = M3
1200 FOR I = 1 TO M1
1210    LET C(I) = 0
1220 NEXT I
1230 LET Q(1) = 0
1240 LET D(1) = 0
1250 LET D(2) = 0
1260 LET A(1) = 1
1270 LET D2 = 0
1280 LET P1 = 0
1290 LET S1 = 0
1300 LET G1 = 0
1310 LET I1 = 0
1320 LET S2 = W(1)
1330 REM FIND THE MAXIMUM AND MINIMUM X & Y
1340 LET X9 = X(1)
1350 LET X1 = X(1)
1360 LET Y9 = Y(1)
1370 LET Y1 = Y(1)
1380 FOR I = 2 TO N
1390    IF X(I) <= X9 THEN 1410
1400    LET X9 = X(I)
1410    IF X(I) >= X1 THEN 1430
1420    LET X1 = X(I)
1430    IF Y(I) <= Y9 THEN 1450
1440    LET Y9 = Y(I)
1450    IF Y(I) >= Y1 THEN 1470
1460    LET Y1 = Y(I)
1470    LET S2 = S2 + W(I)
1480 NEXT I
1490 REM CHECK THAT SUM OF WEIGHTS IS NOT ZERO
1500 IF S2 = 0 THEN 2740
1510 LET Y3 = (Y9 + Y1) / 2
1520 LET Y4 = (Y9 - Y1) / 2
1530 IF Y4 > 0 THEN 1580
1540 LET F(1) = Y(1)
1550 LET N2 = 0
1560 GOTO 2720
1570 REM SCALE Y TERMS INTO THE RANGE +1 TO -1
1580 FOR I = 1 TO N
1590    LET V(I) = (Y(I) - Y3) / Y4
1600    LET D2 = D2 + W(I) * V(I) ^ 2
1610    LET P(I) = 1
1620    LET T(I) = 0
1630    LET P1 = P1 + W(I) * V(I)
1640    LET S1 = S1 + W(I)
1650 NEXT I
1660 LET S(1) = P1 / S1
1670 LET C(1) = S(1)
1680 LET D2 = D2 - S(1) * P1
1690 LET G(1) = ABS(D2 / (N - 1))
1700 LET A1 = 4 / (X9 - X1)
1710 LET B1 = -2 - A1 * X1
1720 REM SCALE X TERMS INTO THE RANGE +2 TO -2
1730 FOR I = 1 TO N
1740    LET U(I) = A1 * X(I) + B1
1750 NEXT I
```

214

```
1760 REM START LOOP FOR EACH ORDER
1770 FOR I = 1 TO M3
1780    LET D1 = 0
1790    FOR J = 1 TO N
1800       LET D1 = D1 + W(J) * U(J) * P(J) ^ 2
1810    NEXT J
1820    REM L IS FORSYTHES ALPHA
1830    LET L(I + 1) = D1 / S1
1840    LET W2 = S1
1850    LET S1 = 0
1860    LET P1 = 0
1870    REM STORE VALUE OF CURRENT ORTHOGONAL POLYNOMIAL IN P( )
1880    REM AND OF PREVIOUS ORTHOGONAL POLYNOMIAL IN T( )
1890    FOR J = 1 TO N
1900       LET D1 = Q(I) * T(J)
1910       LET T(J) = P(J)
1920       LET P(J) = (U(J) - L(I + 1)) * P(J) - D1
1930       LET S1 = S1 + W(J) * P(J) ^ 2
1940       LET P1 = P1 + W(J) * V(J) * P(J)
1950    NEXT J
1960    REM Q IS FORSYTHES BETA
1970    LET Q(I + 1) = S1 / W2
1980    LET S(I + 1) = P1 / S1
1990    LET D2 = D2 - S(I + 1) * P1
2000    LET G(I + 1) = ABS(D2 / (N - I - 1))
2010    IF L > 0 THEN 2180
2020    REM ENTER IF PROGRAM HAS TO DECIDE ON BEST ORDER (L = 0)
2030    IF I1 = 1 THEN 2130
2040    IF G(I + 1) < G(I) THEN 2180
2050    REM ENTER IF A MINIMUM DETECTED
2060    LET N2 = I - 1
2070    LET I1 = 1
2080    LET G1 = G(I)
2090    FOR J = 1 TO M1
2100       LET B(J) = C(J)
2110    NEXT J
2120    GOTO 2180
2130    IF G(I + 1) >= 0.6 * G1 THEN 2180
2140    LET I1 = 0
2150    LET N2 = M3
2160    REM BUILD COEFFICIENTS OF J TH ORDER TERM IN A( ) & SUM TO FORM
2170    REM EXPLICIT POWER SERIES IN C( )
2180    FOR J = 1 TO I
2190       LET D1 = D(J + 1) * Q(I)
2200       LET D(J + 1) = A(J)
2210       LET A(J) = D(J) - L(I + 1) * A(J) - D1
2220       LET C(J) = C(J) + S(I + 1) * A(J)
2230    NEXT J
2240    LET C(I + 1) = S(I + 1)
2250    LET A(I + 1) = 1
2260    LET D(I + 2) = 0
2270    IF I1 = 0 THEN 2320
2280    IF I <> M3 THEN 2320
2290    FOR J = 1 TO M1
2300       LET C(J) = B(J)
2310    NEXT J
2320 NEXT I
2330 LET D(1) = 1
2340 LET B(1) = 1
2350 LET F(1) = C(1)
2360 FOR I = 2 TO M1
```

```
2370    LET D(I) = 1
2380    LET B(I) = B1 * B(I - 1)
2390    LET F(1) = F(1) + C(I) * B(I)
2400    REM WORK OUT EXPLICIT POWER SERIES IN UNSCALED X, & ADD
2410    REM INTO THE COEFFICIENTS F( ) THE RELEVANT CONTRIBUTIONS
2420 NEXT I
2430 FOR J = 2 TO M1
2440    LET D(1) = D(1) * A1
2450    LET F(J) = C(J) * D(1)
2460    LET K1 = 2.
2470    LET J1 = J + 1
2480    IF J1 > M1 THEN 2560
2490    FOR I = J1 TO M1
2500       LET D(K1) = A1 * D(K1) + D(K1 - 1)
2510       LET F(J) = F(J) + C(I) * D(K1) * B(K1)
2520       LET K1 = K1 + 1
2530    NEXT I
2540 NEXT J
2550 REM CALCULATE YCALC & RESIDUAL FOR EACH POINT (ON ORIGINAL SCALE).
2560 FOR I = 1 TO N
2570    LET J = N2 + 1
2580    LET Y5 = F(J)
2590    IF N2 = 0 THEN 2630
2600    FOR K = 1 TO N2
2610       LET Y5 = F(J - 1) + (X(I) * Y5)
2620       LET J = J - 1
2630    NEXT K
2640    LET Z(I) = Y5 * Y4 + Y3
2650    LET R(I) = (V(I) - Y5) * Y4
2660 NEXT I
2670 REM CONVERT COEFF ARRAY F( ) BACK TO ORIGINAL SCALE
2680 LET F(1) = (F(1) * Y4) + Y3
2690 FOR I = 2 TO M1
2700    LET F(I) = F(I) * Y4
2710 NEXT I
2720 RETURN
2730 REM ENTER IF ERRORS DETECTED
2740 PRINT "JOB TERMINATED BY PROGRAM BECAUSE SUM OF WEIGHTS = 0"
2750 STOP
2760 REM SUBROUTINE TO CHECK THAT DATA ARE CORRECT & ALTER IF NECESSARY
2770 PRINT "ARE THE DATA VALUES ENTERED CORRECT?";
2780 REM A4 SHOULD BE SET TO THE NUMBER OF LINES ON THE VDU
2790 LET A4 = 20
2800 GOSUB 3590
2810 IF Q$ = "YES" THEN 3570
2820 PRINT "HERE IS A LIST OF THE CURRENT DATA"
2830 PRINT "LINE NUMBER", "X", "Y", "WEIGHT"
2840 FOR I = 1 TO N
2850    PRINT I, X(I), Y(I), W(I)
2860    IF INT(I / (A4 - 1)) * (A4 - 1) <> I THEN 2900
2870    PRINT "WOULD YOU LIKE TO CONTINUE LISTING";
2880    GOSUB 3590
2890    IF Q$ = "NO" THEN 2910
2900 NEXT I
2910 PRINT "TYPE R TO REPLACE";
2920 IF I$ = "NO" THEN 2940
2930 PRINT " AN EXISTING LINE OF DATA"
2940 IF N = 100 THEN 2990
2950 PRINT TAB(5); " A TO ADD";
2960 IF I$ = "NO" THEN 2980
2970 PRINT " AN EXTRA LINE"
```

216

```
2980 IF N = 1 THEN 3020
2990 PRINT TAB(5); " D TO DELETE";
3000 IF I$ = "NO" THEN 3020
3010 PRINT " AN EXISTING LINE"
3020 PRINT TAB(5); " L TO LIST";
3030 IF I$ = "NO" THEN 3050
3040 PRINT " THE DATA"
3050 PRINT "   OR C TO CONTINUE";
3060 IF I$ = "NO" THEN 3080
3070 PRINT " THE CALCULATION"
3080 INPUT Q$
3090 IF Q$ = "R" THEN 3190
3100 IF N = 100 THEN 3130
3110 IF Q$ = "A" THEN 3330
3120 IF N = 1 THEN 3140
3130 IF Q$ = "D" THEN 3430
3140 IF Q$ = "L" THEN 2820
3150 IF Q$ = "C" THEN 3570
3160 PRINT "REPLY '"; Q$; "' NOT UNDERSTOOD."
3170 GOTO 2910
3180 REM REPLACE LINE
3190 PRINT "TYPE THE LINENUMBER OF THE LINE TO BE REPLACED";
3200 INPUT I
3210 IF I <> INT(I) THEN 3230
3220 IF (I - 1) * (I - N) < = 0 THEN 3260
3230 PRINT "LINENUMBER MUST BE AN INTEGER IN THE RANGE 1 -"; N
3240 PRINT "RE-";
3250 GOTO 3190
3260 PRINT "TYPE THE CORRECT LINE TO REPLACE THE ONE WHICH IS WRONG:"
3270 PRINT "X, Y, WEIGHT"
3280 INPUT X(I), Y(I), W(I)
3290 IF W(I) >= 0 THEN 3400
3300 PRINT "NEGATIVE WEIGHTS ARE IMPOSSIBLE - LAST LINE REJECTED"
3310 GOTO 3260
3320 REM ADD A NEW LINE
3330 LET N = N + 1
3340 PRINT "TYPE THE ADDITIONAL LINE OF DATA AS SHOWN:"
3350 PRINT "X, Y, WEIGHT"
3360 INPUT X(N), Y(N), W(N)
3370 IF W(N) >= 0 THEN 3400
3380 PRINT "NEGATIVE WEIGHTS ARE IMPOSSIBLE - LAST LINE REJECTED"
3390 GOTO 3340
3400 PRINT "OK"
3410 GOTO 2910
3420 REM DELETE A LINE
3430 PRINT "TYPE THE LINENUMBER OF THE LINE TO BE DELETED"
3440 INPUT J
3450 IF (J - 1) * (J - N) >0 THEN 3470
3460 IF J = INT(J) THEN 3490
3470 PRINT "LINENUMBER MUST BE AN INTEGER IN THE RANGE 1 -"; N
3480 GOTO 3430
3490 FOR I = J + 1 TO N
3500    LET X(I - 1) = X(I)
3510    LET Y(I - 1) = Y(I)
3520 NEXT I
3530 LET N = N - 1
3540 PRINT "OK"
3550 IF J > N THEN 2910
3560 GOTO 2820
3570 RETURN
3580 REM SUBROUTINE TO CHECK REPLIES
```

```
3590 IF I$ = "NO" THEN 3610
3600 PRINT " TYPE YES OR NO & PRESS RETURN."
3610 PRINT
3620 INPUT Q$
3630 IF Q$ = "YES" THEN 3670
3640 IF Q$ = "NO" THEN 3670
3650 PRINT "REPLY '"; Q$; "' NOT UNDERSTOOD.";
3660 GOTO 3600
3670 RETURN
3680 END
```

but otherwise a subroutine (lines 1160–2720) is called (line 620) to fit the polynomial.

The polynomial fitting subroutine is based of Forsythe's method of orthogonal polynomials using α and β. The main steps are as follows:

(i) A number of initial values are set (lines 1180–1320).

(ii) The maximum and minimum values of x and y are found (lines 1330–1480).

(iii) A check is performed to make sure that all of the weights are not zero (lines 1490–1500).

(iv) If all of the y values are the same, a zero order polynomial is returned (lines 1530–1560).

(v) The y values are scaled to lie in the range $+1$ to -1 (lines 1570–1650), and the x values are scaled into the range $+2$ to -2 (lines 1700–1750).

(vi) A loop (lines 1760–2320) performs a number of calculations for each order of polynomial tested:

(a) Forsythe's α values are calculated (lines 1780–1830).

(b) The current orthogonal polynomial is calculated (lines 1870–1950) and the previous orthogonal polynomial is stored.

(c) Forsythe's β values are calculated (lines 1960–1970).

(d) The sum of the residuals squared and the goodness of fit are calculated (lines 1990–2000).

(e) If the program is selecting the 'best' polynomial the lines from 2010 to 2150 are used. (The method used is a development from Algorithm 296 by G. J. Makinson in *Communications of the ACM 1967*, **10**, 2, 87–88 with slight improvement.) This calculates the goodness of fit for each of the polynomials from zero order to the maximum order, and selects the order for which the goodness of fit is a local minimum. If there is more than one local minimum, then the later one (higher order) must be better than the first by an empirical factor of 0.6 for it to be selected.

(f) The coefficients c_0, c_1, $c_2, \ldots, c_m$ for the orthogonal polynomials are evaluated (lines 2160–2230) for the current order of polynomial being calculated by the main loop (lines 1770–2320). If the program is choosing the best order, the coefficients for the orthogonal polynomials corresponding to the best polynomial fit are stored in the C array (lines 2270–2310).

The sequence of operations (a), (b), (c), (d), (e) and (f) is repeated for each order of the curve fitting polynomial up to the order specified by the user, or up to the maximum order if the computer is choosing the best polynomial.

(vii) The coefficients $c_0, c_1, c_2, \ldots, c_m$ for the othogonal polynomials are used to calculate the coefficients for the polynomial through the scaled data (lines 2360–2540).

(viii) The scaled residuals are calculated for each data point using the scaled data and the scaled polynomial coefficients. These residuals are then scaled to match the original data (lines 2550–2660).

(ix) Finally the coefficients for the power series polynomial through the scaled data values are converted to give the coefficients a, b, c, d, ... for the polynomial through the original (unscaled) data points (lines 2670–2710).

On returning to the main part of the program from the polynomial fitting subroutine, a message is printed either giving the order of the polynomial specified (line 660) or the maximum order tested for and the best order found (lines 680–690). A table is then printed (lines 710–740) showing the goodness of fit of each of the polynomials from order zero up to either the order specified or the maximum order tested.

The coefficients of the best order polynomial (chosen by the computer) or of the order specified by the user are then printed (lines 760–830). A table of residuals may optionally be printed (lines 860–960).

Finally, the user is asked if another run is required (line 980). The answer which must be YES or NO is checked in a subroutine (lines 3580–3670). Should another run be required, the option is given of re-running with the data values already entered (when the data may be edited or a different order polynomial specified) or alternatively to enter a completely new set of data.

13

Application of Statistics to Biological Assays

Biological assays are used to compare the potency of a test preparation (drug, antibiotic or vitamin) with that of a standard whose potency is known in arbitrary units. The principle is to compare the actions of the test and standard on living material under strictly comparable conditions. Since the biological response to a preparation is inherently variable, the estimated potency of the test material is subject to random errors. It is assumed that systematic errors due to weighing and dilution are small and can be ignored, but it is essential that calculations of the random errors be made. Unless a very large number of observations are made, the estimated errors will themselves be subject to appreciable error, and methods for calculating the fiducial limits are given.

Introduction

A number of crude extracts and preparations are still used in medicine as drugs or vitamins. In many cases the amount of active ingredient cannot be determined by conventional physical or chemical methods for the following reasons:

(i) The active ingredient may not be known.
(ii) There may be no suitable analytical method.
(iii) Other materials in the crude extract may interfere with the analysis, and

separation of the active ingredient may be impracticable or impossible.

For these reasons the activity of crude extracts is measured by biological assay; that is, by measuring the effect of a known dose of the test material on animals, isolated living material or bacteria, and comparing the effect with that produced by known standards. Biological assays are both selective, and hopefully specific, enabling crude extracts to be assayed.

The sort of tests carried out on living material may be to inject a laboratory animal with a preparation such as a hormone or vitamin, and observe the change in sugar or other chemical in the blood, or the change in weight of the whole animal or one of its organs. Irritability tests are performed by injecting preparations into the eyes of animals. Whole animal work of this kind is being replaced by model systems using isolated tissue or other cell or bacterial cultures. For example, the effect of adrenalin or other materials on isolated animal gut or muscle can be studied conveniently, or the growth of a bacterial colony in a culture medium containing a vitamin or an antibiotic provides a simple experimental method.

All assay experiments are based on the principle that if the response of a tissue to a given dose of the preparation under test is the same as that from a known dose of standard, then the two doses contain the same quantity of active ingredient or drug.

In theory it is possible to measure the

response to a 'test' dose and to compare this with the responses from a series of standard doses until the responses match. In practice this is extremely difficult because of the variation in response from apparently identical samples which is inherent in all work using living material. In order to reduce the random variation it is essential to run replicate tests and to average the results. (When this necessitates killing a large number of test animals, the statistical reliability is sometimes outweighed by the financial cost.)

Threshold Dose Assay

In this simple form of assay, suitable dilutions of the unknown preparation under test and the standard preparation are injected into animals. The dose—that is the volume of preparation injected—is varied to find the minimum volume required to produce a given effect in each animal. The experiment is repeated several times. The volumes actually injected are then corrected to allow for any dilution and are divided by the body weight to give a series of X_u and X_s values. These are the undiluted volumes per unit of body weight for the test and standard preparations respectively which produce the threshold response. From these, the mean values $\bar{X}_u$ and $\bar{X}_s$ are calculated, and the potency ratio of the test/standard is given by $\bar{X}_s/\bar{X}_u$ since the potency is inversely proportional to the volume required to produced the response.

The limits of error in the potency ratio are calculated from the standard errors of the means, and the appropriate Student t-value. The method is best illustrated by an example.

Example 1

A digitalis preparation of unknown potency was assayed by measuring the minimum dose required to arrest the heart of six guinea pigs and comparing with the doses of a known standard on a further six guinea pigs. The results shown in Table 13.1 were obtained with a standard of potency 1.251 units/ml.

Table 13.1

Volume of unknown preparation (X_u)	Volume of standard preparation (X_s)
0.71	0.92
0.80	0.85
0.75	0.94
0.82	0.88
0.86	0.81
0.79	0.97
$\Sigma X_u = 4.73$	$\Sigma X_s = 5.37$
Mean, $\bar{X}_u = 0.788$	Mean, $\bar{X}_s = 0.895$

The potency ratio (test/standard)

$$= \frac{\bar{X}_s}{\bar{X}_u} = \frac{0.895}{0.788} = 1.136$$

The unknown preparation is thus 1.136 times as potent as the standard. The absolute potency of the unknown is thus $1.136 \times 1.251 = 1.421$ units per ml.

Limits of Error for the Threshold Dose Assay

The limits of error are calculated in the following way:

(i) The estimated standard deviations of X_u and X_s are calculated.
(ii) The standard errors of $\bar{X}_u$ and $\bar{X}_s$ are estimated.
(iii) From these the standard error in the potency ratio is estimated.
(iv) A t-table is then used to obtain the number of standard errors variation for a given probability level, which is frequently chosen as 95%.
(v) The limits of error in the absolute potency are evaluated from the standard error and the t-value. The *British Pharmacopoeia* expresses the limits of error as a percentage of the absolute potency.

221

(i) The estimated standard deviation of X_u is calculated:

Estimated standard deviation of $X_u =$

$$= \sqrt{\frac{\Sigma(X_u - \bar{X}_u)^2}{(n_u - 1)}}$$

Thus the estimated standard deviation of X_u is

$$\sqrt{\frac{\begin{array}{l}[(0.71 - 0.788)^2 + (0.80 - 0.788)^2 + \\ + (0.75 - 0.788)^2 + (0.82 - 0.788)^2 + \\ + (0.86 - 0.788)^2 + (0.79 - 0.788)^2]\end{array}}{6 - 1}}$$
$$= 0.0527$$

Similarly the estimated standard deviation of X_s is 0.0597.

(ii) The mean values $\bar{X}_u$ and $\bar{X}_s$ are themselves subject to error. This is because a repeat of the experiment would yield slightly different means due to experimental error. Some indication of the variability of the means is required. This is called the standard error of the mean and is calculated as: Standard error of $\bar{X}_u =$

$$= \frac{\text{Estimated standard deviation of } X_u}{\sqrt{n_u}}$$

$$= 0.0215$$

Similarly the standard error of $\bar{X}_s$ is 0.0244.

(iii) Since the mean values $\bar{X}_u$ and $\bar{X}_s$ used to calculate the potency ratio are each subject to error, the potency ratio is also subject to error. The standard error of the quotient of two terms may be expressed as: Standard error of A/B

$$= \frac{\bar{B}}{\bar{A}} \sqrt{\left[\frac{(\text{Standard error of } A)^2}{\bar{A}^2} + \right.}$$

$$\left. + \frac{(\text{Standard error of } B)^2}{\bar{B}^2}\right]$$

(This is discussed in Chapter 1 and derived in Appendix 8.) Thus the standard error of the potency ratio is

$$\frac{\bar{X}_s}{\bar{X}_u} \sqrt{\left[\frac{(\text{Standard error of } \bar{X}_s)^2}{\bar{X}_s^2} + \right.}$$

$$\left. + \frac{(\text{Standard error of } \bar{X}_u)^2}{\bar{X}_u^2}\right] \quad (1)$$

Since the standard error of the potency ratio is to be expressed as a fraction of the absolute potency, Equation 1 becomes:

$$\sqrt{\left[\frac{(\text{Standard error of } \bar{X}_s)^2}{\bar{X}_s^2} + \right.}$$

$$\left. + \frac{(\text{Standard error of } \bar{X}_u)^2}{\bar{X}_u^2}\right]$$

In this case

$$\sqrt{\left(\frac{0.0244^2}{0.0895^2} + \frac{0.0215^2}{0.788^2}\right)}$$

$$= 0.0386$$

(iv) It is never possible to guarantee 100% that the true potency of a preparation will lie within the limits derived from an experiment such as a biological assay. In practice, limits of 95% are commonly used. This means that the true potency may be expected to lie within the calculated limits in 95% of the experiments undertaken. The 95% limit is calculated using the standard error of the potency ratio calculated above and a value from the t-distribution. The t-values are tabulated in Appendix 6 for varying probability levels and number of degrees of freedom. The 95% limit chosen above means that there is a 5% chance of the true potency being outside the limit of error. The 5% two-tailed probabilities should be used from the table, since it does not matter if the true potency is above or below the limits of error. The number of degrees of freedom is the total number of animals tested minus two. In this case for 10 degrees of freedom the 5% two-tailed t-value is 2.228.

(v) The limits of error are calculated as:

$$\pm 2.228 \times 0.0386$$

$$= \pm 0.0859 \text{ as a fraction of the potency}$$

$$= \pm 8.6\%$$

The result shows that there is a 95% chance that the potency of the unknown lies in the range $1.421 \pm 8.6\%$; that is 1.299–1.543 unit/ml.

Description of Threshold Dose Assay Program (Program 13.1)

After printing a heading (lines 20–30), a message (line 50) asks for the potency of the standard. The value typed in is checked to ensure that it is greater than zero. If not, an error message is printed (line 80) and the user is asked to re-input the value.

A message (line 110) asks the user how many animals were tested with the standard. The number typed is checked to make sure that it is an integer, and that it is in the range 2–50 inclusive. The lower limit is imposed by the calculation for fiducial limits and the upper limit is arbitrarily set at 50 by the dimensioning of the arrays in line 10. If the value typed is unacceptable, one of the error messages (line 140 or line 170) is printed and the user asked to retype the correct value.

Next a message (lines 190–200) asks the user to type in the threshold doses of the standard preparation. These are input in a loop from lines 220 to 280, and are checked to ensure that they are greater than zero.

The number of animals tested with the unknown preparation, and the threshold doses of the unknown are input in a similar manner, and are subjected to the same checks in lines 290–460.

The mean volumes of standard and unknown are calculated and printed (lines

Program 13.1   Trial run

```
THRESHOLD DOSE ASSAY
========= ==== =====

TYPE THE POTENCY OF THE STANDARD AND PRESS RETURN
? 1.251

TYPE NUMBER OF ANIMALS TESTED WITH STANDARD AND PRESS RETURN
? 6
TYPE IN THE THRESHOLD DOSES OF STANDARD ONE AT A TIME
AND PRESS RETURN AFTER EACH VALUE
? 0.92
? 0.85
? 0.94
? 0.88
? 0.81
? 0.97
TYPE NUMBER OF ANIMALS TESTED WITH UNKNOWN AND PRESS RETURN
? 6
TYPE IN THE THRESHOLD DOSES OF UNKNOWN ONE AT A TIME
AND PRESS RETURN AFTER EACH VALUE
? 0.71
? 0.80
? 0.75
? 0.82
? 0.86
? 0.79
MEAN VOLUME OF STANDARD = 0.895
MEAN VOLUME OF UNKNOWN  = 0.788333
POTENCY RATIO (TEST/STANDARD) = 1.13531
ABSOLUTE POTENCY OF UNKNOWN = 1.42027

STANDARD ERROR OF STANDARD DOSE = 0.0243242
STANDARD ERROR OF TEST     DOSE = 0.0215123

FIDUCIAL LIMITS FOR ABSOLUTE POTENCY ARE + OR - 8.6 %
THAT IS 1.3 TO 1.54

END OF JOB
```

```
10 DIM U(50), S(50)
20 PRINT TAB(20); "THRESHOLD DOSE ASSAY"
30 PRINT TAB(20); "========= ==== ====="
40 PRINT
50 PRINT "TYPE THE POTENCY OF THE STANDARD AND PRESS RETURN"
60 INPUT P
70 IF P > 0 THEN 100
80 PRINT "VALUE REJECTED - POTENCY MUST BE GREATER THAN ZERO"
90 GOTO 40
100 PRINT
110 PRINT "TYPE NUMBER OF ANIMALS TESTED WITH STANDARD AND PRESS RETURN"
120 INPUT N1
130 IF N1 = INT(N1) THEN 160
140 PRINT "THERE MUST BE A WHOLE NUMBER OF ANIMALS. RE-"
150 GOTO 110
160 IF (N1 - 2) * (N1 - 50) <= 0 THEN 190
170 PRINT "VALUE MUST BE BETWEEN 2 AND 50. RE-"
180 GOTO 110
190 PRINT "TYPE IN THE THRESHOLD DOSES OF STANDARD ONE AT A TIME"
200 PRINT "AND PRESS RETURN AFTER EACH VALUE"
210 LET S1 = 0
220 FOR I = 1 TO N1
230    INPUT S(I)
240    IF S(I) > 0 THEN 270
250    PRINT "DOSE MUST BE POSITIVE - RETYPE CORRECTLY"
260    GOTO 230
270    LET S1 = S1 + S(I)
280 NEXT I
290 PRINT "TYPE NUMBER OF ANIMALS TESTED WITH UNKNOWN AND PRESS RETURN"
300 INPUT N2
310 IF N2 = INT(N2) THEN 340
320 PRINT "THERE MUST BE A WHOLE NUMBER OF ANIMALS"
330 GOTO 300
340 IF (N2 - 2) * (N2 - 50) <= 0 THEN 370
350 PRINT "VALUE MUST BE BETWEEN 2 AND 50. RE-"
360 GOTO 300
370 PRINT "TYPE IN THE THRESHOLD DOSES OF UNKNOWN ONE AT A TIME"
380 PRINT "AND PRESS RETURN AFTER EACH VALUE"
390 LET U1 = 0
400 FOR I = 1 TO N2
410    INPUT U(I)
420    IF U(I) > 0 THEN 450
430    PRINT "DOSE MUST BE POSITIVE - RETYPE CORRECTLY"
440    GOTO 410
450    LET U1 = U1 + U(I)
460 NEXT I
470 LET M1 = S1 / N1
480 LET M2 = U1 / N2
490 PRINT "MEAN VOLUME OF STANDARD ="; M1
500 PRINT "MEAN VOLUME OF UNKNOWN  ="; M2
510 PRINT "POTENCY RATIO (TEST/STANDARD) ="; M1 / M2
520 PRINT "ABSOLUTE POTENCY OF UNKNOWN ="; M1 / M2 * P
530 PRINT
540 REM *** CALCULATE STANDARD ERROR OF STANDARD DOSE
550 LET E1 = 0
560 FOR I = 1 TO N1
570    LET E1 = E1 + (S(I) - M1) * (S(I) - M1)
580 NEXT I
590 LET V1 = E1 / (N1 * (N1 - 1))
600 PRINT "STANDARD ERROR OF STANDARD DOSE ="; SQR(V1)
610 REM *** CALCULATE STANDARD ERROR OF UNKNOWN DOSE
```

224

```
620 LET E2 = 0
630 FOR I = 1 TO N2
640    LET E2 = E2 + (U(I) - M2) * (U(I) - M2)
650 NEXT I
660 LET V2 = E2 / (N2 * (N2 - 1))
670 PRINT "STANDARD ERROR OF TEST     DOSE ="; SQR(V2)
680 PRINT
690 REM *** CALCULATE STANDARD ERROR OF POTENCY RATIO
700 LET E = SQR(V1 / (M1 * M1) + V2 / (M2 * M2))
710 REM CALCULATE CONFIDENCE LIMITS USING INFINITY VALUES
720 LET T = 1.95996
730 REM RE-CALCULATE 95% CONFIDENCE LIMIT IF DEGREES OF FREEDOM <= 30
740 FOR I = 1 TO 30
750    READ T1
760    IF I <> N1 + N2 - 2 THEN 780
770    LET T = T1
780 NEXT I
790 DATA 12.706, 4.303, 3.182, 2.776, 2.571, 2.447, 2.365
800 DATA 2.306, 2.262, 2.228, 2.201, 2.197, 2.160, 2.145, 2.131
810 DATA 2.120, 2.110, 2.101, 2.093, 2.086, 2.080, 2.074, 2.069
820 DATA 2.064, 2.060, 2.056, 2.052, 2.048, 2.045, 2.042
830 LET L =INT(1000 * E * T + 0.5) / 10
840 PRINT "FIDUCIAL LIMITS FOR ABSOLUTE POTENCY ARE + OR -"; L ;"%"
850 LET L1 = INT(100 * M1 / M2 * (1 - E * T) * P + 0.5) / 100
860 LET L2 = INT(100 * M1 / M2 * (1 + E * T) * P + 0.5) / 100
870 PRINT "THAT IS"; L1; "TO"; L2
880 PRINT
890 PRINT "END OF JOB"
900 END
```

470–500). The potency ratio of test/standard is calculated from these means and printed in line 510. Using the known potency of the standard and the potency ratio, the absolute potency of the unknown is calculated and printed (line 520).

The standard error of the mean standard dose is calculated and printed (lines 540–600) and the standard error of the mean unknown dose is calculated and printed (lines 610–670). From these the standard error of the potency ratio is evaluated (lines 690–700).

The value from the *t*-distribution corresponding to a probability of 95% for the appropriate number of degrees of freedom is read from a table of values held in DATA statements (lines 790–820), provided that the number of degrees of freedom does not exceed 30. The number of degrees of freedom is the total number of animals tested minus two. If more animals were used, then the infinity value is used for *t*. The *t*-value and the standard error are combined to give the confidence (fiducial) limits, which are calcu-

lated and printed (lines 830–870). The percentage fiducial limit is rounded to give one decimal figure and the range of values is rounded to give two decimal figures.

Finally, a finishing message is printed (line 890).

Three-Point Assay

Another simple form of assay is known as a 'bracketing' assay. This requires the response from the 'test' sample plus two responses from different doses of a standard sample. One standard must produce a larger response than the 'test' response whilst the other standard must produce a smaller response. Clearly the amount of drug in the 'test' sample lies between the amount in the two standard drug doses. The simplest approximation is to estimate the 'test' concentration as the mean of the two standard concentrations.

Assumptions about the relationship between response and dose are necessary to estimate the quantity of the drug in the 'test' sample to any greater accuracy. For most agonists (drugs which produce a response, as opposed to antagonists which inhibit a response) and tissues, the graph of logarithm (dose) against response is sigmoid in shape. However, the graph is roughly linear in the middle ranges between about 25% and 75% of the maximum response which can be obtained from the tissue (Fig. 13.1).

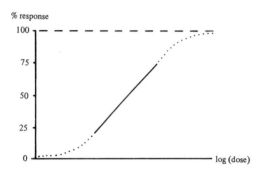

Fig. 13.1

Derivation of the Three-Point Assay

Consider the assay of a mixed extract. Let the smaller dose of standard be s_1 and the response it produces is S_1. Similarly the larger dose of standard is s_2 and the response produced is S_2. The dose of the 'unknown test' is u and the response it produces is U. Finally let the dose of standard equivalent to u be ψ.

It is required to estimate the potency ratio ψ/u from s_1, s_2, S_1, S_2 and U (Fig 13.2).

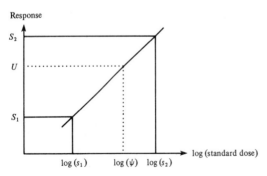

Fig. 13.2

The procedure for estimating the potency ratio is as follows:

(i) Estimate the slope, b, of the log (standard dose)/response line.

(ii) Compare the responses from u and s_1 and obtain the difference in response $U - S_1$.

(iii) Using the difference from (ii) and the slope from (i), the difference in log (standard dose) between s_1 and ψ is estimated.

(iv) Calculate the equivalent standard dose ψ from the difference in log (standard dose) and obtain the potency ratio ψ/u.

Details of these steps are as follows.

(i) The slope b is estimated as the difference in observed response divided by the difference in log (standard dose):

$$\text{Slope } b = \frac{S_2 - S_1}{\log(s_2) - \log(s_1)}$$

$$= \frac{S_2 - S_1}{\log(s_2/s_1)} \qquad (2)$$

(ii) The difference is response between 'test' u and the smaller dose, s_1 is $U - S_1$.

(iii) Dividing the difference by the slope gives the difference in log (standard dose). Hence

$$\log(\psi) - \log(s_1) = \frac{U - S_1}{\left[\dfrac{S_2 - S_1}{\log(s_2/s_1)}\right]}$$

Rearranging,

$$\log(\psi/s_1) = \frac{U - S_1}{S_2 - S_1} \log\left(\frac{s_2}{s_1}\right)$$

Hence

$$\psi/s_1 = \text{antilog}\left\{\frac{U - S_1}{S_2 - S_1} \log\left(\frac{s_2}{s_1}\right)\right\}$$

$$\psi = s_1 \text{ antilog}\left\{\frac{U - S_1}{S_2 - S_1} \log\left(\frac{s_2}{s_1}\right)\right\} \qquad (3)$$

The potency ratio, R, for unknown/standard is:

$$R = \frac{\psi}{u} = \frac{s_1}{u} \text{ antilog}\left\{\frac{U - S_1}{S_2 - S_1} \log\left(\frac{s_2}{s_1}\right)\right\}$$

The absolute potency of the unknown test preparation can then be calculated from the known potency of the standard and the potency ratio R.

Limits of Error for the Three-Point Assay

Although a value has been obtained for the potency, it is subject to the usual random errors associated with biological systems. Because of this it is essential that the magnitude of the errors is estimated in order that limits of error (fiducial limits) may be placed on the calculated potency. The upper and lower limits of error may be calculated from:

$$\text{antilog} \left\{ \frac{M}{(1-g)} \pm \frac{t}{b(1-g)} \right.$$
$$\left. \sqrt{\frac{3}{2} V (1-g) + \frac{2VM^2}{[\log(s_2/s_1)]^2}} \right\} \quad (4)$$

where M is the log (potency ratio) $= \log(R)$, b is the slope of the log (dose)/response line, s_1 and s_2 are the doses of the standard, t is the 95% probability level (taken as 5% in the two tailed t-table in Appendix 6) with $3(n-1)$ degrees of freedom, n is the number of assays performed on each sample, V is the pooled estimate of the variance of the means, given by:

$$V = \frac{\Sigma(S_1 - \bar{S}_1)^2 + \Sigma(S_2 - \bar{S}_2)^2 + \Sigma(U - \bar{U})^2}{3n(n-1)}$$

and g is Finney's significance of regression and is given by:

$$\frac{2Vt^2}{[\log(s_2/s_1)]^2 b^2} = \frac{2Vt^2}{(S_2 - S_1)^2}$$

Note that when g is greater than one the result is of no value. S_1, S_2 and U are the mean responses to doses of s_1, s_2 and u. The use of the three-point assay is generally restricted to applications where high accuracy is not needed, or the test solution is in short supply. The four-point assay described next is more widely used.

Four-Point Assay (2 × 2 Assay)

This assay provides a means of determining the potency of an 'unknown' test solution by comparison of its biological activity with a known standard. Two different doses of standard solution are administered, and the mean response from a number of replicates is obtained for each dose. In a similar fashion, two different doses of test solution are used and their mean responses are calculated. The use of two standards and two 'unknowns' accounts for the name 2 × 2 assay.

The following symbols are used:

n	number of replicates used for each drug
s_1, s_2	smaller and larger doses of standard
S_1, S_2	mean response to doses of standard s_1 and s_2
u_1, u_2	smaller and larger doses of unknown
U_1, U_2	mean response to doses of unknown u_1 and u_2
ψ_1, ψ_2	equivalent standard dose to u_1 and u_2, i.e. dose of standard which would produce the same response as was produced by u_1 and u_2
b	slope of log (dose)/response line
R	potency ratio of test/standard
M	log (potency ratio)

If necessary the standard or the 'unknown' preparation is diluted by a known amount in order to make the responses S_1 and U_1 approximately equal.

To simplify the calculations it is usual to choose the same ratio between the doses u_1 and u_2 as was used for the doses s_1 and s_2. That is

$$\frac{s_2}{s_1} = \frac{u_2}{u_1} \quad (5)$$

This simplification is assumed in the subsequent equations.

In practice it is common to make the doses s_1 and u_1 equal, and consequently s_2 and u_2 will also be equal. Though this simplifies the derivation, the general case is worked out which does *not* make this assumption.

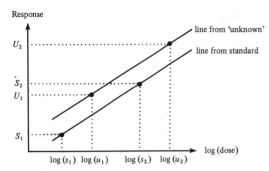

Response

U_2 ⋯⋯⋯⋯⋯⋯⋯⋯ line from 'unknown'

line from standard

S_2 ⋯⋯⋯⋯⋯⋯

U_1 ⋯⋯⋯⋯⋯

S_1 ⋯⋯

log (dose)

log (s_1) log (u_1) log (s_2) log (u_2)

Fig. 13.3   Note that the two slopes are equal and that the differences in response $(U_2 - S_2)$ and $(U_1 - S_1)$ are equal.

Once the values for S_1, S_2, U_1 and U_2 have been measured, the four mean values for the responses are plotted against log (dose). Ideally two parallel lines should result (Fig. 13.3).

The potency ratio may be defined as the dose of standard divided by the dose of test which produce the same response. Ideally the log (potency ratio) can be obtained directly from the graph (Fig. 13.4).

In practice the slopes of the two lines are not usually parallel (Fig. 13.5).

The analysis required to obtain the potency ratio is more complicated. The result is given in Equation 12, which may be derived as follows:

(i) The slope of log (dose)/response line is estimated from the standards in exactly the

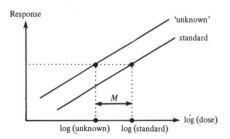

Response

'unknown'

standard

M

log (dose)

log (unknown)   log (standard)

Fig. 13.4 M is log (standard) − log (unknown) = log (dose of standard/dose of unknown) = log (potency ratio).

same manner as in the three-point assay, using the values of s_1, s_2, S_1 and S_2.

(ii) The slope of the log (dose)/response line is estimated from the 'test' using the values of u_1, u_2, U_1 and U_2.

(iii) The two slopes should be the same except for experimental error, since the responses are caused by the same drug in the same ratio of doses. The best estimate for the slope of the line is the average of the values obtained in (i) and (ii).

(iv) The responses from doses u_1 and s_1 are compared, and the difference in response $U_1 - S_1$ is obtained.

(v) The responses from doses u_2 and s_2 are compared, and the difference in response $U_2 - S_2$ is obtained.

(vi) The differences in response obtained in (iv) and (v) should be the same, but due to experimental error they will probably differ. The best estimate for the difference is therefore the mean of the two values. A theoretical justification for the two differences being equal is given below. Since

$$\psi_2/\psi_1 = u_2/u_1$$

Equation 5 becomes

$$\frac{s_2}{s_1} = \frac{\psi_2}{\psi_1}$$

Hence

$$\frac{\psi_1}{s_1} = \frac{\psi_2}{s_2}$$

Taking logarithms,

$$\log (\psi_1) - \log (s_1) = \log (\psi_2) - \log (s_2)$$

Finally, since response should be proportional to log (dose), $(U_1 - S_1)$ should be equal to $(U_2 - S_2)$. In practice, these two differences in response may be different (see Fig. 13.5).

(vii) The difference in log (dose) between s_1 and ψ_1 is calculated using the mean difference in response from (vi) and the mean slope from (iii).

(viii) The value of ψ_1 is then estimated using the result from (vii) and s_1.

(ix) The potency ratio (test/standard) is calculated as ψ_1/u_1.

228

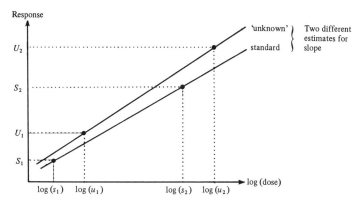

Fig. 13.5

Details of these steps are as follows.

(i) The slope is estimated from the standards as

$$\frac{S_2 - S_1}{\log(s_2/s_1)} \qquad (6)$$

It can be seen that Equation 6 is identical to Equation 2.

(ii) Similarly the slope is estimated from the 'tests' as

$$\frac{U_2 - U_1}{\log(u_2/u_1)} \qquad (7)$$

(iii) The best estimate for the slope b of the line is the average of Equations 6 and 7. Best estimate for slope

$$b = \frac{1}{2}\left[\frac{S_2 - S_1}{\log(s_2/s_1)} + \frac{U_2 - U_1}{\log(u_2/u_1)}\right]$$

$$= \frac{1}{2}(S_2 - S_1 + U_2 - U_1)/\log(s_2/s_1) \qquad (8)$$

since $s_2/s_1 = u_2/u_1$.

(iv) The difference in response is

$$U_1 - S_1. \qquad (9)$$

(v) The difference in response is

$$U_2 - S_2. \qquad (10)$$

(vi) The best estimate for the difference in response between 'test' u and standard s is the average of Expressions 9 and 10

$$= \frac{1}{2}(U_1 - S_1 + U_2 - S_2) \qquad (11)$$

(vii) The estimate for the difference in log (dose) between ψ_1 and s_1 (and also between ψ_2 and s_2) is the difference in response from Equation 11 divided by the slope from Equation 8.

$$\log(\psi_1) - \log(s_1) =$$

$$= \frac{\frac{1}{2}(U_1 - S_1 + U_2 - S_2)}{\frac{1}{2}(S_2 - S_1 + U_2 - U_1)/\log(s_2/s_1)}$$

Hence

$$\log(\psi_1/s_1) =$$

$$= \frac{U_1 - S_1 + U_2 - S_2}{S_2 - S_1 + U_2 - U_1}\log\left(\frac{s_2}{s_1}\right) \qquad (12)$$

(viii)

$$\frac{\psi_1}{s_1} =$$

$$= \text{antilog}\left\{\frac{U_1 - S_1 + U_2 - S_2}{S_2 - S_1 + U_2 - U_1}\log\left(\frac{s_2}{s_1}\right)\right\}$$

$$\qquad (13)$$

Equation 1 also evaluates ψ_2/s_2.

Equation 1 also evaluates ψ_2/s_2. Thus $\psi_1 =$

$$s_1\, \text{antilog}\left\{\frac{U_1 - S_1 + U_2 - S_2}{S_2 - S_1 + U_2 - U_1}\log\left(\frac{s_2}{s_1}\right)\right\}$$

and $\psi_2 =$

$$s_2\, \text{antilog}\left\{\frac{U_1 - S_1 + U_2 - S_2}{S_2 - S_1 + U_2 - U_1}\log\left(\frac{s_2}{s_1}\right)\right\}$$

229

(ix) The potency ratio (test/standard) is calculated as ψ_1/u_1. Potency ratio R =

$$\frac{s_1}{u_1} \text{ antilog} \left\{ \frac{U_1 - S_1 + U_2 - S_2}{S_2 - S_1 + U_2 - U_1} \log \left(\frac{s_2}{s_1}\right) \right\} \quad (14)$$

It is common practice, but not essential, to choose the doses such that $s_1 = u_1$. This has two advantages—firstly, the test and standard solutions may have been diluted with saline or Ringer's solution, and injecting the same volume ensures that the same amount of extraneous material is added in each case; secondly, the arithmetic is simplified in each case and the term s_1/u_1 can be omitted from the calculation of the potency ratio in Equation 14, and Equation 12 becomes the log (potency ratio) M:

$$M = \frac{U_1 - S_1 + U_2 - S_2}{S_2 - S_1 + U_2 - U_1} \log \left(\frac{s_2}{s_1}\right) \quad (15)$$

Advantages of the Four-Point Assay

1. Since two estimates are made for the slope, and these are averaged, a higher degree of accuracy is obtained.
2. The activity of the 'test' solution is based on two different doses of unknown which is more reliable than the single dose used in the three-point assay.

Test of Validity of the Four-Point Assay (Parallelism)

The derivation of the four-point assay is based on the assumption that slopes of the test and standard lines are equal within the limits of experimental error. If the slopes of the two lines differ significantly, this assumption is incorrect and consequently the result of the assay is not valid. The two slopes can be compared by means of a *t*-test to determine whether they differ significantly. The steps are outlined below:

(i) The difference in slopes of the test and standard lines is calculated.

(ii) The standard error of the difference in slope is calculated.

(iii) The difference is divided by the standard error to give a *t*-value, which is looked up in a *t*-significance table.

Details of these steps are as follows.

(i) The slope of the standard line

$$= (S_2 - S_1)/\log(s_2/s_1)$$

The slope of the test line

$$= (U_2 - U_1)/\log(u_2/u_1)$$

$$= (U_2 - U_1)/\log(s_2/s_1)$$

since the ratio of the doses was chosen to be the same (see Equation 5). The difference in slopes is

$$\frac{(S_2 - S_1)}{\log(s_2/s_1)} - \frac{(U_2 - U_1)}{\log(s_2/s_1)} =$$

$$= \frac{(S_2 - S_1 - U_2 + U_1)}{\log(s_2/s_1)} \quad (16)$$

(ii) The standard error of the difference in slopes is calculated as follows. First it should be noted that the errors in the numerator of Equation 16 are due to variations in the animal or tissue, whereas errors in the denominator arise from measuring errors in administering the doses. The latter errors are negligible in comparison with the former errors. The standard error of the difference in slopes is thus:

$$\frac{\text{Standard error of } S_2 - S_1 - U_2 + U_1}{\log(s_2/s_1)} \quad (17)$$

The standard error of $S_2 - S_1 - U_2 + U_1$ cannot be calculated directly. However, in Appendix 8 it is shown that

Variance $(x + y) =$
    Variance (x) + Variance (y)

and

Variance $(x - y) =$
    Variance (x) + Variance (y)

Hence the variance of $S_2 - S_1 - U_2 + U_1$ equals

the variance of S_2 plus the variance of S_1 plus the variance of U_2 plus the variance of U_1.

However, variance is standard error squared, so, if e_1, e_2, e_3 and e_4 are the standard errors of S_2, S_1, U_2 and U_1, the standard error in $S_2 - S_1 - U_2 + U_1$ is:

$$\sqrt{(e_1{}^2 + e_2{}^2 + e_3{}^2 + e_4{}^2)}$$

Equation 17 for the standard error in the difference in slopes can therefore be rewritten

$$\frac{\sqrt{(e_1{}^2 + e_2{}^2 + e_3{}^2 + e_4{}^2)}}{\log(s_2/s_1)} \qquad (18)$$

(In the *British Pharmacopoeia*,

$$e_1{}^2 + e_2{}^2 + e_3{}^2 + e_4{}^2$$

is referred to as $4V$ where V is the average variance of the mean responses.)

(iii) The t-value is calculated as difference in slopes/standard error of difference which is Equation 16 divided by Equation 18

$$t = \frac{\dfrac{S_2 - S_1 - U_2 + U_1}{\log(s_2/s_1)}}{\dfrac{\sqrt{(e_1{}^2 + e_2{}^2 + e_3{}^2 + e_4{}^2)}}{\log(s_2/s_1)}}$$

$$t = \frac{S_2 - S_1 - U_2 + U_1}{\sqrt{(e_1{}^2 + e_2{}^2 + e_3{}^2 + e_4{}^2)}}$$

or

$$\frac{S_2 - S_1 - U_2 + U_1}{\sqrt{(4V)}} \qquad (19)$$

The value of t is looked up in a two-tailed t-table (Appendix 6) with $(4n - 4)$ degrees of freedom to give a probability. A large value of t corresponds to a large difference in slopes and gives a small probability that such a t-value has occurred by chance. If the probability is less than 5% then the difference in slopes invalidates the assay. The value of n is the number of assays performed with each dose.

Limits of Error (Fiducial Limits) for the Four-Point Assay

It is essential to estimate the magnitude of the errors in the potency ratio so that fiducial

limits can be given. The upper and lower fiducial limits may be calculated from

$$\text{antilog} \left\{ \frac{M}{(1-g)} \pm \frac{t}{b(1-g)} \right.$$
$$\left. \sqrt{V(1-g) + \frac{VM^2}{[\log(s_2/s_1)]^2}} \right\} \qquad (20)$$

where M is the log (potency ratio), b is the mean slope of the log (dose)/response lines, s_1 and s_2 are the doses of the standard, t is the 95% probability level (taken as 5% in the two-tailed t-table in Appendix 6) with $4(n-1)$ degrees of freedom, n is the number of assays performed on each sample, V is the pooled estimate of the variance of the means, given by:

$$V = \frac{\begin{array}{c}\Sigma(S_1 - \bar{S}_1)^2 + \Sigma(S_2 - \bar{S}_2)^2 + \\ + \Sigma(U_1 - \bar{U}_1)^2 + \Sigma(U_2 - \bar{U}_2)^2\end{array}}{4n(n-1)} \qquad (21)$$

and g is Finney's significance of regression, given by:

$$\frac{Vt^2}{[\log(s_2/s_1)]^2 \, b^2} =$$
$$= \frac{Vt^2}{\frac{1}{4}(S_2 - S_1 + U_2 - U_1)^2} \qquad (22)$$

If g is greater than 1, the result of the assay is valueless. S_1, S_2, U_1 and U_2 are the mean responses to doses of s_1, s_2, u_1 and u_2.

Derivation of Fiducial Limits

The fiducial limits for M (log potency ratio) are:

$$M + t \cdot (\text{Standard error of } M) \qquad (23)$$

where t is the 95% t-table value of the standard error of M for $4(n-1)$ degrees of freedom. The standard error of M is the standard error of:

$$\frac{U_1 - S_1 + U_2 - S_2}{S_2 - S_1 + U_2 - U_1}$$

since any measuring errors in s_1, s_2, u_1 or u_2 are negligible compared with the large random errors in response. Using the equations for combining variances given in Appendix 8 it follows that:

$$\left(\frac{\text{Standard error of } M}{M}\right)^2 =$$

$$= \left(\frac{\text{Standard error of } U_1 - S_1 + U_2 - S_2}{U_1 - S_1 + U_2 - S_2}\right)^2 +$$

$$+ \left(\frac{\text{Standard error of } S_2 - S_1 + U_2 - U_1}{S_2 - S_1 + U_2 - U_1}\right)^2$$

If e_1, e_2, e_3 and e_4 are the standard errors of S_2, S_1, U_2 and U_1 then Standard error of $U_1 - S_1 + U_2 - S_2$

$$= \sqrt{(e_4{}^2 + e_2{}^2 + e_3{}^2 + e_1{}^2)}$$

and Standard error of $S_2 - S_1 + U_2 - U_1$

$$= \sqrt{(e_1{}^2 + e_2{}^2 + e_3{}^2 + e_4{}^2)}$$

(The *British Pharmacopoeia* uses $4V$ in place of $e_1{}^2 + e_2{}^2 + e_3{}^2 + e_4{}^2$.) Substituting for the standard errors,

$$\left(\frac{\text{Standard error of } M}{M}\right)^2 =$$

$$= \left(\frac{\sqrt{(4V)}}{U_1 - S_1 + U_2 - S_2}\right)^2 +$$

$$+ \left(\frac{\sqrt{(4V)}}{S_2 - S_1 + U_2 - U_1}\right)^2 \quad (24)$$

Reference to Equation 8 shows that $S_2 - S_1 + U_2 - U_1$ is equal to

$$2b \log(s_2/s_1) \quad (25)$$

Further $U_1 - S_1 + U_2 - S_2$ is twice the mean difference in response and is therefore equal to twice the slope times log potency ratio

$$= 2\,bM \quad (26)$$

Substituting Equations 25 and 26 into Equation 24,

$$\left(\frac{\text{Standard error of } M}{M}\right)^2 =$$

$$= \frac{4V}{(2\,bM)^2} + \frac{4V}{[2b \log(s_2/s_1)]^2}$$

$$\left(\frac{\text{Standard error of } M}{M}\right)^2 =$$

$$= \frac{V}{b^2 M^2} + \frac{V}{b^2[\log(s_2/s_1)]^2}$$

Standard error of $M =$

$$= \sqrt{\left\{\frac{V}{b^2} + \frac{VM^2}{b^2[\log(s_2/s_1)]^2}\right\}}$$

Standard error of M =

$$= \frac{1}{b}\sqrt{\left\{V + \frac{VM^2}{[\log(s_2/s_1)]^2}\right\}} \quad (27)$$

Substituting Equation 27 into Equation 23 gives the fiducial limits for M the log (potency ratio) as:

$$M + \frac{t}{b}\sqrt{\left\{V + \frac{VM^2}{[\log(s_2/s_1)]^2}\right\}}$$

hence the fiducial limits of the potency ratio are

$$\text{antilog}\left(M + \frac{t}{b}\sqrt{\left\{V + \frac{VM^2}{[\log(s_2/s_1)]^2}\right\}}\right) \quad (28)$$

It is necessary to correct this equation if the variance V is large. The correction involves Finney's significance of regression g, and if this is included, Equation 28 becomes Equation 20. If the value of g is between 0.1 and 1, Equation 28 may be used to calculate the fiducial limits. If g is between 0.1 and 1, Equation 20 should be used, and if g is 1 or more the assay is valueless.

An Alternative Formula for Calculating Fiducial Limits for 2 × 2 Assay

A number of publications including the *European Pharmacopoeia* and the *Pharmacopoeia of the United States* quote a somewhat different looking equation from Equation 20 for evaluating fiducial limits. This is fiducial limits of potency ratio

$$= \text{antilog}\left(CM \pm\right.$$

$$\left.\pm \sqrt{(C-1)\{CM^2 + [\log(s_2/s_1)]^2\}}\right) \quad (29)$$

where M is the log (potency ratio), s_1 and s_2 are

the smaller and larger standard doses, and C is a measure of the significance of regression and is given by $1/(1-g)$, where g is Finney's significance of regression.

Equation 29 may be derived from Equation 20 as follows. Fiducial limit of potency ratio

$$= \text{antilog}\left\{\frac{M}{(1-g)} \pm \frac{t}{b(1-g)}\right.$$

$$\left.\sqrt{V(1-g)+\frac{VM^2}{[\log(s_2/s_1)]^2}}\right\} \quad (20)$$

$$= \text{antilog}\left\{CM \pm\right.$$

$$\left.\pm \frac{Ct}{b}\sqrt{\frac{V}{C}+\frac{VM^2}{[\log(s_2/s_1)]^2}}\right\}$$

$$= \text{antilog}\left\{CM \pm\right.$$

$$\left.\pm\sqrt{\frac{C^2Vt^2}{Cb^2}+\frac{C^2M^2Vt^2}{[\log(s_2/s_1)]^2b^2}}\right\}$$

substituting g for $Vt^2/(\log(s_2/s_1))^2b^2$ and $g(\log(s_2/s_1))^2$ for Vt^2/b^2

$$= \text{antilog}\left\{CM \pm\right.$$

$$\left.\pm\sqrt{Cg[\log(s_2/s_1)]^2+C^2M^2g}\right\}$$

$$= \text{antilog}\left\{CM \pm\right.$$

$$\left.\pm\sqrt{gC[[\log(s_2/s_1)]^2+CM^2]}\right\} \quad (30)$$

however, $C=1/(1-g)$. Hence

$$C(1-g) = 1 = C-gC$$

Therefore

$$gC = C-1$$

Substituting this into Equation 30, fiducial limit of potency ratio

$$= \text{antilog}\left(CM \pm\right.$$

$$\left.\pm\sqrt{(C-1)\{[\log(s_2/s_1)]^2+CM^2\}}\right) \quad (31)$$

It can be seen that Equation 31 is equivalent to Equation 29.

Example 2

The results shown in Table 13.2 were obtained by a student from a 2×2 assay for histamine using a standard preparation containing 20 μg ml^{-1}.

From Equation 8, the best estimate for the slope b is:

$$b = \tfrac{1}{2}(S_2-S_1+U_2-U_1)/\log(s_2/s_1)$$

$$= \tfrac{1}{2}(7.08-4.1+7.45-4.6)/$$
$$\log(0.10/0.05)$$

$$= 9.68$$

From Equation 11, the best estimate for the difference in response

$$= \tfrac{1}{2}(U_1-S_1+U_2-S_2)$$

$$= \tfrac{1}{2}(4.6-4.1+7.45-7.08)$$

$$= 0.435$$

The best estimate for the difference in log (dose) is

$$0.435/9.68 = 0.0449$$

Table 13.2

	Dose/ml	Responses/cm				Mean
Smaller standard	$s_1=0.05$	$S_1=3.0$	4.5	3.9	5.0	4.1
Larger standard	$s_2=0.10$	$S_2=6.1$	7.4	7.4	7.4	7.08
Smaller 'unknown'	$u_1=0.10$	$U_1=4.7$	4.6	5.1	4.0	4.6
Larger 'unknown'	$u_2=0.20$	$U_2=6.6$	7.4	7.8	8.0	7.45

This is the term inside { } in Equation 13, hence

$$\frac{\psi_1}{s_1} = \text{antilog } \{0.0449\}$$

$$= 1.11$$

Hence ψ_1 is the equivalent standard dose to $u_1 = 1.11 \times 0.05 = 0.0555$. The potency ratio (test/standard) R is $\psi_1/u_1 = 0.0555/0.1 = 0.555$. Alternatively this result can be obtained directly and more simply by evaluating Equation 14:

$$R = \frac{0.05}{0.10} \text{ antilog}$$

$$\left[\frac{4.6 - 4.1 + 7.45 - 7.08}{7.08 - 4.1 + 7.45 - 4.6} \log \left(\frac{0.10}{0.05} \right) \right]$$

$$= 0.555$$

(Thus log potency ratio, $M = \log R = -0.256$.) The absolute potency $= 0.555 \times 20$ $\mu g \ ml^{-1} = 11.1 \ \mu g^{-1}$.

Test for Parallelism

First the standard errors e_1, e_2, e_3 and e_4 of S_2, S_1, U_2 and U_1 are calculated. For example,

$$e_1 = \sqrt{ \left(\frac{\begin{pmatrix} \Sigma \text{ Difference between individual} \\ \text{response and mean response } S_2)^2 \end{pmatrix}}{n(n-1)} \right) }$$

$$e_1 = \sqrt{ \left((6.1 - 7.08)^2 + (7.4 - 7.08)^2 + \frac{+ (7.4 - 7.08)^2 + (7.4 - 7.08)^2}{4(4-1)} \right) }$$

$$= 0.325$$

Similarly, $e_2 = 0.430$, $e_3 = 0.310$ and $e_4 = 0.227$. The average variance

$$V = \tfrac{1}{4}(e_1^2 + e_2^2 + e_3^2 + e_4^2) = 0.110$$

The t-value is obtained from Equation 19 as:

$$t = \frac{7.08 - 4.1 - 7.45 + 4.6}{\sqrt{(4 \times 0.110)}}$$

$$= 0.196$$

The calculated value of $t = 0.196$ is compared with 5% two-tailed t-value given in Appendix 6

234

for $4(4-1) = 12$ degrees of freedom. The value of 0.196 is much smaller than the table value of 2.179, hence there is no significant difference between the slopes for the test and standard lines.

Fiducial Limits

First V is calculated from Equation 21 giving a value of 0.110. (This is identical to the value obtained from the e^2 values previously.) Next, g is evaluated using Equation 22, which requires the table value for t of 2.179 obtained previously.

$$g = \frac{0.110 \times 2.179^2}{\tfrac{1}{4}(7.08 - 4.1 + 7.45 - 4.6)^2}$$

$$= 0.0615$$

For values of g less than 0.1, it is permissible to use the simplified equation for fiducial limits given in Equation 28. Fiducial limit of potency ratio

$$= \text{antilog} \left(-0.256 \pm \frac{2.179}{9.68} \right.$$

$$\left. \sqrt{ \left\{ 0.110 + \frac{0.110(-0.256)^2}{[\log(0.10/0.05)]^2} \right\} } \right)$$

$$= \text{antilog } \{-0.256 \pm 0.098\}$$

$$= 0.443 \text{ to } 0.695$$

Multiplying these values by the absolute potency of the standard (20 $\mu g \ ml^{-1}$) gives fiducial limits for the absolute potency of the unknown as 8.86 to 13.9 $\mu g \ ml^{-1}$. The *British Pharmacopoeia* expresses these limits as percentages of the calculated absolute potency of 11.1 $\mu g \ ml^{-1}$; that is, the limits are -20% and $+25\%$ of the absolute potency. (It may seem strange that the two percentages are different. The two limits on the log (potency) are the same, and this difference in magnitude occurs when antilogarithms are taken.)

For comparison, the fiducial limits are evaluated using the more general formula given in Equation 20, though this is not necessary since g is less than 0.1. Fiducial limit of potency ratio

$$= \text{antilog}\left(\frac{-0.256}{1-0.0615} \pm \frac{2.179}{9.68(1-0.0615)} \cdot \right.$$
$$\left. \sqrt{\left\{0.110(1-0.0615) + \frac{0.110(-0.256)^2}{[\log(0.10/0.05)]^2}\right\}}\right)$$
$$= 0.421 \text{ to } 0.676$$

These values differ only slightly from those calculated using the approximate formula, since g is small.

Description of Program for Three- and Four-Point Assays

First the program (Program 13.2) prints a heading (lines 20–30), and then asks for the potency of the standard solution (line 50). The value typed is checked to ensure that it is greater than zero (lines 70–90). Next a message asks whether a three- or four-point assay was performed (lines 100–110), and the reply is checked to ensure that it is either three or four in lines 130–160. A message (line 170) asks for the number of replicates n of each dose, and the value typed is checked (lines 190–220) to ensure that it is an integer, and in the range 2–50 inclusive. The lower limit is imposed by the calculation of fiducial limits, and the upper limit is set by the DIMension of X(50) in line 10.

A message (line 230) requests the value for the smaller standard dose s_1. This is input, and checked to ensure that the value is greater than zero in lines 250–270. Then a message (line 280) asks for each of the individual responses to the smaller standard dose. These values are input using a subroutine (lines 1610–1790). The subroutine checks that each of the values is greater than zero, and calculates and prints the mean response (lines 1710–1720).

Program 13.2   Trial run.

```
3 AND 4 POINT ASSAY
= === = ===== =====

TYPE THE POTENCY OF THE STANDARD SOLUTION
? 20
DID YOU PERFORM A THREE POINT OR A FOUR POINT ASSAY
TYPE 3 OR 4 AND PRESS RETURN
? 4
TYPE IN THE NUMBER OF REPLICATES FOR EACH DOSE
? 4
TYPE THE SMALLER STANDARD DOSE
? 0.05
TYPE IN THE 4 RESPONSES TO THE SMALLER STANDARD DOSE
ONE AT A TIME AND PRESS RETURN AFTER EACH RESPONSE
? 3
? 4.5
? 3.9
? 5
MEAN RESPONSE = 4.1

TYPE THE LARGER STANDARD DOSE
? 0.1
TYPE IN THE 4 RESPONSES TO THE LARGER STANDARD DOSE
ONE AT A TIME AND PRESS RETURN AFTER EACH RESPONSE
? 6.1
? 7.4
? 7.4
? 7.4
MEAN RESPONSE = 7.075
```

```
TYPE IN THE SMALLER TEST DOSE
? 0.1
TYPE IN THE 4 RESPONSES TO THE SMALLER TEST DOSE
ONE AT A TIME AND PRESS RETURN AFTER EACH RESPONSE
? 4.7
? 4.6
? 5.1
? 4.0
MEAN RESPONSE = 4.6

TYPE IN THE LARGER TEST DOSE
? 0.2
TYPE IN THE 4 RESPONSES TO THE LARGER TEST DOSE
ONE AT A TIME AND PRESS RETURN AFTER EACH RESPONSE
? 6.6
? 7.4
? 7.8
? 8.0
MEAN RESPONSE = 7.45

EQUIVALENT DOSE OF STANDARD FOR SMALLER TEST = 0.055
EQUIVALENT DOSE OF STANDARD FOR LARGER  TEST = 0.111
POTENCY RATIO (TEST/STANDARD) = 0.555
ABSOLUTE POTENCY OF TEST = 11.097
FIDUCIAL LIMITS FOR POTENCY ARE 8.41 AND 13.54

THE DIFFERENCE IN SLOPES BETWEEN THE TEST & STANDARD LINES
IS SMALL ENOUGH TO HAVE OCCURRED BY RANDOM CHANCE

WOULD YOU LIKE ANOTHER RUN  TYPE YES OR NO AND PRESS RETURN
? NO

END OF JOB


10 DIM Q$(10), X(50)
20 PRINT TAB(25); "3 AND 4 POINT ASSAY"
30 PRINT TAB(25); "= === = ===== ====="
40 PRINT
50 PRINT "TYPE THE POTENCY OF THE STANDARD SOLUTION"
60 INPUT P
70 IF P > 0 THEN 100
80 PRINT "POTENCY OF STANDARD SOLUTION MUST BE GREATER THAN ZERO.   RE"
90 GOTO 50
100 PRINT "DID YOU PERFORM A THREE POINT OR A FOUR POINT ASSAY"
110 PRINT "TYPE 3 OR 4 AND PRESS RETURN"
120 INPUT P$
130 IF P$ = "3" THEN 170
140 IF P$ = "4" THEN 170
150 PRINT "REPLY '"; P$; "' NOT UNDERSTOOD. RE-";
160 GOTO 110
170 PRINT "TYPE IN THE NUMBER OF REPLICATES FOR EACH DOSE"
180 INPUT N
190 IF N <> INT(N) THEN 210
200 IF (N - 2) * (N - 50) <= 0 THEN 230
210 PRINT "NUMBER OF REPLICATES MUST BE A WHOLE NUMBER BETWEEN 2 & 50"
220 GOTO 170
230 PRINT "TYPE THE SMALLER STANDARD DOSE"
240 INPUT R1
250 IF R1 > 0 THEN 280
260 PRINT "THE SMALLER STANDARD DOSE MUST BE GREATER THAN ZERO.   RETYPE"
```

```
270  GOTO 240
280  PRINT "TYPE IN THE"; N; "RESPONSES TO THE SMALLER STANDARD DOSE"
290  GOSUB 1620
300  LET S1 = M
310  LET V1 = S
320  PRINT "TYPE THE LARGER STANDARD DOSE"
330  INPUT R2
340  IF R2 > R1 THEN 380
350  PRINT "THE LARGER STANDARD DOSE MUST BE GREATER THAN THE SMALLER"
360  PRINT "STANDARD DOSE.  RETYPE THE LARGER STANDARD DOSE CORRECTLY"
370  GOTO 330
380  LET I = LOG(R2 / R1)
390  PRINT "TYPE IN THE"; N; "RESPONSES TO THE LARGER STANDARD DOSE"
400  GOSUB 1620
410  LET S2 = M
420  LET V2 = S
430  IF S2 > S1 THEN 480
440  PRINT "THE MEAN RESPONSE FROM THE LARGER STANDARD MUST BE GREATER"
450  PRINT "THAN THE MEAN RESPONSE FROM THE SMALLER STANDARD."
460  PRINT "RE-";
470  GOTO 380
480  IF P$ = "4" THEN 870
490  REM *** THREE POINT ASSAY
500  PRINT "TYPE IN THE TEST DOSE"
510  INPUT T
520  IF T > 0 THEN 550
530  PRINT "THE DOSE OF TEST MUST BE GREATER THAN ZERO.  RETYPE"
540  GOTO 510
550  PRINT "TYPE IN THE"; N; "RESPONSES TO THE TEST DOSE"
560  GOSUB 1620
570  LET U = M
580  LET V3 = S
590  IF (S1 - U) * (S2 - U) <= 0 THEN 630
600  PRINT "THE MEAN TEST RESPONSE MUST BE BETWEEN THE MEAN RESPONSES"
610  PRINT "FROM THE SMALLER AND LARGER STANDARD DOSES.  RE-";
620  GOTO 550
630  LET B = (S2 - S1) / I
640  LET K = R1 * EXP((U - S1) / B)
650  PRINT "EQUIVALENT DOSE OF STANDARD FOR TEST ="; INT(1000*K+0.5)/1000
660  LET R = K / T
670  PRINT "POTENCY RATIO (TEST/STANDARD) ="; INT(1000 * R + 0.5) / 1000
680  PRINT "ABSOLUTE POTENCY OF TEST ="; INT(1000 * R * P + 0.5) / 1000
690  LET M = LOG(R)
700  REM CALCULATE POOLED ESTIMATE OF THE VARIANCE OF THE MEANS, V
710  LET V = (V1 + V2 + V3) / 3
720  REM GET T VALUE
730  LET D = 3 * (N - 1)
740  GOSUB 1810
750  REM CALCULATE SIGNIFICANCE OF REGRESSION - FINNEY
760  LET G = 2 * V * T * T / (I * I * B * B)
770  IF G >= 1 THEN 1480
780  REM CALCULATE LIMITS OF ERROR
790  LET L = 1.5 * V * (1 - G) + 2 * V * M * M / (I * I)
800  LET L = T / (B * (1 - G)) * SQR(L)
810  LET F1 = P * EXP(M / (1 - G) - L)
820  LET F2 = P * EXP(M / (1 - G) + L)
830  PRINT "FIDUCIAL LIMITS FOR POTENCY ARE"; INT(100 * F1 + 0.5) / 100
840  PRINT "AND"; INT(100 * F2 + 0.5) / 100
850  GOTO 1500
860  REM *** FOUR POINT ASSAY
870  PRINT "TYPE IN THE SMALLER TEST DOSE"
```

```
880 INPUT T1
890 IF T1 > 0 THEN 920
900 PRINT "SMALLER TEST DOSE MUST BE GREATER THEN ZERO.  RETYPE"
910 GOTO 880
920 PRINT "TYPE IN THE"; N; "RESPONSES TO THE SMALLER TEST DOSE"
930 GOSUB 1620
940 LET U1 = M
950 LET V3 = S
960 PRINT "TYPE IN THE LARGER TEST DOSE"
970 INPUT T2
980 IF T2 > T1 THEN 1020
990 PRINT "LARGER TEST DOSE MUST BE GREATER THEN THE SMALLER TEST"
1000 PRINT "DOSE.  RE-";
1010 GOTO 870
1020 IF T2 / T1 = R2 / R1 THEN 1060
1030 PRINT "RATIO OF TEST DOSES MUST EQUAL THE RATIO OF STANDARD DOSES"
1040 PRINT "REFER TO THE DERIVATION IN CHAPTER 13"
1050 GOTO 1490
1060 PRINT "TYPE IN THE"; N; "RESPONSES TO THE LARGER TEST DOSE"
1070 GOSUB 1620
1080 LET U2 = M
1090 LET V4 = S
1100 IF U2 > U1 THEN 1130
1110 PRINT "MEAN RESPONSE MUST BE LARGER THAN FOR THE SMALLER DOSE"
1120 GOTO 1490
1130 LET B = 0.5 * (S2 - S1 + U2 - U1) / I
1140 LET K = 0.5 * (U1 - S1 + U2 - S2) / B
1150 LET K1 = R1 * EXP(K)
1160 LET K2 = R2 * EXP(K)
1170 PRINT "EQUIVALENT DOSE OF STANDARD FOR SMALLER TEST =";
1180 PRINT INT(1000 * K1 + 0.5) / 1000
1190 PRINT "EQUIVALENT DOSE OF STANDARD FOR LARGER  TEST =";
1200 PRINT INT(1000 * K2 + 0.5) / 1000
1210 LET R = K1 / T1
1220 PRINT "POTENCY RATIO (TEST/STANDARD) ="; INT(1000 * R + 0.5) / 1000
1230 PRINT "ABSOLUTE POTENCY OF TEST ="; INT(1000 * R * P + 0.5) / 1000
1240 LET M = LOG(R)
1250 REM CALCULATE POOLED ESTIMATE OF THE VARIANCE OF THE MEANS, V
1260 LET V = (V1 + V2 + V3 + V4) / 4
1270 REM GET T VALUE
1280 LET D = 4 * (N - 1)
1290 GOSUB 1810
1300 REM CALCULATE SIGNIFICANCE OF REGRESSION - FINNEY
1310 LET G = V * T * T / (I * I * B * B)
1320 IF G >= 1 THEN 1480
1330 REM CALCULATE LIMITS OF ERROR
1340 LET L = V * (1 - G) + V * M * M / (I * I)
1350 LET L = T / (B * (1 - G)) * SQR(L)
1360 LET F1 = P * EXP(M / (1 - G) - L)
1370 LET F2 = P * EXP(M / (1 - G) + L)
1380 PRINT "FIDUCIAL LIMITS FOR POTENCY ARE"; INT(100*F1 + 0.5)/100;
1390 PRINT "AND"; INT(100 * F2 + 0.5) / 100
1400 PRINT
1410 LET T1 = (U2 - U1 - S2 + S1) / SQR(4 * V)
1420 PRINT "THE DIFFERENCE IN SLOPES BETWEEN THE TEST & STANDARD LINES"
1430 IF ABS(T1) < T THEN 1460
1440 PRINT "IS SO LARGE THAT THE CALCULATION OF POTENCY IS NOT VALID"
1450 GOTO 1500
1460 PRINT "IS SMALL ENOUGH TO HAVE OCCURRED BY RANDOM CHANCE"
1470 GOTO 1500
1480 PRINT "THE ASSAY IS VALUELESS SINCE G IS GREATER THAN ONE: G ="; G
```

```
1490 PRINT "RUN ABANDONED ON THIS DATA"
1500 PRINT
1510 PRINT "WOULD YOU LIKE ANOTHER RUN ";
1520 PRINT "TYPE YES OR NO AND PRESS RETURN"
1530 INPUT Q$
1540 RESTORE
1550 IF Q$ = "YES" THEN 40
1560 IF Q$ = "NO" THEN 1590
1570 PRINT "REPLY '"; Q$; "' NOT UNDERSTOOD.  RE-";
1580 GOTO 1520
1590 PRINT "END OF JOB"
1600 GOTO 1930
1610 REM *** SUBROUTINE TO INPUT N VALUES & CALCULATE MEAN & STAND DEV
1620 PRINT "ONE AT A TIME AND PRESS RETURN AFTER EACH RESPONSE"
1630 S = 0
1640 FOR J = 1 TO N
1650    INPUT X(J)
1660    IF X(J) > 0 THEN 1690
1670    PRINT "THE RESPONSES MUST BE POSITIVE.  RETYPE CORRECTLY"
1680    GOTO 1650
1690    LET S = S + X(J)
1700 NEXT J
1710 LET M = S / N
1720 PRINT "MEAN RESPONSE ="; M
1730 PRINT
1740 LET S = 0
1750 FOR J = 1 TO N
1760    LET S = S + (X(J) - M) * (X(J) - M)
1770 NEXT J
1780 LET S = S / (N * (N - 1))
1790 RETURN
1800 REM *** SUBROUTINE TO RETURN 95% T VALUE FOR D DEGREES OF FREEDOM
1810 LET T = 1.95996
1820 REM RE-CALCULATE 95% CONFIDENCE LIMIT IF DEGREES OF FREEDOM <= 30
1830 FOR J = 1 TO 30
1840 READ A
1850 IF J <> D THEN 1870
1860 LET T = A
1870 NEXT J
1880 DATA 12.706, 4.303, 3.182, 2.776, 2.571, 2.447, 2.365
1890 DATA 2.306, 2.262, 2.228, 2.201, 2.197, 2.160, 2.145, 2.131
1900 DATA 2.120, 2.110, 2.101, 2.093, 2.086, 2.080, 2.074, 2.069
1910 DATA 2.064, 2.060, 2.056, 2.052, 2.048, 2.045, 2.042
1920 RETURN
1930 END
```

A message (line 320) requests the user to type the value for the larger standard dose s_2. This is input and checked to ensure that the value is larger than the smaller standard dose (lines 330–370). The value of $\log(s_2/s_1)$ is calculated and stored as I (line 380) for use in later calculations. Next a message (line 390) requests the user to type the individual responses to the larger standard dose. The values are input and checked in a subroutine (lines 1610–1790) as in the previous paragraph. A further check is carried out (lines 430–470) to ensure that the mean response from the larger standard dose is greater than the mean response from the smaller standard dose. At this point (line 480) a test is performed to see if a three-point or a four-point assay was requested. Lines 490–850 deal with a three-point assay, while lines 860–1470 calculate the four-point assay.

Three-Point Assay

A message (line 500) asks for the test dose, u. The value is input and checked to ensure that it is greater than zero (lines 510–540). Next the individual responses to the test dose are requested (line 550), and the values are input and checked by a subroutine (lines 1610–1790) as in the previous two paragraphs. A check is then performed (lines 590–620) to ensure that the mean response from the test is between the mean responses from the smaller and larger standard doses.

The slope of the log (dose)/response line, b, is calculated (line 630) and the equivalent dose of standard ψ to the test is calculated and printed in lines 640–650. The potency ratio, R, is calculated and printed in lines 660–670, followed by the absolute potency in line 680. The pooled estimate of the variance of the means is calculated (lines 700–710). Lines 720–740 get the appropriate 95% two-tailed t-value by setting the number of degrees of freedom to $3(n-1)$ and calling the t-subroutine (lines 1800–1920). This first sets t to the value for an infinite number of degrees of freedom and then substitutes an appropriate value read from the DATA statements if there are 30 or fewer degrees of freedom. Finney's significance of regression, g, is calculated (lines 750–760), and, if g is greater than or equal to 1, the run is abandoned (line 770) and a message is printed (lines 1480–1490). Otherwise, the fiducial limits are calculated and printed suitably rounded (lines 750–840). The program then asks if another run is required (lines 1500–1520).

Four-Point Assay

The smaller and larger test doses and the individual responses produced by them are input and checked in lines 870–1120, using the same input subroutine (lines 1610–1790) as was used for the standard doses and responses. One additional check is performed (lines 1020–1050) to ensure that the ratio of the two test doses is equal to the ratio of the two standard doses. The mean slope b is calculated in line 1130 using Equation 8. Then the equivalent standard doses ψ_1 and ψ_2 of the smaller and larger test doses are calculated and printed suitably rounded (lines 1140–1200). The potency ratio and absolute potency are calculated and printed suitably rounded in lines 1210–1230. The pooled estimate of the variance of the means is calculated (lines 1250–1260), and the appropriate t-value is obtained (lines 1270–1290) using the t-subroutine as before (lines 1800–1920). Finney's significance of regression g is calculated using Equation 22 in lines 1300–1310, and if g is greater than or equal to 1 the run is abandoned (line 1320) with a suitable message (lines 1480–1490). Otherwise the fiducial limits of error are calculated, rounded and printed (lines 1330–1390). The t-value for the difference in slopes is evaluated using Equation 19 and an appropriate message printed (lines 1410–1470).

Finally a message (lines 1510–1520) asks is another run is required, and the reply is checked (lines 1530–1580) to ensure that it is YES or NO.

Appendix 1

Significance Table for Pearson's Correlation Coefficient *r*

Number of degrees of freedom = number of *x*, *y* pairs minus two.

Two-tailed	Probability of *r* occurring by chance				
	10%	5%	2%	1%	0·1%
No. of degrees of freedom					
1	.987 69	.996 92	.999 507	.999 877	.999 998 8
2	.900 00	.950 00	.980 00	.990 000	.999 00
3	.805 4	.878 3	.934 33	.958 73	.991 16
4	.729 3	.811 4	.882 2	.917 20	.974 06
5	.669 4	.754 5	.832 9	.874 5	.950 74
6	.621 5	.706 7	.788 7	.834 3	.924 93
7	.582 2	.666 4	.749 8	.797 7	.898 2
8	.549 4	.631 9	.715 5	.764 6	.872 1
9	.521 4	.602 1	.685 1	.734 8	.847 1
10	.497 3	.576 0	.658 1	.707 9	.823 3
11	.476 2	.552 9	.633 9	.683 5	.801 0
12	.457 5	.532 4	.612 0	.661 4	.780 0
13	.440 9	.513 9	.592 3	.641 1	.760 3
14	.425 9	.497 3	.574 2	.622 6	.742 0
15	.412 4	.482 1	.557 7	.605 5	.724 6
16	.400 0	.468 3	.542 5	.589 7	.708 4
17	.388 7	.455 5	.528 5	.575 1	.693 2
18	.378 3	.443 8	.515 5	.561 4	.678 7
19	.368 7	.432 9	.503 4	.548 7	.665 2
20	.359 8	.422 7	.492 1	.536 8	.652 4
25	.323 3	.380 9	.445 1	.486 9	.597 4
30	.296 0	.349 4	.409 3	.448 7	.554 1
35	.274 6	.324 6	.381 0	.418 2	.518 9
40	.257 3	.304 4	.357 8	.393 2	.489 6
45	.242 8	.287 5	.338 4	.372 1	.464 8
50	.230 6	.273 2	.321 8	.354 1	.443 3
60	.210 8	.250 0	.294 8	.324 8	.407 8
70	.195 4	.231 9	.273 7	.301 7	.379 9
80	.182 9	.217 2	.256 5	.283 0	.356 8
90	.172 6	.205 0	.242 2	.267 3	.337 5
100	.163 8	.194 6	.230 1	.254 0	.321 1
One-tailed	5%	2·5%	1%	0·5%	0·05%

241

Appendix 2

Significance Table for Spearman's Rank Correlation Coefficient ρ

		Probability of rho occurring by chance					
Two-tailed		20%	10%	5%	2%	1%	0.2%
Number of (x,y) pairs n	4	.8000	.8000				
	5	.7000	.8000	.9000	.9000		
	6	.6000	.7714	.8286	.8857	.9429	
	7	.5357	.6786	.7450	.8571	.8929	.9643
	8	.5000	.6190	.7143	.8095	.8571	.9286
	9	.4667	.5833	.6833	.7667	.8167	.9000
	10	.4424	.5515	.6364	.7333	.7818	.8667
	11	.4182	.5273	.6091	.7000	.7455	.8364
	12	.3986	.4965	.5804	.6713	.7273	.8182
	13	.3791	.4780	.5549	.6429	.6978	.7912
	14	.3626	.4593	.5341	.6220	.6747	.7670
	15	.3500	.4429	.5179	.6000	.6536	.7464
	16	.3382	.4265	.5000	.5824	.6324	.7265
	17	.3260	.4118	.4853	.5637	.6152	.7083
	18	.3148	.3994	.4716	.5480	.5975	.6904
	19	.3070	.3895	.4579	.5333	.5825	.6737
	20	.2977	.3789	.4451	.5203	.5684	.6586
	21	.2909	.3688	.4351	.5078	.5545	.6455
	22	.2829	.3597	.4241	.4963	.5426	.6318
	23	.2767	.3518	.4150	.4852	.5306	.6186
	24	.2704	.3435	.4061	.4748	.5200	.6070
	25	.2646	.3362	.3977	.4654	.5100	.5962
	26	.2588	.3299	.3894	.4564	.5002	.5856
	27	.2540	.3236	.3822	.4481	.4915	.5757
	28	.2490	.3175	.3749	.4401	.4828	.5660
	29	.2443	.3113	.3685	.4320	.4744	.5567
	30	.2400	.3059	.3620	.4251	.4665	.5479
	35	.2198	.2821	.3361	.3990	.4417	.5300
	40	.2052	.2634	.3139	.3725	.4125	.4948
	45	.1932	.2480	.2955	.3507	.3883	.4659
	50	.1831	.2350	.2800	.3323	.3680	.4415
	55	.1744	.2238	.2667	.3166	.3505	.4205
	60	.1669	.2141	.2552	.3029	.3353	.4023
	65	.1602	.2056	.2450	.2908	.3220	.3863
	70	.1543	.1980	.2360	.2801	.3101	.3720
	75	.1490	.1912	.2278	.2704	.2994	.3592
	80	.1442	.1851	.2205	.2617	.2898	.3477
	85	.1398	.1795	.2139	.2538	.2810	.3372
	90	.1358	.1744	.2078	.2466	.2730	.3276
	95	.1322	.1697	.2022	.2399	.2657	.3187
	100	.1288	.1653	.1970	.2338	.2589	.3106
One-tailed		10%	5%	2.5%	1%	0.5%	0.1%

The top of the table has been calculated exactly, whereas a slight approximation that the distribution of rho is normal with variance (n-1) has been used for n > 30.

Appendix 3

Significance Table for Kendall's Rank Correlation Coefficient τ

n	20%	10%	5%	2%	1%	0.2%
4	1.0000	1.0000				
5	0.8000	0.8000	1.0000	1.0000		
6	0.6000	0.7333	0.8667	0.8667	1.0000	
7	0.5238	0.6190	0.7143	0.8095	0.9048	1.0000
8	0.4286	0.5714	0.6429	0.7143	0.7857	0.8571
9	0.3889	0.5000	0.5556	0.6667	0.7222	0.8333
10	0.3778	0.4667	0.5111	0.6000	0.6444	0.7778
11	0.3455	0.4182	0.4909	0.5636	0.6000	0.7091
12	0.3030	0.3939	0.4545	0.5455	0.5758	0.6667
13	0.3077	0.3590	0.4359	0.5128	0.5641	0.6410
14	0.2747	0.3626	0.4066	0.4725	0.5165	0.6044
15	0.2762	0.3333	0.3905	0.4667	0.5048	0.5810
16	0.2500	0.3167	0.3833	0.4333	0.4833	0.5667
17	0.2500	0.3088	0.3676	0.4265	0.4706	0.5441
18	0.2418	0.2941	0.3464	0.4118	0.4510	0.5294
19	0.2281	0.2865	0.3333	0.3918	0.4386	0.5088
20	0.2211	0.2737	0.3263	0.3789	0.4211	0.4947
21	0.2095	0.2667	0.3143	0.3714	0.4095	0.4857
22	0.2035	0.2641	0.3074	0.3593	0.3939	0.4719
23	0.2016	0.2569	0.2964	0.3518	0.3913	0.4545
24	0.1957	0.2464	0.2899	0.3406	0.3768	0.4493
25	0.1933	0.2400	0.2867	0.3333	0.3667	0.4400
26	0.1877	0.2369	0.2800	0.3292	0.3600	0.4277
27	0.1795	0.2308	0.2707	0.3219	0.3561	0.4188
28	0.1799	0.2275	0.2646	0.3122	0.3439	0.4127
29	0.1724	0.2217	0.2611	0.3103	0.3399	0.4039
30	0.1724	0.2184	0.2552	0.3011	0.3333	0.3931
31	0.1656	0.2129	0.2516	0.2946	0.3247	0.3892
32	0.1653	0.2097	0.2460	0.2903	0.3226	0.3790
33	0.1629	0.2045	0.2424	0.2879	0.3144	0.3750
34	0.1586	0.2014	0.2371	0.2799	0.3119	0.3690
35	0.1563	0.1966	0.2336	0.2773	0.3042	0.3613
36	0.1524	0.1937	0.2317	0.2730	0.3016	0.3587
37	0.1502	0.1922	0.2282	0.2673	0.2973	0.3514
38	0.1494	0.1892	0.2233	0.2632	0.2916	0.3457
39	0.1471	0.1876	0.2200	0.2605	0.2874	0.3414
40	0.1436	0.1846	0.2179	0.2564	0.2846	0.3385
41	0.1415	0.1805	0.2146	0.2537	0.2805	0.3341
42	0.1405	0.1777	0.2125	0.2497	0.2753	0.3287
43	0.1384	0.1761	0.2093	0.2470	0.2735	0.3245
44	0.1374	0.1734	0.2072	0.2431	0.2685	0.3214
45	0.1354	0.1717	0.2040	0.2404	0.2667	0.3172
46	0.1324	0.1691	0.2019	0.2386	0.2638	0.3140
47	0.1323	0.1674	0.1989	0.2359	0.2599	0.3099
48	0.1294	0.1667	0.1968	0.2323	0.2571	0.3067
49	0.1293	0.1633	0.1956	0.2296	0.2534	0.3027
50	0.1265	0.1624	0.1918	0.2278	0.2506	0.2996
51	0.1263	0.1608	0.1906	0.2251	0.2486	0.2973
52	0.1237	0.1584	0.1885	0.2232	0.2459	0.2941
53	0.1234	0.1567	0.1872	0.2206	0.2438	0.2903
54	0.1223	0.1558	0.1852	0.2187	0.2411	0.2872
55	0.1205	0.1542	0.1825	0.2162	0.2391	0.2848
56	0.1195	0.1519	0.1805	0.2143	0.2364	0.2818
57	0.1178	0.1516	0.1792	0.2118	0.2343	0.2794
58	0.1168	0.1494	0.1773	0.2099	0.2317	0.2765
59	0.1163	0.1479	0.1759	0.2086	0.2297	0.2741
60	0.1153	0.1469	0.1740	0.2068	0.2282	0.2723

Appendix 4

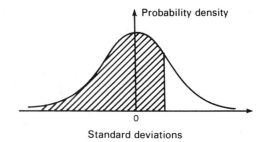

Area Under Normal Curve Table

NUMBER OF STANDARD DEVIATIONS	0.00	0.01	0.02	0.03	0.04	0.05	0.06	0.07	0.08	0.09
0.0	.5000	.5040	.5080	.5120	.5160	.5199	.5239	.5279	.5319	.5359
0.1	.5398	.5438	.5478	.5517	.5557	.5596	.5636	.5675	.5714	.5753
0.2	.5793	.5832	.5871	.5910	.5948	.5987	.6026	.6064	.6103	.6141
0.3	.6179	.6217	.6255	.6293	.6331	.6368	.6406	.6443	.6480	.6517
0.4	.6554	.6591	.6628	.6664	.6700	.6736	.6772	.6808	.6844	.6879
0.5	.6915	.6950	.6985	.7019	.7054	.7088	.7123	.7157	.7190	.7224
0.6	.7257	.7291	.7324	.7357	.7389	.7422	.7454	.7486	.7517	.7549
0.7	.7580	.7611	.7642	.7673	.7704	.7734	.7764	.7794	.7823	.7852
0.8	.7881	.7910	.7939	.7967	.7995	.8023	.8051	.8078	.8106	.8133
0.9	.8159	.8186	.8212	.8238	.8264	.8289	.8315	.8340	.8365	.8389
1.0	.8413	.8438	.8461	.8485	.8508	.8531	.8554	.8577	.8599	.8621
1.1	.8643	.8665	.8686	.8708	.8729	.8749	.8770	.8790	.8810	.8830
1.2	.8849	.8869	.8888	.8907	.8925	.8944	.8962	.8980	.8997	.9015
1.3	.9032	.9049	.9066	.9082	.9099	.9115	.9131	.9147	.9162	.9177
1.4	.9192	.9207	.9222	.9236	.9251	.9265	.9279	.9292	.9306	.9319
1.5	.9332	.9345	.9357	.9370	.9382	.9394	.9406	.9418	.9429	.9441
1.6	.9452	.9463	.9474	.9484	.9495	.9505	.9515	.9525	.9535	.9545
1.7	.9554	.9564	.9573	.9582	.9591	.9599	.9608	.9616	.9625	.9633
1.8	.9641	.9649	.9656	.9664	.9671	.9678	.9686	.9693	.9699	.9706
1.9	.9713	.9719	.9726	.9732	.9738	.9744	.9750	.9756	.9761	.9767
2.0	.9772	.9778	.9783	.9788	.9793	.9798	.9803	.9808	.9812	.9817
2.1	.9821	.9826	.9830	.9834	.9838	.9842	.9846	.9850	.9854	.9857
2.2	.9861	.9864	.9868	.9871	.9875	.9878	.9881	.9884	.9887	.9890
2.3	.9893	.9896	.9898	.9901	.9904	.9906	.9909	.9911	.9913	.9916
2.4	.9918	.9920	.9922	.9925	.9927	.9929	.9931	.9932	.9934	.9936
2.5	.9938	.9940	.9941	.9943	.9945	.9946	.9948	.9949	.9951	.9952
2.6	.9953	.9955	.9956	.9957	.9959	.9960	.9961	.9962	.9963	.9964
2.7	.9965	.9966	.9967	.9968	.9969	.9970	.9971	.9972	.9973	.9974
2.8	.9974	.9975	.9976	.9977	.9977	.9978	.9979	.9979	.9980	.9981
2.9	.9981	.9982	.9982	.9983	.9984	.9984	.9985	.9985	.9986	.9986
3.0	.9987	.9987	.9987	.9988	.9988	.9989	.9989	.9989	.9990	.9990
3.1	.9990	.9991	.9991	.9991	.9992	.9992	.9992	.9992	.9993	.9993
3.2	.9993	.9993	.9994	.9994	.9994	.9994	.9994	.9995	.9995	.9995
3.3	.9995	.9995	.9995	.9996	.9996	.9996	.9996	.9996	.9996	.9997
3.4	.9997	.9997	.9997	.9997	.9997	.9997	.9997	.9997	.9997	.9998
3.5	.9998	.9998	.9998	.9998	.9998	.9998	.9998	.9998	.9998	.9998

Program to Produce Area Under Normal Curve Table

```
10  PRINT TAB(26); "AREA UNDER NORMAL CURVE TABLE"
20  PRINT TAB(26); "==== ===== ====== ===== ====="
30  PRINT
40  PRINT "NUMBER OF"
50  PRINT "STANDARD"
60  PRINT "DEVIATIONS 0.00  0.01  0.02  0.03  0.04";
70  PRINT TAB(42); "0.05  0.06  0.07  0.08  0.09"
80  PRINT
90  FOR I = 0 TO 35
100    PRINT TAB(5); I / 10;
110    FOR J = 0 TO 9
120      LET X = I / 10 + J / 100
130      GOSUB 270
140      PRINT TAB(12 + 6 * J); INT((1 - F) * 10000 + .5) / 10000;
150    NEXT J
160    PRINT
170    IF INT(I / 5) * 5 <> I THEN 190
180    PRINT
190  NEXT I
200  PRINT
210  PRINT
220  PRINT "TABLE FINISHED"
230  GOTO 420
240  REM CALC CUMULATIVE AREA UNDER NORMAL CURVE
250  REM CONSTANTS SET FOR 8 FIGURE ACCURACY
260  REM USE ONLY FOR -4.5 TO 9 STANDARD DEVIATIONS
270  LET X9 = -X * .707107
280  LET T = 1 - 7.5 / (ABS(X9) + 3.75)
290  LET Y = 0
300  FOR I1 = 1 TO 12
310    READ C
320    LET Y = Y * T + C
330  NEXT I1
340  RESTORE
350  DATA 3.14753E-05, -.000138746, -6.41279E-06, .00178663
360  DATA -.00823169, .0241519, -.0547992, .102602
370  DATA -.163572, .226008, -.273422, .14559
380  LET F = 0.5 * EXP(-X9 * X9) * Y
390  IF X9 <= 0 THEN 410
400  LET F = 1 - F
410  RETURN
420  END
```

Appendix 5
Chi-Squared Significance Table

For larger v, $\sqrt{(2\chi^2)}$ is approximately normally distributed with mean $\sqrt{(2v-1)}$ and $\sigma^2 = 1$.

| | Probability of chi-squared χ^2 occurring by chance | | | | | | | | | | | | | |
No. of degrees of freedom v	99%	98%	95%	90%	80%	70%	50%	30%	20%	10%	5%	2%	1%	0.1%
1	.000157	.000628	.00393	.0158	.0642	.148	.455	1.074	1.642	2.706	3.841	5.412	6.635	10.827
2	.0201	.0404	.103	.211	.446	.713	1.386	2.408	3.219	4.605	5.991	7.824	9.210	13.815
3	.115	.185	.352	.584	1.005	1.424	2.366	3.665	4.642	6.251	7.815	9.837	11.345	16.266
4	.297	.429	.711	1.064	1.649	2.195	3.357	4.878	5.989	7.779	9.488	11.668	13.277	18.467
5	.554	.752	1.145	1.610	2.343	3.000	4.351	6.064	7.289	9.236	11.070	13.388	15.086	20.515
6	.872	1.134	1.635	2.204	3.070	3.828	5.348	7.231	8.558	10.645	12.592	15.033	16.812	22.457
7	1.239	1.564	2.167	2.833	3.822	4.671	6.346	8.383	9.803	12.017	14.067	16.622	18.475	24.322
8	1.646	2.032	2.733	3.490	4.594	5.527	7.344	9.524	11.030	13.362	15.507	18.168	20.090	26.125
9	2.088	2.532	3.325	4.168	5.380	6.393	8.343	10.656	12.242	14.684	16.919	19.679	21.666	27.877
10	2.558	3.059	3.940	4.865	6.179	7.267	9.342	11.781	13.442	15.987	18.307	21.161	23.209	29.588
11	3.053	3.609	4.575	5.578	6.989	8.148	10.341	12.899	14.631	17.275	19.675	22.618	24.725	31.264
12	3.571	4.178	5.226	6.304	7.807	9.034	11.340	14.011	15.812	18.549	21.026	24.054	26.217	32.909
13	4.107	4.765	5.892	7.042	8.634	9.926	12.340	15.119	16.985	19.812	22.362	25.472	27.688	34.528
14	4.660	5.368	6.571	7.790	9.467	10.821	13.339	16.222	18.151	21.064	23.685	26.873	29.141	36.123
15	5.229	5.985	7.261	8.547	10.307	11.721	14.339	17.322	19.311	22.307	24.996	28.259	30.578	37.697
16	5.812	6.614	7.962	9.312	11.152	12.624	15.338	18.418	20.465	23.542	26.296	29.633	32.000	39.252
17	6.408	7.255	8.672	10.085	12.002	13.531	16.338	19.511	21.615	24.769	27.587	30.995	33.409	40.790
18	7.015	7.906	9.390	10.865	12.857	14.440	17.338	20.601	22.760	25.989	28.869	32.346	34.805	42.312
19	7.633	8.567	10.117	11.651	13.716	15.352	18.338	21.689	23.900	27.204	30.144	33.687	36.191	43.820
20	8.260	9.237	10.851	12.443	14.578	16.266	19.337	22.775	25.038	28.412	31.410	35.020	37.566	45.315
21	8.897	9.915	11.591	13.240	15.445	17.182	20.337	23.858	26.171	29.615	32.671	36.343	38.932	46.797
22	9.542	10.600	12.338	14.041	16.314	18.101	21.337	24.939	27.301	30.813	33.924	37.659	40.289	48.268
23	10.196	11.293	13.091	14.848	17.187	19.021	22.337	26.018	28.429	32.007	35.172	38.968	41.638	49.728
24	10.856	11.992	13.848	15.659	18.062	19.943	23.337	27.096	29.553	33.196	36.415	40.270	42.980	51.179
25	11.524	12.697	14.611	16.473	18.940	20.867	24.337	28.172	30.675	34.382	37.652	41.566	44.314	52.620
26	12.198	13.409	15.379	17.292	19.820	21.792	25.336	29.246	31.795	35.563	38.885	42.856	45.642	54.052
27	12.879	14.125	16.151	18.114	20.703	22.719	26.336	30.319	32.912	36.741	40.113	44.140	46.963	55.476
28	13.565	14.847	16.928	18.939	21.588	23.647	27.336	31.391	34.027	37.916	41.337	45.419	48.278	56.893
29	14.256	15.574	17.708	19.768	22.475	24.577	28.336	32.461	35.139	39.087	42.557	46.693	49.588	58.302
30	14.953	16.306	18.493	20.599	23.364	25.508	29.336	33.530	36.250	40.256	43.773	47.962	50.892	59.703
32	16.362	17.783	20.072	22.271	25.148	27.373	31.336	35.665	38.466	42.585	46.194	50.487	53.486	62.487
34	17.789	19.275	21.664	23.952	26.938	29.242	33.336	37.795	40.676	44.903	48.602	52.995	56.061	65.247
36	19.233	20.783	23.269	25.643	28.735	31.115	35.336	39.922	42.879	47.212	50.999	55.489	58.619	67.985
38	20.691	22.304	24.884	27.343	30.537	32.992	37.335	42.045	45.076	49.513	53.384	57.969	61.162	70.703
40	22.164	23.838	26.509	29.051	32.345	34.872	39.335	44.165	47.269	51.805	55.759	60.436	63.691	73.402
42	23.650	25.383	28.144	30.765	34.157	36.755	41.335	46.282	49.456	54.090	58.124	62.892	66.206	76.084
44	25.148	26.939	29.787	32.487	35.974	38.641	43.335	48.396	51.639	56.369	60.481	65.337	68.710	78.750
46	26.657	28.504	31.439	34.215	37.795	40.529	45.335	50.507	53.818	58.641	62.830	67.771	71.201	81.400
48	28.177	30.080	33.098	35.949	39.621	42.420	47.335	52.616	55.993	60.907	65.171	70.197	73.683	84.037
50	29.707	31.664	34.764	37.689	41.449	44.313	49.335	54.723	58.164	63.167	67.505	72.613	76.154	86.661
52	31.246	33.256	36.437	39.433	43.281	46.209	51.335	56.827	60.332	65.422	69.832	75.021	78.616	89.272
54	32.793	34.856	38.116	41.183	45.117	48.106	53.335	58.930	62.496	67.673	72.153	77.422	81.069	91.872
56	34.350	36.464	39.801	42.937	46.955	50.005	55.335	61.031	64.658	69.919	74.468	79.815	83.513	94.461
58	35.913	38.078	41.492	44.696	48.797	51.906	57.335	63.129	66.816	72.160	76.778	82.201	85.950	97.039
60	37.485	39.699	43.188	46.459	50.641	53.809	59.335	65.227	68.972	74.397	79.082	84.580	88.379	99.607
62	39.063	41.327	44.889	48.226	52.487	55.714	61.335	67.322	71.125	76.630	81.381	86.953	90.802	102.166
64	40.649	42.960	46.595	49.996	54.336	57.620	63.335	69.416	73.276	78.860	83.675	89.320	93.217	104.716
66	42.240	44.599	48.305	51.770	56.188	59.527	65.335	71.508	75.424	81.085	85.965	91.681	94.626	107.258
68	43.838	46.244	50.020	53.548	58.042	61.436	67.335	73.600	77.571	83.308	88.250	94.037	98.028	109.791
70	45.442	47.893	51.739	55.329	59.898	63.346	69.334	75.689	79.715	85.527	90.531	96.388	100.425	112.317

Appendix 6
t-Distribution Table

Two-tailed	Probability of *t* occurring by chance												
	90%	80%	70%	60%	50%	40%	30%	20%	10%	5%	2%	1%	0.1%
No of degrees of freedom v													
1	.158	.325	.510	.727	1.000	1.376	1.963	3.078	6.314	12.706	31.821	63.657	636.619
2	.142	.289	.445	.617	.816	1.061	1.386	1.886	2.920	4.303	6.965	9.925	31.598
3	.137	.277	.424	.584	.765	.978	1.250	1.638	2.353	3.182	4.541	5.841	12.924
4	.134	.271	.414	.569	.741	.941	1.190	1.533	2.132	2.776	3.747	4.604	8.610
5	.132	.267	.408	.559	.727	.920	1.156	1.476	2.015	2.571	3.365	4.032	6.869
6	.131	.265	.404	.553	.718	.906	1.134	1.440	1.943	2.447	3.143	3.707	5.959
7	.130	.263	.402	.549	.711	.896	1.119	1.415	1.895	2.365	2.998	3.499	5.408
8	.130	.262	.399	.546	.706	.889	1.108	1.397	1.860	2.306	2.896	3.355	5.041
9	.129	.261	.398	.543	.703	.883	1.100	1.383	1.833	2.262	2.821	3.250	4.781
10	.129	.260	.397	.542	.700	.879	1.093	1.372	1.812	2.228	2.764	3.169	4.587
11	.129	.260	.396	.540	.697	.876	1.088	1.363	1.796	2.201	2.718	3.106	4.437
12	.128	.259	.395	.539	.695	.873	1.083	1.356	1.782	2.179	2.681	3.055	4.318
13	.128	.259	.394	.538	.694	.870	1.079	1.350	1.771	2.160	2.650	3.012	4.221
14	.128	.258	.393	.537	.692	.868	1.076	1.345	1.761	2.145	2.624	2.977	4.140
15	.128	.258	.393	.536	.691	.866	1.074	1.341	1.753	2.131	2.602	2.947	4.073
16	.128	.258	.392	.535	.690	.865	1.071	1.337	1.746	2.120	2.583	2.921	4.015
17	.128	.257	.392	.534	.689	.863	1.069	1.333	1.740	2.110	2.567	2.898	3.965
18	.127	.257	.392	.534	.688	.862	1.067	1.330	1.734	2.101	2.552	2.878	3.922
19	.127	.257	.391	.533	.688	.861	1.066	1.328	1.729	2.093	2.539	2.861	3.883
20	.127	.257	.391	.533	.687	.860	1.064	1.325	1.725	2.086	2.528	2.845	3.850
21	.127	.257	.391	.532	.686	.859	1.063	1.323	1.721	2.080	2.518	2.831	3.819
22	.127	.256	.390	.532	.686	.858	1.061	1.321	1.717	2.074	2.508	2.819	3.792
23	.127	.256	.390	.532	.685	.858	1.060	1.319	1.714	2.069	2.500	2.807	3.767
24	.127	.256	.390	.531	.685	.857	1.059	1.318	1.711	2.064	2.492	2.797	3.745
25	.127	.256	.390	.531	.684	.856	1.058	1.316	1.708	2.060	2.485	2.787	3.725
26	.127	.256	.390	.531	.684	.856	1.058	1.315	1.706	2.056	2.479	2.779	3.707
27	.127	.256	.389	.531	.684	.855	1.057	1.314	1.703	2.052	2.473	2.771	3.690
28	.127	.256	.389	.530	.683	.855	1.056	1.313	1.701	2.048	2.467	2.763	3.674
29	.127	.256	.389	.530	.683	.854	1.055	1.311	1.699	2.045	2.462	2.756	3.659
30	.127	.256	.389	.530	.683	.854	1.055	1.310	1.697	2.042	2.457	2.750	3.646
40	.126	.255	.388	.529	.681	.851	1.050	1.303	1.684	2.021	2.423	2.704	3.551
60	.126	.254	.387	.527	.679	.848	1.046	1.296	1.671	2.000	2.390	2.660	3.460
120	.126	.254	.386	.526	.677	.845	1.041	1.289	1.658	1.980	2.358	2.617	3.373
Normal $= \infty$	.126	.253	.385	.524	.674	.842	1.036	1.282	1.645	1.960	2.326	2.576	3.291
One-tailed	45%	40%	35%	30%	25%	20%	15%	10%	5%	2.5%	1%	0.5%	0.05%

Appendix 7

Significance Table for F-Test

Probability 10%

No. of degrees of freedom v_1	1	2	3	4	5	6	8	12	24	∞
No. of degrees of freedom v_2										
1	39.86	49.50	53.59	55.83	57.24	58.20	59.44	60.70	62.00	63.33
2	8.53	9.00	9.16	9.24	9.29	9.33	9.37	9.41	9.45	9.49
3	5.54	5.46	5.39	5.34	5.31	5.28	5.25	5.22	5.18	5.13
4	4.54	4.32	4.19	4.11	4.05	4.01	3.95	3.90	3.83	3.76
5	4.06	3.78	3.62	3.52	3.45	3.40	3.34	3.27	3.19	3.10
6	3.78	3.46	3.29	3.18	3.11	3.05	2.98	2.90	2.82	2.72
7	3.59	3.26	3.07	2.96	2.88	2.83	2.75	2.67	2.58	2.47
8	3.46	3.11	2.92	2.81	2.73	2.67	2.59	2.50	2.40	2.29
9	3.36	3.01	2.81	2.69	2.61	2.55	2.47	2.38	2.28	2.16
10	3.28	2.92	2.73	2.61	2.52	2.46	2.38	2.28	2.18	2.06
11	3.23	2.86	2.66	2.54	2.45	2.39	2.30	2.21	2.10	1.97
12	3.18	2.81	2.61	2.48	2.39	2.33	2.24	2.15	2.04	1.90
13	3.14	2.76	2.56	2.43	2.35	2.28	2.20	2.10	1.98	1.85
14	3.10	2.73	2.52	2.39	2.31	2.24	2.15	2.05	1.94	1.80
15	3.07	2.70	2.49	2.36	2.27	2.21	2.12	2.02	1.90	1.76
16	3.05	2.67	2.46	2.33	2.24	2.18	2.09	1.99	1.87	1.72
17	3.03	2.64	2.44	2.31	2.22	2.15	2.06	1.96	1.84	1.69
18	3.01	2.62	2.42	2.29	2.20	2.13	2.04	1.93	1.79	1.66
19	2.99	2.61	2.40	2.27	2.18	2.11	2.02	1.91	1.79	1.63
20	2.97	2.59	2.38	2.25	2.16	2.09	2.00	1.89	1.77	1.61
21	2.96	2.57	2.36	2.23	2.14	2.08	1.98	1.88	1.75	1.59
22	2.95	2.56	2.35	2.22	2.13	2.06	1.97	1.86	1.73	1.57
23	2.94	2.55	2.34	2.21	2.11	2.05	1.95	1.84	1.72	1.55
24	2.93	2.54	2.33	2.19	2.10	2.04	1.94	1.83	1.70	1.53
25	2.92	2.53	2.32	2.18	2.09	2.02	1.93	1.82	1.69	1.52
26	2.91	2.52	2.31	2.17	2.08	2.01	1.92	1.81	1.68	1.50
27	2.90	2.51	2.30	2.17	2.07	2.00	1.91	1.80	1.67	1.49
28	2.89	2.50	2.29	2.16	2.06	2.00	1.90	1.79	1.66	1.48
29	2.89	2.50	2.28	2.15	2.06	1.99	1.89	1.78	1.65	1.47
30	2.88	2.49	2.28	2.14	2.05	1.98	1.88	1.77	1.64	1.46
40	2.84	2.44	2.23	2.09	2.00	1.93	1.83	1.71	1.57	1.38
60	2.79	2.39	2.18	2.04	1.95	1.87	1.77	1.66	1.51	1.29
120	2.75	2.35	2.13	1.99	1.90	1.82	1.72	1.60	1.45	1.19
∞	2.71	2.30	2.08	1.94	1.85	1.77	1.67	1.55	1.38	1.00

Probability 5%

No. of degrees of freedom v_1	1	2	3	4	5	6	8	12	24	∞
No. of degrees of freedom v_2										
1	161.4	199.5	215.7	224.6	230.2	234.0	238.9	243.9	249.0	254.3
2	18.51	19.00	19.16	19.25	19.30	19.33	19.37	19.41	19.45	19.50
3	10.13	9.55	9.28	9.12	9.01	8.94	8.84	8.74	8.64	8.53
4	7.71	6.94	6.59	6.39	6.26	6.16	6.04	5.91	5.77	5.63
5	6.61	5.79	5.41	5.19	5.05	4.95	4.82	4.68	4.53	4.36
6	5.99	5.14	4.76	4.53	4.39	4.28	4.15	4.00	3.84	3.67
7	5.59	4.74	4.35	4.12	3.97	3.87	3.73	3.57	3.41	3.23
8	5.32	4.46	4.07	3.84	3.69	3.58	3.44	3.28	3.12	2.93
9	5.12	4.26	3.86	3.63	3.48	3.37	3.23	3.07	2.90	2.71
10	4.96	4.10	3.71	3.48	3.33	3.22	3.07	2.91	2.74	2.54
11	4.84	3.98	3.59	3.36	3.20	3.09	2.95	2.79	2.61	2.40
12	4.75	3.88	3.49	3.26	3.11	3.00	2.85	2.69	2.50	2.30
13	4.67	3.80	3.41	3.18	3.02	2.92	2.77	2.60	2.42	2.21
14	4.60	3.74	3.34	3.11	2.96	2.85	2.70	2.53	2.35	2.13
15	4.54	3.68	3.29	3.06	2.90	2.79	2.64	2.48	2.29	2.07
16	4.49	3.63	3.24	3.01	2.85	2.74	2.59	2.42	2.24	2.01
17	4.45	3.59	3.20	2.96	2.81	2.70	2.55	2.38	2.19	1.96
18	4.41	3.55	3.16	2.93	2.77	2.66	2.51	2.34	2.15	1.92
19	4.38	3.52	3.13	2.90	2.74	2.63	2.48	2.31	2.11	1.88
20	4.35	3.49	3.10	2.87	2.71	2.60	2.45	2.28	2.08	1.84
21	4.32	3.47	3.07	2.84	2.68	2.57	2.42	2.25	2.05	1.81
22	4.30	3.44	3.05	2.82	2.66	2.55	2.40	2.23	2.03	1.78
23	4.28	3.42	3.03	2.80	2.64	2.53	2.38	2.20	2.00	1.76
24	4.26	3.40	3.01	2.78	2.62	2.51	2.36	2.18	1.98	1.73
25	4.24	3.38	2.99	2.76	2.60	2.49	2.34	2.16	1.96	1.71
26	4.22	3.37	2.98	2.74	2.59	2.47	2.32	2.15	1.95	1.69
27	4.21	3.35	2.96	2.73	2.57	2.46	2.30	2.13	1.93	1.67
28	4.20	3.34	2.95	2.71	2.56	2.44	2.29	2.12	1.91	1.65
29	4.18	3.33	2.93	2.70	2.54	2.43	2.28	2.10	1.90	1.64
30	4.17	3.32	2.92	2.69	2.53	2.42	2.27	2.09	1.89	1.62
40	4.08	3.23	2.84	2.61	2.45	2.34	2.18	2.00	2.79	1.51
60	4.00	3.15	2.76	2.52	2.37	2.25	2.10	1.92	1.70	2.39
120	3.92	3.07	2.68	2.45	2.29	2.17	2.02	1.83	1.61	1.25
∞	3.84	2.99	2.60	2.37	2.21	2.10	1.94	1.75	1.52	1.00

Probability 1%

No. of degrees of freedom v_1	1	2	3	4	5	6	8	12	24	∞
No. of degrees of freedom v_2										
1	4052.	4999.	5403.	5625.	5764.	5859.	5982.	6106.	6234.	6366.
2	98.50	99.00	99.17	99.25	99.30	99.33	99.37	99.42	99.46	99.50
3	34.12	30.82	29.46	28.71	28.24	27.91	27.49	27.05	26.60	26.12
4	21.20	18.00	16.69	15.98	15.52	15.21	14.80	14.37	13.93	13.46
5	16.26	13.27	12.06	11.39	10.97	10.67	10.29	9.89	9.47	9.02
6	13.74	10.92	9.78	9.15	8.75	8.47	8.10	7.72	7.31	6.88
7	12.25	9.55	8.45	7.85	7.46	7.19	6.84	6.47	6.07	5.65
8	11.26	8.65	7.59	7.01	6.63	6.37	6.03	5.67	5.28	4.86
9	10.56	8.02	6.99	6.42	6.06	5.80	5.47	5.11	4.73	4.31
10	10.04	7.56	6.55	5.99	5.64	5.39	5.06	4.71	4.33	3.91
11	9.65	7.20	6.22	5.67	5.32	5.07	4.74	4.40	4.02	3.60
12	9.33	6.93	5.95	5.41	5.06	4.82	4.50	4.16	3.78	3.36
13	9.07	6.70	5.74	5.20	4.86	4.62	4.30	3.96	3.59	3.16
14	8.86	6.51	5.56	5.03	4.69	4.46	4.14	3.80	3.43	3.00
15	8.68	6.36	5.42	4.89	4.56	4.32	4.00	3.67	3.29	2.87
16	8.53	6.23	5.29	4.77	4.44	4.20	3.89	3.55	3.18	2.75
17	8.40	6.11	5.18	4.67	4.34	4.10	3.79	3.45	3.08	2.65
18	8.28	6.01	5.09	4.58	4.25	4.01	3.71	3.37	3.00	2.57
19	8.18	5.93	5.01	4.50	4.17	3.94	3.63	3.30	2.92	2.49
20	8.10	5.85	4.94	4.43	4.10	3.87	3.56	3.23	2.86	2.42
21	8.02	5.78	4.87	4.37	4.04	3.81	3.51	3.17	2.80	2.36
22	7.94	5.72	4.82	4.31	3.99	3.76	3.45	3.12	2.75	2.31
23	7.88	5.66	4.76	4.26	3.94	3.71	3.41	3.07	2.70	2.26
24	7.82	5.61	4.72	4.22	3.90	3.67	3.36	3.03	2.66	2.21
25	7.77	5.57	4.68	4.18	3.86	3.63	3.32	2.99	2.62	2.17
26	7.72	5.53	4.64	4.14	3.82	3.59	3.29	2.96	2.58	2.13
27	7.68	5.49	4.60	4.11	3.78	3.56	3.26	2.93	2.55	2.10
28	7.64	5.45	4.57	4.07	3.75	3.53	3.23	2.90	2.52	2.06
29	7.60	5.42	4.54	4.04	3.73	3.50	3.20	2.87	2.49	2.03
30	7.56	5.39	4.51	4.02	3.70	3.47	3.17	2.84	2.47	2.01
40	7.31	5.18	4.31	3.83	3.51	3.29	2.99	2.66	2.29	1.80
60	7.08	4.98	4.13	3.65	3.34	3.12	2.82	2.50	2.12	1.60
120	6.85	4.79	3.95	3.48	3.17	2.96	2.66	2.34	1.95	1.38
∞	6.64	4.60	3.78	3.32	3.02	2.80	2.51	2.18	1.79	1.00

Probability 0.1%

No. of degrees of freedom v_1	1	2	3	4	5	6	8	12	24	∞
No. of degrees of freedom v_2										
1	405284	500000	540379	562500	576405	585937	598144	610667	623497	636619
2	998.5	999.0	999.2	999.2	999.3	999.3	999.4	999.4	999.5	999.5
3	167.0	148.5	141.1	137.1	134.6	132.8	130.6	128.3	125.9	123.5
4	74.14	61.25	56.18	53.44	51.71	50.53	49.00	47.41	45.77	44.05
5	47.18	37.12	33.20	31.09	29.75	28.84	27.64	26.42	25.14	23.78
6	35.51	27.00	23.70	21.92	20.81	20.03	19.03	17.99	16.89	15.75
7	29.25	21.69	18.77	17.19	16.21	15.52	14.63	13.71	12.73	11.69
8	25.42	18.49	15.83	14.39	13.49	12.86	12.04	11.19	10.30	9.34
9	22.86	16.39	13.90	12.56	11.71	11.13	10.37	9.57	8.72	7.81
10	21.04	14.91	12.55	11.28	10.48	9.92	9.20	8.45	7.64	6.76
11	19.69	13.81	11.56	10.35	9.58	9.05	8.35	7.63	6.85	6.00
12	18.64	12.97	10.80	9.63	8.89	8.38	7.71	7.00	6.25	5.42
13	17.81	12.31	10.21	9.07	8.35	7.86	7.21	6.52	5.78	4.97
14	17.14	11.78	9.73	8.62	7.92	7.43	6.80	6.13	5.41	4.60
15	16.59	11.34	9.34	8.25	7.57	7.09	6.47	5.81	5.10	4.31
16	16.12	10.97	9.00	7.94	7.27	6.81	6.19	5.55	4.85	4.06
17	15.72	10.66	8.73	7.68	7.02	6.56	5.96	5.32	4.63	3.85
18	15.38	10.39	8.49	7.46	6.81	6.35	5.76	5.13	4.45	3.67
19	15.08	10.16	8.28	7.26	6.62	6.18	5.59	4.97	4.29	3.52
20	14.82	9.95	8.10	7.10	6.46	6.02	5.44	4.82	4.15	3.38
21	14.59	9.77	7.94	6.95	6.32	5.88	5.31	4.70	4.03	3.26
22	14.38	9.61	7.80	6.81	6.19	5.76	5.19	4.58	3.92	3.15
23	14.19	9.47	7.67	6.69	6.08	5.65	5.09	4.48	3.82	3.05
24	14.03	9.34	7.55	6.59	5.98	5.55	4.99	4.39	3.74	2.97
25	13.88	9.22	7.45	6.49	5.88	5.46	4.91	4.31	3.66	2.89
26	13.74	9.12	7.36	6.41	5.80	5.38	4.83	4.24	3.59	2.82
27	13.61	9.02	7.27	6.33	5.73	5.31	4.76	4.17	3.52	2.75
28	13.50	8.93	7.19	6.25	5.66	5.24	4.69	4.11	3.46	2.70
29	13.39	8.85	7.12	6.19	5.59	5.18	4.64	4.05	3.41	2.64
30	13.29	8.77	7.05	6.12	5.53	5.12	4.58	4.00	3.36	2.59
40	12.61	8.25	6.60	5.70	5.13	4.73	4.21	3.64	3.01	2.23
60	11.97	7.76	6.17	5.31	4.76	4.37	3.87	3.31	2.69	1.90
120	11.38	7.32	5.79	4.95	4.42	4.04	3.55	3.02	2.40	1.54
∞	10.83	6.91	5.42	4.62	4.10	3.74	3.27	2.74	2.13	1.00

Appendix 8

Derivation of Formulae for Combining Variances

Consider two variables x and y and let z be a known function of x and y

$$\bar{z} = f(\bar{x}, \bar{y})$$

the mean values of x and y are $\bar{x}$ and $\bar{y}$, thus,

$$\bar{z} = f(\bar{x}, \bar{y})$$

provided that x and y are independent (unrelated).

Take a pair of x, y values where x is an amount Δx from $\bar{x}$ and y is an amount Δy from $\bar{y}$. The difference Δz between z and $\bar{z}$ may be estimated:

$$\Delta z \simeq \frac{\partial z}{\partial x} \Delta x + \frac{\partial z}{\partial y} \Delta y \qquad (1)$$

If a large number n of x, y pairs is taken then the variance of z may be calculated as:

$$\text{Variance of } z = \sigma_z^2 = \Sigma (\Delta z)^2 / n \qquad (2)$$

$$= \left(\frac{\partial z}{\partial x}\right)^2 \frac{(\Delta x)^2}{n} + \left(\frac{\partial z}{\partial y}\right)^2 \frac{(\Delta y)^2}{n} +$$

$$+ 2 \frac{\partial z}{\partial x} \frac{\partial z}{\partial y} \frac{\Delta x \, \Delta y}{n}$$

but

$(\Delta x)^2/n$ is the variance of $x = \sigma_x^2$

$(\Delta y)^2/n$ is the variance of $y = \sigma_y^2$

and

$\Delta x \, \Delta y / n$ is the covariance of x and y which equals σ_{xy}^2. (This is the same as s_{xy} used in Chapter 2.)

If x and y are independent (that is unrelated) then the covariance of x and y is zero, and hence the variance of z becomes:

$$\sigma_z^2 = \left(\frac{\partial z}{\partial x}\right)^2 \sigma_x^2 + \left(\frac{\partial z}{\partial y}\right)^2 \sigma_y^2 \qquad (3)$$

Four functions of x and y are now considered, namely addition, subtraction, multiplication and division.

(i) *Addition* $z = x + y$
Partially differentiating
$$\frac{\partial z}{\partial x} = 1 \text{ and } \frac{\partial z}{\partial y} = 1$$
Substituting into Equation 3,
$$\sigma_z^2 = \sigma_x^2 + \sigma_y^2 \qquad (4)$$

(ii) *Subtraction* $z = x - y$
Partially differentiating
$$\frac{\partial z}{\partial x} = 1 \text{ and } \frac{\partial z}{\partial y} = -1$$
Substituting into Equation 3
$$\sigma_z^2 = \sigma_x^2 + \sigma_y^2 \qquad (5)$$

(iii) *Multiplication* $z = xy$
Partially differentiating
$$\frac{\partial z}{\partial x} = y \text{ and } \frac{\partial z}{\partial y} = x$$
Substituting into Equation 3
$$\sigma_z^2 = y^2 \sigma_x^2 + x^2 \sigma_y^2$$
this is sometimes expressed as
$$\frac{\sigma_z^2}{x^2 y^2} = \frac{\sigma_x^2}{x^2} + \frac{\sigma_y^2}{y^2}$$
or
$$\frac{\sigma_z^2}{z^2} = \frac{\sigma_x^2}{x^2} + \frac{\sigma_y^2}{y^2} \qquad (6)$$

(iv) *Division* $z = x/y$
Partially differentiating

$$\frac{\partial z}{\partial x} = \frac{1}{y} \text{ and } \frac{\partial z}{\partial y} = \frac{-x}{y^2}$$

Substituting into Equation 3

$$\sigma_z^2 = \frac{1}{y^2} \sigma_x^2 + \frac{x^2}{y^4} \sigma_y^2$$

$$\frac{\sigma_z^2}{x^2/y^2} = \frac{1}{x^2} \sigma_x^2 + \frac{1}{y^2} \sigma_y^2$$

Hence

$$\frac{\sigma_z^2}{z^2} = \frac{\sigma_x^2}{x^2} + \frac{\sigma_y^2}{y^2} \qquad (7)$$

The final equations for combining variances or standard deviations are the same regardless of whether n or $(n-1)$ is used in the denominator.

Appendix 9

List of Symbols

A	area under curve	$\bar{x}$	mean value of x_i
a, b, c, d	coefficients for quadratic and cubic equations	X_i	$(x_i - \bar{x})$
		y_i	set of y values
c	intercept on y axis	$\bar{y}$	mean value of y_i
D	difference in ranks	Y_i	$(y_i - \bar{y})$
E	expected frequency (theoretical frequency)	z	standardised normal deviate
		Δ	large difference
F	variance ratio (F-test)	∂	partial derivative
$f(\ )$	frequency of	μ	mean of parent population
m	slope of straight line or mean of sample	ν	number of degrees of freedom
m_1, m_2	means of samples 1 and 2	π	3.141 592 65
$n!$	factorial n, i.e. $n \cdot (n-1) \cdot (n-2) \ldots 3.2.1$	ρ	Spearman's rank correlation coefficient
n	number of values	Σ	sum of
O	observed frequency	σ	standard deviation of parent population
r	Pearson's correlation coefficient	σ^2	variance of parent populatition
s, s_1, s_2, s_x, s_y	standard deviations	τ	Kendall's rank correlation coefficient
s_{xy}	covariance of x and y		
t	Student's t value	χ^2	chi-squared
$T(+), T(-)$	rank sums for Wilcoxon's test	$\int$	integral of
T	Wilcoxon's test criterion		
x_i	set of x values		

Bibliography

General Statistics

Bajpai, A.C., Calus, I.M. and Fairley, J.A., *Statistical Methods for Engineers and Scientists*, Wiley, London (1979).

Clarke, G.M. and Cooke, D., *A Basic Course in Statistics*, Arnold, London (1978).

Eckschlager, K., *Errors, Measurement and Results in Chemical Analysis*, Van Nostrand Reinhold, London (1969).

Loveday, R., *A Second Course in Statistics, 2nd. ed.*, Cambridge University Press, Cambridge (1975).

Snedecor, G.W. and Cochran, W.G., *Statistical Methods, 6th. ed.*, Iowa State College Press, Iowa (1963).

Topping, J., *Errors of Observation and their Treatment, 3rd. ed.*, Chapman and Hall, London (1962).

Biological and Medical Statistics

Armitage, P.A., *Statistical Methods of Medical Research*, Blackwell, Oxford (1971).

Bailey, N.T.J., *Statistical Methods in Biology*, English Universities Press, London (1959).

Bourke, G.J. and McGilvray, J., *Interpretation and Uses of Medical Statistics*, Blackwell, Oxford (1975).

Campbell, R.C., *Statistics for Biologists, 2nd. ed.*, Cambridge University Press, London (1974).

Causton, D.R., *A Biologist's Mathematics*, Arnold, London (1977).

Daniel, W.W., *Biostatistics: A Foundation for Analysis in the Health Sciences*, Wiley, London (1974).

Fisher, R.A., *Statistical Methods for Research Workers, 11th. ed.*, Oliver and Boyd, Edinburgh (1950).

Heath, O.V.S., *Investigation by Experiment* (Studies in Biology No. 23), Arnold, London (1970).

Mather, K., *Statistical Analysis in Biology*, Methuen, London (1949).

Mather, K., *The Elements of Biometry*, Methuen, London (1967).

Parker, R.E., *Introductory Statistics for Biology*, Arnold, London (1973).

Pearce, S.C., *Biological Statistics—An Introduction*, McGraw-Hill, London (1965).

Saunders, L. and Fleming, R., *Mathematics and Statistics for Use in Pharmacy, Biology and Chemistry, revised edition*, Pharmaceutical Press, London (1966).

Smart, J.V., *Elements of Medical Statistics*, Staples Press (1970).

Sokal, R.R. and Rohlf, F.J., *Biometry*, Freeman, London (1969).

Sokal, R.R. and Rohlf, F.J., *Introduction to Biostatistics*, Freeman, London (1973).

Experimental Design

Clarke, G.M., *Statistics and Experimental Design, 2nd. ed.*, Arnold, London (1980).

Cochran, W.G. and Cox, G.M., *Experimental Designs, 2nd. ed.*, Wiley, Chichester (1957).

Edwards, A.L., *Experimental Design in Psychological Research*, Holt Rinehart and Winston, New York (1968).

Non-parametric Statistics

Conover, W.J., *Practical Non-parametric Statistics, 2nd. ed.*, Wiley, London (1980).

Siegel, S., *Non-parametric Statistics for the Behavioural Sciences*, McGraw-Hill, Maidenhead (1956).

Statistical Tables

Fisher, R.A. and Yates, F., *Statistical Tables for Biological, Medical and Agricultural Research, 6th. ed.*, Longman, London (1974).

Lindley, D.V. and Miller, J.C.P., *Cambridge Elementary Statistical Table*, Cambridge University Press, London (1968).

Pearson, E.S. and Hartley, H.O., *Biometrika Tables for Statisticians, Vol. 1*, Cambridge University Press, London (1962).

Rohlf, F.J. and Sokal, R.R., *Statistical Tables*, Freeman, London (1969).

Answers and Solutions

1.1 (b) 0.5 matches, 7.0 cm

1.2 Density $=$ mass/volume
$$= 3.8251/(4/3 \times 3.1416 \times 0.563^3)$$
$$= 5.117 \text{ g cm}^{-3}$$

Relative error in answer
$$= \frac{0.0001}{3.8251} + \frac{0.0005}{1.126} \times 3 = 0.001\,36$$

Absolute error
$$= 0.001\,36 \times 5.117 = 0.0070$$
Hence result is 5.117 ± 0.007 g cm^{-3}

1.3 Refer to Fig. 1.1
(a) 15.87% below mean corresponds to one standard deviation, hence machine should be set to 255 g.
(b) 260 g
(c) 265 g

2.1 Sample mean $= 60$ m.p.h.; standard deviation of sample $= 7$ m.p.h.
Estimated mean for all cars $= 60$ m.p.h.; estimated standard deviation for all cars $= 7.15$ m.p.h. (For the last result the divisor was $(n-1) = 23$ rather than $n = 24$.)

2.2 Mean $= 160$ cm; standard deviation $= 6.63$ cm with Yates's correction

2.3 (a) Mean $= 2\mu$; standard deviation $= \sqrt{(\sigma^2 + \sigma^2)} = \sqrt{(2 \cdot \sigma)}$
(b) Mean $= 0$; standard deviation $= \sqrt{(2 \cdot \sigma)}$
(c) Mean $= \mu$; standard deviation $= \frac{1}{2}\sqrt{(2 \cdot \sigma)} = \sigma/\sqrt{2}$
(d) Mean $= 0$; standard deviation $= \sigma/\sqrt{2}$

2.4 (a) Mean $= 110$; standard deviation $= 14.1$
(b) Mean $= 0$; standard deviation $= 0$
(c) Mean $= 55$, Standard deviation $= 7.07$

2.5 Mean $=$ £75. Median is in the range £65–75. 104 people have wages $<$£65 and 168 have wages $<$£75. Assuming that the wages of the 64 people in this group are evenly distributed, the wage of the 150.5th. person (median wage) is estimated as £72.3.
Mean-mode approximately equals (mean-median) $\times$ 3, hence mode approximately equals £66.9.
Standard deviation $=$ £20.1 hence skewness $= +0.403$

3.1 Mean catch on one day $= 650$ lbs
Mean weight in one box $= 26$ lbs
Estimated standard deviation for catch on one day $= 112$ lbs
Estimated standard deviation for one box $= 22.4$ lbs

3.2 The sample of 13 people will have a mean weight of $13 \times 70 = 910$ kg and a standard deviation of $\sqrt{13} \times 10 = 36.1$ kg. To overload the lift requires a weight of 90 kg above the mean which is 2.5 standard deviations. Since by central limit theorem the sample means are approximately normally distributed, the chance of this occurring is $1 - 0.9938 = 0.0062 = 1/161$. Thus there is a 1/161 chance of overloading the lift with 13 people. Without being superstitious, this is not very safe!

3.3 $4925 - (75 \times 64) = 125$ kg

$125/(8 \times \sqrt{64}) = 1.953$ standard deviations

From normal distribution table (Appendix 4) probability of a weight less than 1.953 standard deviations is 97.46% hence probability of being overweight = 2.54%

3.4 Mean thickness of 100 washers
= 249.82 mm

Standard deviation of 100 washers
= 2.42

Estimated value for mean thickness of 1 washer = 2.50 mm

Estimated value for standard deviation of 1 washer = $2.42/\sqrt{100} = 0.242$ mm

4.1 15 apples = 10%. From normal table
= 1.282σ

30 apples = 20%. From normal table
= 0.842σ.

Mean $- 1.282\sigma = 71$

Mean $+ 0.842\sigma = 103$

Hence mean = 90.3 g and $\sigma = 15.1$ g.

53 apples above 96 g

4.2 Hint! How many standard deviations are needed for 25% and 75% of a normal population?

4.3 See Table S1. Probability of being in range 100–110 is $0.7340 - 0.5000 = 0.2340$ (since the curve is symmetrical, this is also the value for the range 90–100).

Table S1

Value	Number of standard deviations from mean	Area under normal curve up to this point
100	0	0.5000
110	0.625	0.7340
120	1.25	0.8944
130	1.875	0.9696

Similarly for 110–120 and 80–90
probability = 0.1604

Similarly for 120–130 and 70–80
probability = 0.0752

Similarly for > 130 and < 70
probability = 0.0304.

4.4 Mean = 1.01, variance = 0.93, Poisson distribution:

Number of houses struck			
0	1	2	3 or more
Number of years			
36.4	36.8	18.6	8.2

The data appear to fit a Poisson distribution. This is suggested by the fact that the mean and variance are approximately equal. (A chi-squared test could be used to confirm that the data do not differ significantly from Poisson. Chi-squared = 0.54.)

5.1 (a) Chi-squared = 8.88, probability of worst chi-squared from a fair die = 11%, hence there are no grounds to suspect the die.

(b) Chi-squared = 8.21, probability of worst chi-squared from a fair die = 0.4%, hence die almost certainly unfair.

These results are not contradictory! The die is unfair if one tests the frequency of 3's specifically, but not sufficiently unfair to show in a general test.

5.2 Frequency of digits

7	9	14	8	12	8	12	9	12	9
Digits									
0	1	2	3	4	5	6	7	8	9

chi-squared = 4.8, probability of larger chi-squared from random digits = 85%.

5.3 Chi-squared = 15.36, probability of worse results = 3.2% which suggests either that the theory is wrong or that the data are inaccurate or fabricated.

5.4 If there is no preference then the number of people on each beach will be as follows:

	A	B	C
No. males	72	72	96
No. females	78	78	104

Chi-squared = 4.32, 2 degrees of freedom, probability of a larger discrepancy arising by chance is 12%, hence there is no evidence that the proportion of females differs significantly.

5.5 Chi-squared = 10.4, probability of worse results by chance if the wheel is fair is 0.1%. It is highly likely that the wheel is biased. With 35 zeros, chi-squared = 2.14 and the probability of worse results by chance is 14%, hence there is no evidence for a biased wheel.

5.6 Chi-squared = 2.54, probability of worse results by chance is 11%. There is no evidence that the proportions of animals suffering vitamin deficiency differ between the two diets.

5.7 Expected Poisson distribution

Number of deaths	0	1	2	3	4	5 or more
Number of days	203.4	251.4	155.3	64.0	19.8	6.1

Chi-squared = 3.06, probability of more different results arising by chance = 55%, hence the data do not differ significantly from a Poisson distribution.

5.8 Chi-squared = 0.05 (remember the Yates's correction). Probability of worse results = 82%. The data agree with the theory, and are not so good that their authenticity is questioned.

5.9 Expected frequencies are shown in Table S2.

Table S2

Energy intake	Physical activity in leisure time		
	High	Medium	Low
High	4.99	9.66	9.35
Medium	4.99	9.66	9.35
Low	6.03	11.68	11.30

Strictly groups with a frequency of less than 5 should be pooled, but the values of 4.99 in this case are close enough to 5 to be treated as separate groups. Chi-squared is calculated as 11.4.

With 4 degrees of freedom this is highly significant, the hypothesis that there is no relationship between energy intake and activity is rejected, hence a relationship between these two factors is established for the sample of men tested.

6.1 Difference in means = 24 hours
Standard error of difference in means = 11.8 hours.
Difference in means = 2.03 standard errors.

Using a normal table the probability of A appearing better than B by this amount is $1 - 0.9788 = 0.0212 = 2.12\%$ if the bulbs are both as good. This provides moderately strong evidence that bulbs A last longer than bulbs B.

6.2 Difference in means = 24 hours
Standard error of difference in means = 12.3 hours
Difference in means = 1.95 standard errors which is insignificant at the 5% level.

6.3 Difference in means = 2.76 standard errors, which is significant at the 1% level, hence there is strong evidence for an increase in mean height.

6.4 Estimated standard deviation for the combined sample =

$$\sqrt{\left[\frac{(11-1) \times 1.2^2 + (8-1) \times 1.4^2}{(11-1)+(8-1)}\right]}$$

$$= 1.29 \text{ metres}$$

Standard error of difference in means $= 1.29 \sqrt{(1/11 + 1/8)} = 0.598$ metres.
Difference in means $= 5.3 - 4.2 = 1.1$ metres $= 1.84$ standard erros. Using a t-table (Appendix 6) with $(11-1)+(8-1) = 17$ degrees of freedom, the probability of a t-value of 1.84 occurring by chance is between 5% and $2\frac{1}{2}$% (one-tailed). If the missiles are equally good/bad then there is more than a $2\frac{1}{2}$% chance of getting results which show B to be better by so much, and a $2\frac{1}{2}$% chance of getting results which show A to be better by the same amount. There is no statistical evidence that the missiles differ significantly.

6.5 Road A mean = 30.0 m.p.h.; estimated standard deviation = 0.616 m.p.h.
Road B mean = 31.0 m.p.h.; estimated standard deviation = 0.990 m.p.h.
Estimated standard deviation for the combined group = 0.798 m.p.h.
Difference in means = 1 m.p.h.
Standard error of difference in means = 0.402 m.p.h.
Difference in means = 2.49 standard errors
Using a t-table (Appendix 6) with $(9-1)+(7-1) = 14$ degrees of freedom, the probability of a t-value of 2.49 arising by chance if the mean speed of all cars (rather than the small samples) is the same is between 5% and 2% (two tailed). Since there is less than a 5% probability of the observed result occurring by chance if the mean speed of all cars is the same, it is reasonable to conclude that the mean speeds of cars on the two roads are significantly different.
$F = 2.58$
Using the F tables (Appendix 7) with

$v_1 = (7-1)$ and $v_2 = (9-1)$ it is found that the calculated F value is less than the 10% significance value of 2.67, hence there is more than a 10% probability of standard deviations as different arising by chance if the standard deviation of all cars (as opposed to the samples) is the same. There is no statistical evidence that the standard deviations of speeds of cars on the two roads are different, thus the result of the t-test is valid.

6.6 Ignoring the two people who weighed the same, there are $3+$ (gained weight) and $10-$ (lost weight) from 13 comparisons. The total number of permutations is $2^{13} = 8192$
  Probability of 0 plus signs = 1/8192
  Probability of 1 plus signs = 13/8192
  Probability of 2 plus signs = 78/8192
  Probability of 3 plus signs = 286/8192
Probability of getting up to and including three plus signs (two tailed) = $2 \times 378/8192 = 9.23\%$. This exceeds 5% so the data do not provide evidence for a difference due to the diet. Note that for 13 data values Table 6.3 shows that it is impossible to reach the 5% significance level with 3 signs different.

6.7 Discard the result which gave no difference, leaving 13 comparisons ($11+$ and $2-$). Probability of getting up to and including 2 signs different out of 13 is $2 \times 92/8192 = 2.25\%$ (two-tailed). This is less than 5% and is therefore significant, hence there is evidence for a difference between the two tranquillizers.

6.8 *Sign test*—ignore zero difference leaving $9+$ and $2-$. Two-tailed probability of getting up to and including two signs different = $2 \times 67/2048 = 6.54\%$. This exceeds 5% so the data do not provide statistical evidence for a difference between the response times when rested and when tired.

Wilcoxon—ignore zero difference, sum of positive ranks $= 3$, sum of negative ranks $= -63$, 11 comparisons used. Number of combinations giving $T \leqslant 3$ is 5, total number of permutations is $2^{11} = 2048$, hence probability $= 5/2048 = 0.49\%$. This is less than 1% and thus provides very strong evidence for a difference between response times from rested and tired subjects. This illustrates the greater sensitivity of Wilcoxon's test compared with the sign test.

6.9 *Wilcoxon*—sum of positive ranks $= 24.5$, sum of negative ranks $= -95.5$, number of comparisons $= 15$. Number of combinations giving $T \leqslant 24.5$ is 762, total number of permutations is $2^{15} = 32768$, two-tailed probability $= 4.65\%$. This is less than 5%, hence the data provide statistical evidence for a difference in growth between self-fertilized and cross-fertilized plants.

t-test—self-fertilized: mean $= 17.575$, sum of errors squared $= 58.93$, standard deviation $= 2.05$. Cross-fertilized: mean $= 20.13$, sum of errors squared $= 180.22$, standard deviation $= 3.59$. Calculated value for $t = 2.39$, and the probability that such a difference in means could occur by chance if the parent populations have the same mean and variance is 2.38%. This is less than 5% and would normally suggest a significant difference between self- and cross-fertilized plants. Note, however, that the *t*-test is only valid if the standard deviations of the two groups do not differ significantly. An *F*-test gives $F = 0.327$, and the probability that such a difference in variances could occur by chance if the parent populations have the same variance is 2.25%. This is less than 5% and is significant, so the *t*-test is invalid.

6.10 Ignore the pair which give a zero difference. Sum of positive ranks $= 7$, sum of negative ranks $= -21$, number of comparisons (excluding ties) $= 7$. Number of combinations giving $T \leqslant 7$ is 19, total number of permutations is $2^7 = 128$, and the two-tailed probability of such a *T*-value is 29.7%. This is so large that results as different could arise by chance 29 times out of 100, and the data provide no statistical evidence for a difference between the two drugs.

6.11 *Wilcoxon*: Sum of positive ranks $= 8$, sum of negative ranks $= -47$, number of comparisons $= 10$. Number of combinations giving $T \leqslant 8$ is 25, total number of permutations is $2^{10} = 1024$. Two-tailed probability of such a *T* value is 4.88%. This is less than 5%, so there is statistical evidence for a difference in performance of athletes at sea level and at high altitude.

t-test: sea level: mean $= 3.84$, sum of errors squared $= 0.142$, standard deviation $= 0.125$. Altitude: mean $= 3.95$, sum of errors squared $= 0.268$, standard deviation $= 0.172$. Calculated value for $t = 1.64$, and the probability that such a result could occur by chance is 11.9%. This suggests that there is no evidence for a difference in means between the two groups. Note that the *t*-test is valid only if the standard deviations or variances of the two groups are the same. Calculated $F = 0.529$, and the probability that the variances could differ so much by chance is 17.9%. There is no significant difference in variances, hence the *t*-test stands. Note the better resolution in this case of the Wilcoxon test treating the data as paired (1 athlete at both sea level and high altitude), over the *t*-test which treats the mean of all 10 athletes at sea level and at high altitude.

7.1 Mean square for within samples variance $= 58.68$. Mean square for between samples variance = 158.7. $F = 158.7/58.68 = 2.705$. $v_1 = 5$; $v_2 = 18$.

Probability that such a difference could occur by chance if the variances of the parent populations are the same is 5.4%. This value falls just outside the 5% limit. There is no evidence for a significant difference in growth between the six varieties.

7.2 Mean square for within samples variance = 66.91. Mean square for between samples variance = 153.7. $F = 153.7/66.91 = 2.297$. $v_1 = 2$; $v_2 = 15$.

Probability that such a difference could occur by chance if the variances of the parent populations are the same is 13.5%. The data provide no evidence that the different fertilizer treatments produce significantly different crop yields.

7.3 Mean square for within sample variance = 5·440. Mean square for between samples variance = 20.15. $F = 20.15/5.440 = 3.70$. $v_1 = 3$; $v_2 = 14$.

Probability that such a difference could occur by chance if the variances of the parent populations are the same is 3.76%. Since this is less than 5% there is a significant difference between the varieties of wheat.

7.4 Mean square for within samples variance = 0.0003044. Mean square for between samples variance = 0.0005056. $F = 0.0005056/0.0003044 = 1.661$. $v_1 = 2$; $v_2 = 15$.

Probability that such a difference could occur by chance if the variances of the parent populations are the same is 22.3%.

This value is so large that one concludes that there is no evidence that the analytical methods give significantly different results.

7.5 Treatment totals: 784.3, 791.7, 803.8, 845.7. Block totals: 637.7, 637.7, 612.0, 647.6, 690.5. (See Table S3.) F value for treatment = 3.09. Degrees of freedom: $v_1 = 3$, $v_2 = 12$.

Table S3

	Sum of squares	Mean square
Total	1852.76	—
Treatment	451.15	150.4
Block	818.14	204.5
Residual	583.47	48.62

Probability that mean squares could differ so much by chance if the parent populations have the same variance is 6.8%. The difference in mean squares is insignificant at the 5% level, hence there is no evidence for a difference between the varieties of barley.

F value for blocks = 4.21, which is significant at the 5% level. Thus it was worth while using blocks for the experiment.

7.6 Treatment totals: 20.05, 18.41, 17.74. Block totals: 11.35, 10.50, 11.35, 11.97, 11.03. (See Table S4.) F value for treatment = 22.10. Degrees of freedom, $v_1 = 2$, $v_2 = 8$.

Table S4

	Sum of squares	Mean square
Total	1.050	—
Treatment	0.565	0.282
Block	0.383	0.0957
Residual	0.102	0.0128

Probability that mean squares could differ so much by chance if the parent populations have the same variance is 0.055%—that is, well within the 0.1% probability, hence the density of planting is highly significant.

The F value for blocks = 7.49, which is significant at the 1% level. Thus it was worth using blocks for the experiment.

Table S5

	Sum of squares	Mean square
Total	709.8	—
Treatment	477.3	159.1
Block	176.3	58.75
Residual	56.2	6.25

7.7   Treatment totals: 57, 75, 56, 110. Block totals: 96, 69, 60, 73. (See Table S5.) F value for treatment = 25.45. Degrees of freedom $v_1 = 3$, $v_2 = 9$.

Probability that such a difference in mean squares could occur by chance is 0.0099; that is, well within the 0.1% level. Differences due to the treatments are therefore highly significant. Since one of the treatments is a blank (control), it follows that at least one of the sprays has a highly significant effect.

F value for blocks = 9.4, which is significant at the 1% level. Thus it was worth using blocks in the experiment.

t-values for treatments:

1: control = 9.99
2: control = 6.60
3: control = 10.18

Degrees of freedom = 12.

All the t-values are significant at the 0.1% level, hence all the sprays have a highly significant effect compared with the control. Analysis of variance omitting the control: Treatment totals—57, 75, 56, Block totals—62, 43, 37, 46. (See Table S6.) F value for treatment = 3.21. Degrees of freedom $v_1 = 2$, $v_2 = 6$.

Table S6

	Sum of squares	Mean square
Total	224.7	—
Treatment	57.17	28.58
Block	114.0	38.00
Residual	53.5	8.92

Probability that such a difference in mean squares could occur by chance is 11.3%. This is outside the 5% level, hence there is no evidence for a difference between the three different formulations of copper-containing sprays.

The F value for blocks is 4.26, which is not significant at the 5% level, hence the block design has not eliminated extraneous factors.

7.8   Treatment totals: 26, 32, 24, 42. Block totals: 30, 17, 23, 28, 26. (See Table S7.) F value for treatment = 4.57. Degrees of freedom: $v_1 = 3$, $v_2 = 12$.

Table S7

	Sum of squares	Mean square
Total	99.20	—
Treatment	39.20	13.07
Block	25.70	6.43
Residual	34.30	2.86

Probability that such a difference in mean squares could arise by chance = 2.34%. Since this is less than 5% the difference is significant, and the seed treatments are different.

The F value for blocks = 2.25, which

is not significant at the 5% level. Thus the choice of blocks has not eliminated all extraneous factors, but the difference in seed treatments is still a valid difference.

The 5% least significant difference $= 2.12 \times 3.704 = 7.85$.

The differences in means between 1: control, 2: control and 3: control are all greater than this and are significant at the 5% level.

Alternatively comparison of treatments 1, 2 and 3 with control:

1: control $t = 4.32$
2: control $t = 2.70$
3: control $t = 4.86$

With 16 degrees of freedom, treatments 1 and 3 are within the 0.1% significance level (difference from control highly significant), whilst treatment 2 is within the 5% level (significant).

8.1 (a) $r = 0$ (no linear correlation)
(b) $r = 0.996$ (high degree of linear correlation).

8.2 $r = 0.831$. Using the table in Appendix 1 with 8 degrees of freedom the 1% significance value is 0.765, hence the correlation is very significant.

8.3 $r = 0.742$. 5% significance value $= 0.707$, hence result is significant at this level.

8.4 Observer 1 $\rho = 0.6$, Observer 2 $\rho = 0.7$, hence Observer 2 was the most self-consistent.

8.5 $\rho = 0.9$. Using the table in Appendix 2 with $n = 5$ the result is significant at the 5% level.

8.6 $\tau = 0.714$. Using Appendix 3 with $n = 8$ shows that agreement is significant at the 5% level.

8.7 Pearson's $r = 0.0651$. There is no evidence for a direct relationship.

8.8 Pearson's $r = 0.723$. One-tailed probability $= 1\%$ hence strong evidence for a direct relationship.

8.9 Pearson's $r = 0.664$. One-tailed probability $= 2\%$. There is some evidence for a direct relationship.

8.10 Pearson's $r = 0.721$. One-tailed probability $= 1\%$. There is strong evidence for direct relationship.

9.1 x on y $\quad x = -0.5y + 12.75$
y on x $\quad y = -0.125x + 6.75$

9.2 (b) 41.25, 52.5, 56.25. 68.75, 98.75, 106.25
(c) age (x), pocket money (y), slope 13.61, intercept -113.1.

9.3 Slope -0.0163, intercept 53.4, cost £45.20

9.4 Slope 2.3, intercept 4.3, bill £11.20

9.5 x on y $\quad$ slope 0.696, intercept 4.42, age of wife 21.8 years
y on x $\quad$ slope 0.814, intercept 8.49, age of husband 28.8 years

9.6 Population (x), Civil Servants (y), slope 1.34, intercept 1.85, Civil Servants 22.

9.7 Slope $= 5.39$, Pearson's $r = 0.994$.

9.8 Slope $= 4.00$.

Index

266